Applied Directional Statistics
Modern Methods and Case Studies

Edited by
Christophe Ley
Thomas Verdebout

T0325963

CRC Press
Taylor & Francis Group
Boca Raton London New York

CRC Press is an imprint of the
Taylor & Francis Group, an **informa** business
A CHAPMAN & HALL BOOK

CRC Press
Taylor & Francis Group
6000 Broken Sound Parkway NW, Suite 300
Boca Raton, FL 33487-2742

First issued in paperback 2020

© 2019 by Taylor & Francis Group, LLC
CRC Press is an imprint of Taylor & Francis Group, an Informa business

No claim to original U.S. Government works

ISBN-13: 978-1-138-62643-0 (hbk)
ISBN-13: 978-0-367-73345-2 (pbk)

Library of Congress Cataloging-in-Publication Data

Names: Ley, Christophe, author. | Verdebout, Thomas, author.
Title: Applied directional statistics / Christophe Ley and Thomas Verdebout.
Description: Boca Raton : CRC Press, 2018.
Identifiers: LCCN 2018010549 | ISBN 9781138626430 (hardback)
Subjects: LCSH: Mathematical statistics. | Circular data. | Spherical data.
Classification: LCC QA276 .L43 2018 | DDC 519.5--dc23
LC record available at https://lccn.loc.gov/2018010549

Visit the Taylor & Francis Web site at
http://www.taylorandfrancis.com

and the CRC Press Web site at
http://www.crcpress.com

Contents

Contributors   xi

Editors   xiii

Introduction   xv

1 Directional Statistics in Protein Bioinformatics   1
Kanti V. Mardia, Jesper Illemann Foldager, and Jes Frellsen
1.1   Introduction   1
1.2   Protein Structure   2
1.3   Protein Geometry   4
1.4   Structure Determination and Prediction   6
     1.4.1   Markov Chain Monte Carlo Simulations of Proteins .   8
1.5   Generative Models for the Polypeptide Backbone   10
     1.5.1   Bivariate Angular Distributions   11
         1.5.1.1   Bivariate von Mises   11
         1.5.1.2   Histograms and Fourier Series   11
         1.5.1.3   Mixture of von Mises   12
     1.5.2   A Dynamical Bayesian Network Model: TorusDBN . .   12
1.6   Generative Models for Amino Acids Side Chains   15
1.7   Discussion   17

2 Orientations of Symmetrical Objects   25
Richard Arnold and Peter Jupp
2.1   Ambiguous Rotations   25
     2.1.1   Symmetry Groups   31
     2.1.2   Symmetric Frames   32
2.2   From Symmetric Frames to Symmetric Arrays   33
2.3   Summary Statistics   35
2.4   Testing Uniformity   36
2.5   Distributions of Ambiguous Rotations   38
     2.5.1   A General Class of Distributions on $SO(3)/K$ 38
     2.5.2   Concentrated Distributions   39
2.6   Tests of Location   40
     2.6.1   One-Sample Tests   40
     2.6.2   Two-Sample Tests   41
2.7   Further Developments   41

2.8   Analysis of Examples   42
         2.8.1   Analysis of Example 1   42
         2.8.2   Analysis of Example 2   42
         2.8.3   Analysis of Example 3   42

3   Correlated Cylindrical Data   45
  Francesco Lagona
  3.1   Correlated Cylindrical Data   47
  3.2   Cylindrical Hidden Markov Models   49
         3.2.1   The Abe–Ley Density   49
         3.2.2   Modeling a Cylindrical Time Series   49
         3.2.3   Modelling a Cylindrical Spatial Series   51
  3.3   Identification of Sea Regimes   52
  3.4   Segmentation of Current Fields   54
  3.5   Outline   55

4   Toroidal Diffusions and Protein Structure Evolution   61
  Eduardo García-Portugués, Michael Golden, Michael Sørensen,
  Kanti V. Mardia, Thomas Hamelryck, and Jotun Hein
  4.1   Introduction   62
         4.1.1   Protein Structure   62
         4.1.2   Protein Evolution   64
         4.1.3   Toward a Generative Model of Protein Evolution . . .   66
  4.2   Toroidal Diffusions   67
         4.2.1   Toroidal Ornstein–Uhlenbeck Analogues   70
         4.2.2   Estimation for Toroidal Diffusions   72
         4.2.3   Empirical Performance   75
  4.3   ETDBN: An Evolutionary Model for Protein Pairs   77
         4.3.1   Hidden Markov Model Structure   77
         4.3.2   Site-Classes: Constant Evolution and Jump Events . .   80
         4.3.3   Model Training   81
         4.3.4   Benchmarks   83
  4.4   Case Study: Detection of a Novel Evolutionary Motif   87
  4.5   Conclusions   90

5   Noisy Directional Data   95
  Thanh Mai Pham Ngoc
  5.1   Introduction   95
  5.2   Some Preliminaries about Harmonic Analysis on SO(3) and $\mathbb{S}^2$ 96
  5.3   Model and Assumptions   98
         5.3.1   Null and Alternative Hypotheses   98
         5.3.2   Noise Assumptions   99
  5.4   Test Constructions   100
  5.5   Numerical Illustrations   102
         5.5.1   The Testing Procedures   102

	5.5.2	Alternatives	103
	5.5.3	Simulations	104
	5.5.4	Real Data: Paleomagnetism	107
	5.5.5	Real Data: UHECR	108

6 On Modeling of $SE(3)$ Objects **111**
Louis-Paul Rivest and Karim Oualkacha
6.1		Introduction	111
6.2		The One Axis Model in $SE(3)$	113
	6.2.1	Rotation Matrices and Cardan Angles in $SO(3)$	113
	6.2.2	A Geometric Construction of the One Axis Model	113
6.3		Modeling Data from $SE(3)$	115
6.4		Estimation of the Parameters	116
	6.4.1	The Rotation Only Estimator of the Rotation Axis A_3 and B_3	116
	6.4.2	The Translation Only Estimator of the Parameters	117
	6.4.3	The Rotation-Translation Estimator of the Parameters	118
6.5		Numerical Examples	119
	6.5.1	Simulations	119
	6.5.2	Data Analysis	121
6.6		Discussion	123

7 Spatial and Spatio-Temporal Circular Processes with Application to Wave Directions **129**
Giovanna Jona-Lasinio, Alan E. Gelfand, and Gianluca Mastrantonio
7.1		Introduction	130
7.2		The Wrapped Spatial and Spatio-Temporal Process	131
	7.2.1	Wrapped Spatial Gaussian Process	132
	7.2.2	Wrapped Spatio-Temporal Process	133
	7.2.3	Kriging and Forecasting	133
	7.2.4	Wave Data for the Examples	134
	7.2.5	Wrapped Skewed Gaussian Process	136
	7.2.5.1	Space-Time Analysis Using the Wrapped Skewed Gaussian Process	140
7.3		The Projected Gaussian Process	141
	7.3.1	Univariate Projected Normal Distribution	141
	7.3.2	Projected Gaussian Spatial Processes	142
	7.3.3	Model Fitting and Inference	145
	7.3.4	Kriging with the Projected Gaussian Processes	146
	7.3.4.1	An Example Using the Projected Gaussian Process	147
	7.3.5	The Space-Time Projected Gaussian Process	149
	7.3.6	A Separable Space-Time Wave Direction Data Example	149
7.4		Space-Time Comparison of the WN and PN Models	151
7.5		Joint Modeling of Wave Height and Direction	153

7.6   Concluding Remarks   157

**8 Cylindrical Distributions and Their Applications to
    Biological Data   163**
Toshihiro Abe and Ichiro Ken Shimatani
8.1   Introduction   164
8.2   Example: Commonly Observed Patterns for Cylindrical Data   165
8.3   A Brief Review of the Univariate Probability Distribution . .   165
      8.3.1   Probability Distributions on $[0, \infty)$ 165
      8.3.2   Circular Distributions   167
      8.3.3   Sine-Skewed Perturbation to the Symmetric Circular
              Distributions   168
8.4   Cylindrical Distributions   168
      8.4.1   The Johnson–Wehrly Distribution   168
      8.4.2   The Weibull–von Mises Distribution   169
      8.4.3   Gamma–von Mises Distribution   170
      8.4.4   Generalized Gamma–von Mises Distribution   171
      8.4.5   Sine-Skewed Weibull–von-Mises Distribution   172
      8.4.6   Parameter Estimation   173
8.5   Application 1: Quantification of the Speed/Turning Angle
      Patterns of a Flying Bird   173
8.6   Application of Cylindrical Distributions 2: How Trees Are
      Expanding Crowns   174
      8.6.1   Crown Asymmetry in Boreal Forests   176
      8.6.2   Crown Asymmetry Model   177
      8.6.3   Results of the Cylindrical Models   180
8.7   Concluding Remarks   184

9 Directional Statistics for Wildfires   187
Jose Ameijeiras–Alonso, Rosa M. Crujeiras, and
Alberto Rodríguez Casal
9.1   Introduction to Wildfire modeling   187
9.2   Fires' Seasonality   188
      9.2.1   Landscape Scale   189
      9.2.2   Global Scale   195
9.3   Fires' Orientation   197
      9.3.1   Main Spread on the Orientation of Fires   199
      9.3.2   Orientation–Size Joint Distribution   200
      9.3.3   Orientation–Size Regression Modeling   205
9.4   Open Problems   206

**10 Bayesian Analysis of Circular Data in Social and Behavioral
    Sciences   211**
Irene Klugkist, Jolien Cremers, and Kees Mulder
10.1 Introduction   212

10.2 Introducing Two Approaches Conceptually 213
    10.2.1 Intrinsic 214
    10.2.2 Embedding 215
10.3 Bayesian Modeling 216
    10.3.1 Intrinsic 217
    10.3.2 Embedding 218
10.4 The Development of Spatial Cognition 218
    10.4.1 The Data 218
    10.4.2 Bayesian Inference 219
    10.4.3 Inequality Constrained Hypotheses 225
10.5 Basic Human Values in the European Social Survey 227
    10.5.1 The Data 228
    10.5.2 The Model 229
    10.5.3 Variable Selection 230
    10.5.4 Bayesian Inference 231
    10.5.5 Comparison of Approaches 234
10.6 Discussion 236

11 Nonparametric Classification for Circular Data   **241**
Marco Di Marzio, Stefania Fensore, Agnese Panzera,
and Charles C. Taylor
11.1 Density Estimation on the Circle 243
11.2 Classification via Density Estimation 245
11.3 Local Logistic Regression 246
    11.3.1 Binary Regression via Density Estimation 246
    11.3.2 Local Polynomial Binary Regression 248
11.4 Numerical Examples 250
11.5 Classification of Earth's Surface 251
11.6 Conclusion 253

12 Directional Statistics in Machine Learning: A Brief Review 259
Suvrit Sra
12.1 Introduction 259
12.2 Basic Directional Distributions 260
    12.2.1 Uniform Distribution 260
    12.2.2 The von Mises-Fisher Distribution 261
    12.2.3 Watson Distribution 261
    12.2.4 Other Distributions 262
12.3 Related Work and Applications 263
12.4 Modeling Directional Data: Maximum-Likelihood Estimation 263
    12.4.1 Maximum-Likelihood Estimation for vMF 264
    12.4.2 Maximum-Likelihood Estimation for Watson 265
12.5 Mixture Models 266
    12.5.1 EM Algorithm 267
    12.5.2 Limiting Versions 268

        12.5.3 Application: Clustering Using movMF   269
        12.5.4 Application: Clustering Using moW   271
    12.6 Conclusion   272

13 Applied Directional Statistics with R: An Overview   277
    Arthur Pewsey
    13.1 Introduction   277
    13.2 The Circular Package   278
    13.3 Packages that Use the Circular Package   280
    13.4 Other Packages for Circular Statistics   281
    13.5 The Directional Package   281
    13.6 Other Packages for Directional Statistics   283
    13.7 Unsupported Directional Statistics Methodologies   284
    13.8 Conclusions   285

Index   291

Contributors

Toshihiro Abe
Nanzan University

Jose Ameijeiras-Alonso
University of Santiago de
  Compostela

Richard Arnold
Victoria University of Wellington

Jolien Cremers
Utrecht University

Rosa M. Crujeiras
University of Santiago de
  Compostela

Marco Di Marzio
Chieti-Pescara University

Stefania Fensore
Chieti-Pescara University

Jesper Foldager
University of Copenhagen

Jes Frellsen
IT University of Copenhagen

Eduardo Garciá-Portugués
Carlos III University of Madrid

Alan Gelfand
Duke University

Michael Golden
University of Oxford

Thomas Hamelryck
University of Copenhagen

Jotun Hein
University of Oxford

Giovanna Jona-Lasinio
Sapienza University of Rome

Peter Jupp
University of St. Andrews

Irene Klugkist
Utrecht University and Twente
  University

Francesco Lagona
University of Roma Tre

Kanti V. Mardia
University of Leeds
University of Oxford

Kees Mulder
Utrecht University

Agnese Panzera
Florence University

Thanh Mai Pham Ngoc
Université Paris-Sud
Université Paris-Saclay

Gianluca Mastrantonio
Polytechnic of Turin

Karim Oualkacha
Université du Québec à Montréal

Arthur Pewsey
University of Extremadura Cáceres

Louis-Paul Rivest
Université Laval

Alberto Rodríguez Casal
University of Santiago de
    Compostela

Ichiro Ken Shimatani
The Institute of Statistical
    Mathematics

Michael Sørensen
University of Copenhagen

Suvrit Sra
Massachussets Institute of
    Technology

Charles C. Taylor
University of Leeds

Editors

Christophe Ley
Ghent University

Thomas Verdebout
Université libre de Bruxelles

Introduction

Aim of the book

Directional statistics are concerned with data that are directions. The typical supports for directional data are the unit circle and unit (hyper-)sphere, or more generally Riemannian manifolds. The nonlinear nature of these manifolds implies that the classical statistical techniques and tools cannot be used to analyze directional data, and this has given rise to the research flow called directional statistics which has been particularly active over the past two decades.

The present book is intended to be a companion book to our manuscript *Modern Directional Statistics* published by Chapman & Hall/CRC Press in 2017. While the latter book mainly covers theoretical aspects of recent developments in the field, the present book is dedicated to methodological advances and treatments of various modern real data applications.

Content of the Companion Book *Modern Directional Statistics*

In a nutshell, we now summarize the material described in *Modern Directional Statistics*. It begins with a very detailed description of the recently proposed probability distributions for data on the circle, sphere, torus, and cylinder. The book then focuses on more inferential aspects such as nonparametric density estimation, quantile-/depth-based inference, order-restricted inference, rank-based inference, tests of uniformity and symmetry, among others. The Le Cam methodology adapted to directional supports is described in detail and theoretical applications presented. Finally, the book deals with high-dimensional inference on hyperspheres.

Content of the present book

Various modern application areas will be described in this book, as well as the new methods that have been developed to analyze the corresponding directional data. These areas include protein bioinformatics (Chapter 1 by Mardia, Foldager and Frellsen, as well as Chapter 4 by García-Portugués, Golden, Sørensen, Mardia, Hamelryck and Hein), the study of sea regimes (Chapter 3 by Lagona and Chapter 7 by Gelfand, Jona Lasinio and Mastrantonio), biology (Chapter 8 by Abe and Shimatani), the study of wildfires (Chapter 9 by Ameijeiras-Alonso, Crujeiras and Rodríguez Casal), social and behavioural sciences (Chapter 10 by Klugkist, Cremers and Mulder), and machine learning (Chapter 12 by Sra). Specific topics with applications in diverse domains have also been addressed: ambiguous rotations (Chapter 2 by Arnold and Jupp), inference under noisy data (Chapter 5 by Pham Ngoc), the modeling of rotation matrices (Chapter 6 by Rivest and Oualkacha), and nonparametric classification (Chapter 11 by Di Marzio, Fensore, Panzera and Taylor). Finally, Chapter 13 by Pewsey provides an overview of existing R packages that are relevant for directional data analysis.

We wish to thank all contributors to the present book which, we hope, will please the reader and provide further motivation to delve into the passionating field of directional statistics.

1

Directional Statistics in Protein Bioinformatics

Kanti V. Mardia

University of Leeds

Jesper Illemann Foldager

University of Copenhagen

Jes Frellsen

IT University of Copenhagen

CONTENTS

1.1	Introduction	1
1.2	Protein Structure	2
1.3	Protein Geometry	4
1.4	Structure Determination and Prediction	6
	1.4.1 Markov Chain Monte Carlo Simulations of Proteins	8
1.5	Generative Models for the Polypeptide Backbone	10
	1.5.1 Bivariate Angular Distributions	11
	1.5.1.1 Bivariate von Mises	11
	1.5.1.2 Histograms and Fourier Series	11
	1.5.1.3 Mixture of von Mises	12
	1.5.2 A Dynamical Bayesian Network Model: TorusDBN	12
1.6	Generative Models for Amino Acids Side Chains	15
1.7	Discussion	17
	Bibliography	18

1.1 Introduction

Directional statistics has been applied in several different branches of biology [2], including the modelling of periodic properties in biological tissues [19], movement of organisms [47], and in the study of circadian rhythms, such as wake-sleep cycles [32]. In this chapter we will outline several usages of directional statistics in protein bioinformatics, which is a field dealing with the modelling and prediction of the three-dimensional structure of proteins. We will first give the biochemical background for studying these biological macromolecules and then we will outline different approaches to molecular modelling making use of methods from directional statistics. We will conclude the chapter with a discussion on related research and open challenges.

1.2   Protein Structure

In this chapter we will be considering proteins, which are a type of biological macromolecule. Proteins are essential to all living cells and they are often called the workhorses of cells, due to their central roles in cellular structures and activities. The functionality of proteins mainly arises through their structure, and in this section we take a closer look at both the terms used to describe the different levels of structure and some general aspects of what constrains the three-dimensional structure of molecules.

Molecules are made up of atoms connected by covalent bonds. The structure of a molecule consists of the three-dimensional positions of its atoms. At biological temperatures the atoms of a molecule do not remain at fixed positions, but evolve over time. The dynamics of most molecules are ergodic and as a consequence the probability of seeing a specific three-dimensional configuration is the same when sampling a slice from a time trajectory of a single molecule, and sampling a single molecule from a pool of molecules at a specific time. Therefore, simulation studies are often done using only a single molecule. Some molecules are more dynamic and can be found in a wide range of different configurations, while other molecules are more constrained and only vibrate around a single three-dimensional configuration. The distribution depends both on the temperature and the chemical properties of the molecule and its surroundings.

Proteins are biopolymers constructed from a linear sequence of monomeric subunits. In proteins these subunits are *amino acids*, which are joined by covalent bonding to form a single macromolecule. There are 20 different amino acids in naturally occurring proteins. They are identical in the part involved in the polymerization forming the *backbone*, but differ by what is called the *side chain*. The amino acids are compactly represented by a single letter code using the letters A, C, D, E, F, G, H, I, K, L, M, N, P, Q, R, S, T, V, W, and Y. For instance, A is the code for Alanine and C is the code for Cysteine. In the following sections we will use glutamic acid as an example, which is abbreviated E. The specific sequence of amino acids differs between different proteins and is called the *primary structure* of the molecule, which can be represented by a string of characters. An example of a primary structure is shown in Figure 1.1(a).

(a) Primary structure

MQIFVKTLTGKTITLEVEPSDTIENVKAKIQDKEGIPPDQQRLIFAGKQLEDGRTLSDYNIQKESTLHLVLR

(b) Secondary structure

CEEEEECCCCCEEEEECCCCCCHHHHHHHHHHHHHHCCCCCEEEEECCEECCCCCCCCCCCCCCCCEEEEEC

(c) Tertiary structure

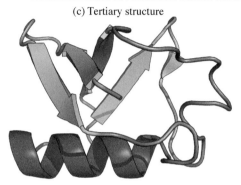

FIGURE 1.1

Protein structure is often divided into four levels, which here is illustrated with human ubiquitin [25]. The *primary structure* (a) is the linear sequence of amino acids, here shown by their one letter code. The *secondary structure* (b) is a classification of amino acids into local structural elements that they are part of. The secondary structure classification typically consists of three or eight classes; here we used the three classes helix (H) and strand (E) and coil (C). The *tertiary structure* (c) is the three-dimensional positions of all the atoms in the protein, which here is shown using a simplified cartoon representation. The cartoon is colored according to secondary structure, such that the helix is red, the strands are green, and coil regions are gray. The *quaternary structure* is the structure formed by multiple protein subunits (not shown here). The figure is reproduced from Frellsen [14].

Proteins typically fold into complex three-dimensional structures, as illustrated in Figure 1.1(c). The three-dimensional structure of a single protein is referred to as the *tertiary structure*. The tertiary structure mainly depends on the primary structure, i.e., the sequence of amino acids in the polymer, as the different amino acids have different biochemical properties. The function of a protein is normally dictated by its three-dimensional structure. However, experimentally determining the three-dimensional structure of a protein is typically a difficult and expensive process, and therefore substantial research efforts have been invested in developing computational methods for predicting the tertiary structure of proteins given their primary structure.

The individual subunits of proteins can be classified into reoccurring local structural patterns; this is called the *secondary structure*. Based on the hydrogen bond patterns in proteins each amino acid is classified into being a member of a helix, a sheet, or a coil as illustrated in Figure 1.1(b).

The covalent bonds in a molecule severely constrain the relative positions of the atoms. First of all, covalent bond lengths vary little, constraining the distance between bound atoms. Further, the geometry of the atoms covalently bound to a central atom is largely fixed, and depends on chemical properties that we will not describe here. In organic chemistry the most occurring geometry is that of a carbon atom with four elements bound, which forms a tetrahedron with the carbon atom at the center. There can be small deviations from a perfect tetrahedron, but this relatively fixed geometry translates to small variations in bond angles. The main degree of freedom in larger molecules, including proteins, comes from rotation around covalent bonds, which can be measured by dihedral angles [5]. The variations in the covalent bond angles and lengths are less significant, and sometimes fixed, ideal values are used when modelling proteins. In the following section (1.3) we will review the geometry and dihedral angles in proteins, and then we will return to the problem of structure prediction in Section 1.4.

1.3   Protein Geometry

The sequence of amino acids in a protein are linked together by *peptide bonds* between the carboxyl group of one amino acid and the amino group of the following, which form the *polypeptide backbone* of the protein. After amino acids form polypeptides the correct term for them is *amino acid residues* as the chemical groups they are named by have been modified. However, they are commonly still referred to as amino acids, which we will also do in this chapter when there is no ambiguity. As illustrated in Figure 1.2, the polypeptide backbone consists of a repeated sequence of three atoms: a nitrogen (N), a carbon (C_α), and another carbon (C). If we index the atoms by their position in the amino acid sequence, we can write the sequence of backbone atoms as

$$\mathrm{N}^{(1)} - C_\alpha^{(1)} - C^{(1)} - \mathrm{N}^{(2)} - C_\alpha^{(2)} - C^{(2)} - \cdots - \mathrm{N}^{(n)} - C_\alpha^{(n)} - C^{(n)}.$$

In this sequence, the peptide bonds are $C^{(i)} - N^{(i+1)}$. The atoms of an amino acid that are not part of the backbone are part of the *side chain*, which is attached to C_α. The backbone is identical across all 20 amino acids, while the side chains are different, and it is this difference that gives the amino acids different biochemical properties. For each amino acid, i, there are three dihedral angles in the backbone:

- $\phi^{(i)}$ formed by $C^{(i-1)} - N^{(i)} - C_\alpha^{(i)} - C^{(i)}$,

- $\psi^{(i)}$ formed by $N^{(i)} - C_\alpha^{(i)} - C^{(i)} - N^{(i+1)}$ and

- $\omega^{(i)}$ formed by $C_\alpha^{(i)} - C^{(i)} - N^{(i+1)} - C_\alpha^{(i+1)}$.

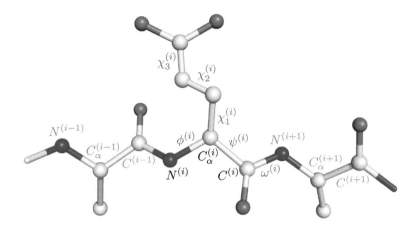

FIGURE 1.2
A protein fragment with selected atom and dihedral angle names. Atoms are colored by element: nitrogen, blue; carbon, gray; oxygen, red; and hydrogens are left out. Backbone atom names are in gray for the noncentral amino acids and black for the central amino acid. The central amino acid is a glutamic acid with dihedral angle names in red. The glutamic acid has the three dihedral angles in the side chain, χ_1, χ_2, and χ_3. The dihedral angles are shown on the covalent bond they specify rotation around. The figure is adapted from Frellsen [14].

The three dihedral angles of the backbone are illustrated in Figure 1.2.

The peptide bonds have a partial double-bond character, which means that the bond is planar and the ω angle is concentrated around two modes. The two modes are denoted *cis* isomer with $\omega \approx 0°$ and *trans* isomers with $\omega \approx 180°$. The main flexibility in the backbone is due to the rotational degrees of freedom of the ϕ and ψ angles. However, due to stereochemical properties of the polypeptide chain, only a limited number of conformations of these angles are energetically allowed. This is typically illustrated in a Ramachandran plot [50], a scatter plot where the ψ angle is plotted against the ϕ angle for all amino acids in a set of proteins. The Ramachandran plot from the original work by Ramachandran et al. [50] is shown in Figure 1.4. The Ramachandran plot is normally divided into energetically favorable and energetically disallowed regions, making it a useful tool in structure quality assessment. If most of the amino acids of a protein are not in the favorable regions, the model is likely in poor agreement with the real protein structure. Figure 1.3 shows the Ramachandran plot in the plan and on the two-dimensional torus.

There are between zero and four free dihedral angles in the side chain depending on the amino acid type. These angles are denoted χ_1, χ_2, χ_3, and χ_4,

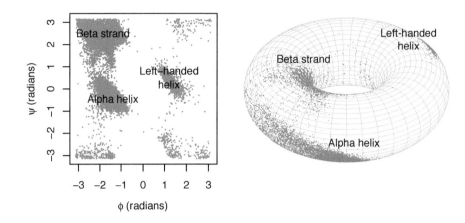

FIGURE 1.3

Two Ramachandran plots constructed from the Top 100 database [58]. The left is a scatter plot in the plane and the right is a scatter plot on the two-dimensional torus. In the torus plot, ϕ is the toroidal angle and ψ is the poloidal angle. Regions for right-handed α-helices, left-handed α-helices, and β-strands are shown on both plots. Figure reproduced from Mardia and Frellsen [38].

and they are also constrained to a limited number of allowed conformations. The side chain angles are shown in Figure 1.2.

1.4   Structure Determination and Prediction

There exist a number of experimental biophysical methods for determining the structure of proteins, where the two most predominantly used methods are X-ray crystallography and NMR spectroscopy. Both methods obtain measurements of physical quantities that can be used to infer the tertiary structure, often at atomic level resolution. Predicting the structure of proteins is a major research challenge in molecular biology, computational chemistry and bioinformatics. The problem can be formulated as: given the primary structure of a molecule (the sequence of amino acids), predict the secondary and the tertiary structure.

   Due to very active research in the past decades, the secondary structure of proteins can be predicted with great accuracy. Traditionally, artificial neural networks have been used for secondary structure prediction [24], and today

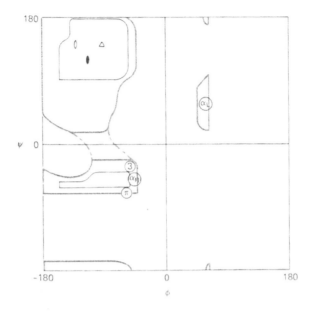

FIGURE 1.4
The original Ramachandran plot reproduced from Ramachandran et al. [50]. The regions are labelled according to secondary structure, including α_R for right-handed helix, α_L for left-handed helix, and the upper left region for strands.

the three classes of secondary structure can be predicted with more than 80% accuracy [57].

Tertiary structure prediction is a more challenging problem. If the molecule has a close sequence homolog (i.e., a molecule with shared ancestry) for which the structure is known, the tertiary structure of the molecule can often be well predicted by using the structure of the homolog as a scaffold. This approach is denoted *homology modelling* or *template-based modelling*, and if the sequence similarity between the two molecules is high the prediction can be quite accurate [59]. If no close sequence homolog exists, tertiary structure prediction is hard and this is the problem addressed by so-called *de novo* structure prediction.

De novo structure prediction methods often make use of a parametrized physical force field, which facilitates calculating the potential energy $U(\mathbf{x})$ of the molecule from the 3D Cartesian coordinates $\mathbf{x} \in \mathbb{R}^{3m}$ of all the m atoms in the molecule. Popular force fields include AMBER [49], CHARMM [7], and OPLS [30]. If we assume that the volume and temperature T of the system are constant, then according to statistical physics [13, 20, 51] the probability of observing a particular configuration $\mathbf{x}$ of the molecule can be expressed by the Boltzmann distribution

$$p(\mathbf{x} \mid \beta) = \frac{\exp(-\beta U(\mathbf{x}))}{Z_\beta}, \tag{1.1}$$

where $Z_\beta = \int_{\mathbb{R}^{3m}} \exp(-\beta U(\mathbf{x})) \, d\mathbf{x}$ is the normalization constant (partition function in physics), $\beta = (k_b T)^{-1}$ is the thermodynamic beta, k_b the Boltzmann constant, and T the temperature. In Equation (1.1) we left out the kinetic contribution, since it is trivial in Cartesian space [13].

The Boltzmann distribution can be used to make inference about the system. For instance, the tertiary structure(s) of the molecule can then be inferred from $p(\mathbf{x} \mid \beta)$ by finding the mode(s) of the distribution. The *Helmholtz free energy* at a given temperature can be calculated for the normalization constant as $F(\beta) = -\beta^{-1} \ln(Z_\beta)$ [13, 51]. Furthermore, given a function $g : \mathbb{R}^{3m} \to \mathbb{R}$, we may be interested in calculating the expectation

$$\mathbb{E}_\beta[g] = \int_{\mathbb{R}^{3m}} g(\mathbf{x}) p(\mathbf{x} \mid \beta) \, d\mathbf{x}. \tag{1.2}$$

For instance we may want to find the mean potential energy $\mathbb{E}_\beta[U]$ at given β-value. This can be used to calculate the *thermodynamic entropy* of the system, $S(\beta) = \beta k_b(\mathbb{E}_\beta[U] - F(\beta))$ [13, 51].

However, for all nontrivial energy functions, inference in $p(\mathbf{x} \mid \beta)$ is analytically intractable. Molecular dynamics (MD) is a simulation-based method for probing $p(\mathbf{x} \mid \beta)$, which assumes that the system follows Newton's laws of motion and performs inference by numerically solving Newton's equations. These simulations assume that $\nabla_{\mathbf{x}} U(\mathbf{x})$ is readily available, which is usually the case. While MD methods have been very successful [53], their main disadvantage is that they are very computationally demanding. In practice, the time step size in an MD simulation is of the magnitude of femtoseconds (10^{-12}s), while proteins fold in the order of magnitudes of microseconds (10^{-6}s) to seconds (s). This means that simulating a single folding event with MD simulations requires millions to trillions of simulation steps.

1.4.1 Markov Chain Monte Carlo Simulations of Proteins

Monte Carlo (MC) based methods do not have the time scale limitation of MD. They work by drawing L samples $\{\mathbf{x}_\ell\}_{\ell=1}^{L}$ from $p(\mathbf{x} \mid \beta)$, and then approximating the integral $\mathbb{E}_\beta[g]$ in equation (1.2) by

$$\hat{I}_L(g) = \frac{1}{L} \sum_{\ell=1}^{L} g(\mathbf{x}_\ell). \tag{1.3}$$

It can be shown that $\hat{I}_L(g)$ is an unbiased estimator for $\mathbb{E}_\beta[g]$ and that it almost surely converges to $\mathbb{E}_\beta[g]$ as $L \to \infty$ [1]. This is known as the *Monte*

Carlo principle. The principle can also be used to obtain estimates of ratios of normalization constants: An unbiased estimate of

$$\frac{Z_{\beta'}}{Z_\beta} = \int_{\mathbb{R}^{3m}} \frac{\exp(-\beta'U(\mathbf{x}))}{Z_\beta} \, d\mathbf{x} = \int_{\mathbb{R}^{3m}} \frac{\exp(-\beta'U(\mathbf{x}))}{\exp(-\beta U(\mathbf{x}))} p(\mathbf{x} \mid \beta) \, d\mathbf{x} \qquad (1.4)$$

can be obtained by $\hat{I}_L \left(\frac{\exp(-\beta'U(\mathbf{x}))}{\exp(-\beta U(\mathbf{x}))} \right)$.

As mentioned before, it is usually not straightforward to efficiently generate samples from $p(\mathbf{x} \mid \beta)$. In Markov chain Monte Carlo (MCMC) this is resolved by constructing a Markov chain that has $P(\mathbf{x} \mid \beta)$ as stationary distribution. In principle this can be done using the Metropolis–Hastings (MH) algorithm [23]. In this algorithm a sequence of states $\{\mathbf{x}_\ell\}_{\ell=1}^L$ is generated one at a time. At the $(\ell+1)^{\text{th}}$ time step, a new state $\mathbf{x}'$ is sampled from a proposal distribution $q(\mathbf{x}' \mid \mathbf{x}_\ell)$ and then either accepted or rejected as the $(\ell+1)^{\text{th}}$ realization of the chain. The probability of accepting the proposed state $\mathbf{x}'$ is

$$\alpha(\mathbf{x}' \mid \mathbf{x}_\ell) = \min \left(1, \frac{p(\mathbf{x}' \mid \beta) \cdot q(\mathbf{x}_\ell \mid \mathbf{x}')}{p(\mathbf{x}_\ell \mid \beta) \cdot q(\mathbf{x}' \mid \mathbf{x}_\ell)} \right). \qquad (1.5)$$

If the proposed state is accepted, then $\mathbf{x}_{\ell+1} = \mathbf{x}'$, otherwise the chain stays in the previous state, $\mathbf{x}_{\ell+1} = \mathbf{x}_\ell$.

A special case of MH is the Metropolis algorithm [41]. In this algorithm a symmetric proposal distribution is used, such that $\frac{q(\mathbf{x}_\ell \mid \mathbf{x}')}{q(\mathbf{x}' \mid \mathbf{x}_\ell)} = 1$, which simplifies Equation (1.5). A proposal is then accepted if $\mathbf{x}'$ is at least as probable as $\mathbf{x}_\ell$, and otherwise $\mathbf{x}'$ is accepted with the probability $\frac{p(\mathbf{x}' \mid \beta)}{p(\mathbf{x}_\ell \mid \beta)}$. This choice of acceptance criterion still ensures the correct stationary distribution. When the proposal distribution is not symmetric, the ratio of proposals in (1.5) can be seen as a correction factor that allows the chain to have the correct stationary distribution no matter the choice of proposal distribution.

For most molecular systems the Markov chain constructed by the MH algorithm will not mix well, i.e., the sample will be highly correlated and the chain will only explore the relevant part of the sample space slowly. So in practice more advanced MCMC methods are used [6, 18], see for example the reviews by Iba [26], Murray [43], or Ferkinghoff-Borg [13].

One of the challenges in MCMC based simulation is designing a good proposal distribution $q(\mathbf{x}' \mid \mathbf{x}_\ell)$. Many methods [6, 27] assume that bond angles and bond lengths of the molecule are constant and represent the molecule by internal dihedral angles $\mathbf{\Phi} \in \mathbb{T}^p$, taking value on the p-dimensional hypertorus $\mathbb{T}^p = [-\pi, \pi)^p$. Changes to the molecule are then proposed as changes in dihedral angles and the proposal distribution takes the form $q(\mathbf{\Phi}' \mid \mathbf{\Phi}_\ell)$.

The most straightforward proposal distributions to use are concentrated Gaussian perturbations [27]. However, a proposal distribution is better when it is closer to the stationary distribution, and to take advantage of this, most proposal distributions incorporate protein structure information. A simple way to achieve this is by proposing angles, or stretches of angles, observed in real proteins. Such methods are very successful, but come with the statistical problem that $q(\mathbf{\Phi}' \mid \mathbf{\Phi}_\ell)$ is not meaningful for continuous $\mathbf{\Phi}$. The solution is often to disregard the term $\frac{q(\mathbf{\Phi}_\ell \mid \mathbf{\Phi}')}{q(\mathbf{\Phi}' \mid \mathbf{\Phi}_\ell)}$ in equation (1.5), and abide that this changes the stationary distribution of the chain and therefore results in biased Monte Carlo estimators, c.f. Equation (1.3). The standard choices for such proposal distributions are to use fragment libraries for backbone angles [28, 29, 54], and rotamer libraries for side chain angles. Rotamer libraries exist in both backbone-independent version [34] and backbone-dependent version [11, 31, 52], where the frequency of each rotamer depends on the backbone angles (ϕ, ψ).

In the following sections we are going to review a number of tractable statistical models for describing the distributions over the dihedral angles in proteins. Ideally we would be interested in a fully tractable model for the conditional distribution of the dihedral angles in a protein $p(\mathbf{\Phi} \mid \mathbf{a})$ given the amino acid sequence $\mathbf{a}$. However, tractable models have only been developed for marginals or conditionals of this distribution.

1.5   Generative Models for the Polypeptide Backbone

Generative models for protein structure draw samples from the joint probability distribution of internal angles. As discussed in Section 1.3, the main degrees of freedom in the protein backbone are the angles ϕ, ψ, and ω. Due to the partial double bond character of the peptide bond, ω can be closely approximated by a discrete two-state variable, leaving most of the variation in the Ramachandran angles (ϕ, ψ). Any generative model either implicitly or explicitly defines the joint probability distribution over all the backbone dihedral angles, $p(\mathbf{\Psi})$, where $\mathbf{\Psi} \in \mathbb{T}^r$ and r is the total number of backbone angles. From a modelling perspective, the full joint distribution is difficult to work with directly without simplifying assumptions, both in terms of functional form and dependency structure. Reducing the problem to independent distributions of Ramachandran angle pairs on $\mathbb{T}^2$ is a good starting point for modelling. A generative model for protein structure must as a minimum recuperate the Ramachandran empirical distribution. However, such a generative model turns out to be very poor for predicting local protein structure, as there is a strong dependency between (ϕ, ψ)-angle pairs at different positions in the protein sequence. A tractable way of introducing dependency structure was introduced by Boomsma et al. [4] using a hidden Markov model, denoted TorusDBN, which encodes dependency along the sequence while any (ϕ, ψ)-pair remain conditionally independent from other angle pairs given the sequence of hidden (latent) variables.

1.5.1 Bivariate Angular Distributions

There are multiple options for the functional form of angle distributions involved in protein structures. Here we focus on bivariate forms, due to the strong dependency between ϕ and ψ observed in Ramachandran plots. Bivariate distributions can be used as a basic building block for more complicated models such as the generative model for protein backbones.

1.5.1.1 Bivariate von Mises

The von Mises distribution can be used for univariate angular data, and has the attractive feature that it is a close approximation to the wrapped normal distribution. By analogy to the normal distribution we would like a bivariate von Mises distribution with five parameters: two mean parameters and three parameters for variance and covariance. However, the "full" bivariate von Mises distribution introduced by Mardia [36] has eight parameters,

$$
\begin{aligned}
p(\phi, \psi \mid \mu, \nu, \kappa_1, \kappa_2, \mathbf{A}) \propto \exp(&\kappa_1 \cos(\phi - \mu) + \kappa_2 \cos(\psi - \nu) \\
&+ \left[\cos(\phi - \mu), \sin(\phi - \mu)\right] \mathbf{A} \left[\cos(\psi - \nu), \sin(\psi - \nu))\right]^{\mathsf{T}}),
\end{aligned} \quad (1.6)
$$

where μ and ν are mean parameters, κ_1 and κ_2 are concentration parameters, and $\mathbf{A} \in \mathbb{R}^{2 \times 2}$ is a two-by-two matrix. Different submodels of Equation (1.6) have been proposed, of which most attention has been given to the sine model by Singh et al. [55]

$$
\begin{aligned}
p_\mathrm{s}(\phi, \psi \mid \mu, \nu, \kappa_1, \kappa_2, \lambda) \\
\propto \exp(\kappa_1 \cos(\phi - \mu) + \kappa_2 \cos(\psi - \nu) + \lambda \sin(\phi - \mu) \sin(\psi - \nu)),
\end{aligned} \quad (1.7)
$$

and the cosine model explored by Mardia et al. [40]

$$
\begin{aligned}
p_\mathrm{c}(\phi, \psi \mid \mu, \nu, \kappa_1, \kappa_2, \kappa_3) \\
\propto \exp(\kappa_1 \cos(\phi - \mu) + \kappa_2 \cos(\psi - \nu) + \kappa_3 \cos(\phi - \mu - \psi + \nu)).
\end{aligned} \quad (1.8)
$$

Both submodels have five parameters and both approximate a normal distribution for higher concentrations. For details on their properties see Mardia et al. [40], Mardia and Jupp [39], Mardia and Frellsen [38], or Ley and Verdebout [33].

1.5.1.2 Histograms and Fourier Series

Initially models for (ϕ, ψ) were histogram based, which involves discretising $\mathbb{T}^2$ into regions and assuming equal density within a region. Histogram methods are very flexible, as the bin size can be arbitrarily small. However, the number of parameters is equal to the number of bins, and grows large for finer meshes. Besides the large number of parameters needed, a histogram using an adequately fine mesh would also suffer from a lack of precision due to the limited number of available protein structures according to Pertsemlidis et al. [48]. Continuous models can have a more compact formulation, and have the inherent advantage of being smooth.

A simple unimodal distribution cannot be used to approximate the quite complicated multimodal Ramachandran empirical distribution, shown in Figure 1.3. Pertsemlidis et al. [48] proposed a parametrized continuous model for the distribution using two-dimensional Fourier series for the log likelihood, which can approximate any density function arbitrarily well by increasing the order of the series. The density across the identity lines was ensured to be continuous by using a period of 2π for the basis functions. A good approximation to the Ramachandran density required 80 parameters [48]. However, the parameters are not easily interpretable in a biological context.

1.5.1.3 Mixture of von Mises

Mardia et al. [40] suggested using a mixture of bivariate von Mises distributions for the Ramachandran empirical distribution, where either Equation (1.8) or (1.7) was the building block for the individual components. The graphical representation of the model is shown in Figure 1.6 and the mixture is defined as

$$p(\phi, \psi \mid \boldsymbol{\mu}, \boldsymbol{\nu}, \boldsymbol{\kappa}_1, \boldsymbol{\kappa}_2, \boldsymbol{\lambda}, \boldsymbol{w}) = \sum_i w^{(i)} p_s(\phi, \psi \mid \mu^{(i)}, \nu^{(i)}, \kappa_1^{(i)}, \kappa_2^{(i)}, \lambda^{(i)}), \quad (1.9)$$

where i denotes the different components, $w^{(i)}$ is a positive weight such that $\sum_i w^{(i)} = 1$, and $p_s(\cdot)$ is the sine based bivariate von Mises distribution (1.7), as it is used by Mardia [37]. A mixture can also be constructed using the cosine model (1.8) as seen in Mardia et al. [40] and in TorusDBN described later [4]. The mixture model needs multiple components to approximate the Ramachandran empirical distribution, but the components are consistent with partitions traditionally assigned to Ramachandran plots [40]. A maximum likelihood fit of the model can be obtained using the expectation-maximization (EM) algorithm [10], and a fitted model using the sine bivariate von Mises is seen in Figure 1.5.

A mixture model is also compatible with how the Ramachandran empirical distribution is thought to arise. In a protein the local context dictates and restricts the possible dihedral angles for a residue. If the possible contexts with reasonable accuracy can be discretised, then each discrete context corresponds to a mixture component, and the distribution over Ramachandran angles is obtained by marginalization. As we will see in the next section, this idea was used in a hidden Markov model, where a discrete hidden state encapsulates the contexts [4].

1.5.2 A Dynamical Bayesian Network Model: TorusDBN

Boomsma et al. [4] proposed a generative model for backbone angles that is similar to a hidden Markov model, but it has more emission nodes. This type

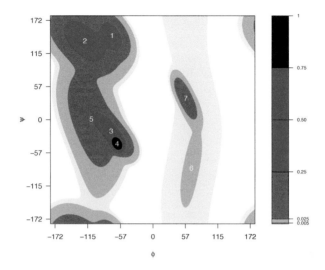

FIGURE 1.5
The Ramachandran plot for a mixture model, Equation (1.9), fitted to a subset
of the Top 500 database [35]. The model is a mixture of seven bivariate von
Mises sine components. Figure reproduced from Mardia [37].

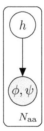

FIGURE 1.6
Graph representation of a mixture model. h is a discrete variable, each state
emitting a distinct distribution for (ϕ, ψ). N_{aa} is the total number of amino
acids.

of model belongs to the broader group of models called dynamical Bayesian
networks [9, 42], however for most purposes it is simpler to consider it a hidden Markov model. The structure of the model can be seen in Figure 1.7.
The model consists of a sequence of hidden nodes, h_i, each connecting to four
emission nodes. The sequence represents the protein chain and N denotes the
sequence length. Each hidden node is a discrete variable and can only take on

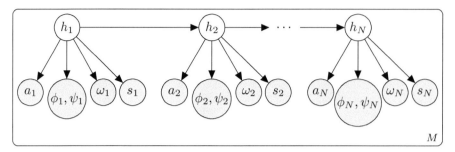

FIGURE 1.7
The independence structure for the protein backbone model TorusDBN by Boomsma et al. [4]. The lack of an arrow between two nodes indicates they are conditionally independent. Hidden nodes are connected to emission nodes for amino acid (a), dihedral angles (ϕ, ψ), cis/trans configuration (ω), and secondary structure (s). M is the total number of proteins and N is the length of individual proteins.

a limited number of states. Each hidden state corresponds to a distinct emission distribution for amino acid (a), Ramachandran angles (ϕ, ψ), cis/trans conformation (ω), and secondary structure (s). The model can be used in multiple ways; it can both be used for evaluating probabilities and generate samples for whole proteins or parts of a protein. When used as a generative model the hidden state sequence is sampled first, followed by sampling of the emission nodes. This method is also applicable for partial resampling, where nodes are resampled conditioned on the remaining fixed part of the sequence.

If some of the emission nodes are known a priori, they can be used to inform the sequence of hidden states. In protein structure prediction the amino acid sequence is known, and using Bayes theorem the hidden state sequence can be sampled conditioned on the observed amino acid sequence. This can be done for any of the emission nodes making the model ideal for generating proposals in MCMC sampling.

The factorization of the joint probability distribution can be read directly from the directed acyclic graph (DAG) in Figure 1.7. Each node contributes with the probability of the node itself conditioned on any input nodes. Starting with the top node h_1 the full factorization reads

$$p(\boldsymbol{a}, \boldsymbol{\phi}, \boldsymbol{\psi}, \boldsymbol{\omega}, \boldsymbol{s}, \boldsymbol{h}) = p(h_1)p(a_1 \mid h_1)p(\phi_1, \psi_1 \mid h_1)p(\omega_1 \mid h_1)p(s_1 \mid h_1)$$
$$p(h_2 \mid h_1)p(a_2 \mid h_2)\ldots p(s_2 \mid h_2)$$
$$\vdots$$
$$p(h_N \mid h_{N-1})p(a_N \mid h_N)\ldots p(s_N \mid h_N).$$

(1.10)

In the TorusDBN model, the transition probabilities $p(h_i \mid h_{i-1})$ are assumed to be a categorical distribution, which means that the parameters can be described by a $\mathbb{R}^{K \times K}$ transition matrix, where K is the number of hidden states.

Similarly, the emission probabilities for the amino acid $p(a_i \mid h_i)$ and the secondary structure $p(s_i \mid h_i)$ are categorical distributions. By assuming that the ω-angle is highly concentrated in the two modes, the cis/trans conformation probability $p(\omega_i \mid h_{i-1})$ can be assumed to be binomial. Finally, the probability of the Ramachandran angles $p(\phi_i, \psi_i \mid h_i)$ is assumed to be a cosine model bivariate von Mises distribution.

As implied by the name, the hidden node sequence is not directly observable, and when evaluating the probability of a protein structure the marginalized distribution is used,

$$
\begin{aligned}
p(\boldsymbol{a}, \boldsymbol{\phi}, \boldsymbol{\psi}, \boldsymbol{\omega}, \boldsymbol{s}) &= \sum_{\boldsymbol{h}} p(\boldsymbol{a} \mid \boldsymbol{h}) p(\boldsymbol{\phi}, \boldsymbol{\psi} \mid \boldsymbol{h}) p(\boldsymbol{\omega} \mid \boldsymbol{h}) p(\boldsymbol{s} \mid \boldsymbol{h}) p(\boldsymbol{h}) \\
&= \sum_{\boldsymbol{h}} \prod_i p(a_i \mid h_i) p(\phi_i, \psi_i \mid h_i) p(\omega_i \mid h_i) p(s_i \mid h_i) p(h_i \mid h_{i-1}),
\end{aligned}
\tag{1.11}
$$

where we conveniently define $p(h_1 \mid h_0) = p(h_1)$ since h_1 has no incoming edges. The calculation scales poorly as $\boldsymbol{h}$ has K^N possible states where K is the number of possible states for each hidden node, and N is the length of the sequence. Fortunately the complexity can be reduced by using the forward-backward algorithm that takes advantage of dynamic programming, see, e.g., Durbin et al. [12]. Similarly, the forward-backtrack algorithm can be used for efficient sampling [8].

The Markov property makes for a simple model but comes at the cost of a short memory. A larger state space could make up for this, but would not scale well. The model can generate protein structures that locally are protein-like which combined with the ability to evaluate exact probabilities makes it a perfect proposal distribution for MCMC sampling.

The parameters of the TorusDBN model can be estimated from data. A maximum likelihood estimate of the parameters can be obtained using the EM algorithm [10] or the stochastic EM algorithm [45].

1.6 Generative Models for Amino Acids Side Chains

In the previous section, we described generative models for the protein backbone. In this section we will review generative models for the amino acids side chains. As mentioned earlier, the main degrees of freedom in the amino acids side chains are the dihedral angles denoted χ_1, χ_2, χ_3, and χ_4. The number of angles varies between amino acids, for example, glutamic acid has three χ-angles, as illustrated in Figure 1.2.

Here we will consider a continuous model for the amino acid side chain angles called BASILISK [22]. The independence assumptions in this model follow an input output hidden Markov model (IOHMM) structure [3, 15, 42], which has previously also been used successfully for modelling the dihedral angles in

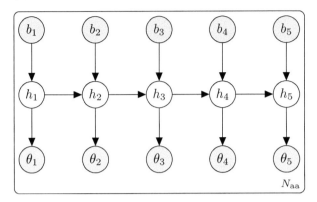

FIGURE 1.8

The general independence structure assumed in the BASILISK model [22] for amino acid side chain angles. In the model $\boldsymbol{\theta}$ are dihedral angles, $\boldsymbol{h}$ are hidden (latent) variables, $\boldsymbol{b}$ are index variables, and N_{aa} is the total number of amino acids. Here the model is unrolled to five slices; the dihedral angles in the first two slices are used for backbone angles, while the dihedral angles in the remaining slices are used for side chain angles. The assumed independence structure is often denoted an input output hidden Markov model [3, 15, 42]

ribonucleic acids (RNA) [15]. The independence assumptions of the IOHMM are illustrated in Figure 1.8 and correspond to the following factorization

$$p(\boldsymbol{\theta}, \boldsymbol{h}, \boldsymbol{b}) = p(b_1)p(h_1 \mid b_1)p(\theta_1 \mid h_1) \prod_{j>1} p(b_j)p(h_j \mid b_j, h_{j-1})p(\theta_j \mid h_j), \quad (1.12)$$

where $\boldsymbol{\theta}$ are dihedral angles, $\boldsymbol{h}$ are hidden (latent) variables, and $\boldsymbol{b}$ are index variables encapsulating information about the amino acid type and the angle number. In this model the conditional distribution of the angles given the hidden variable, $p(\theta_j \mid h_j)$, is univariate von Mises distributed, while the transition probabilities $p(h_j \mid b_j, h_{j-1})$ follow a categorical distribution that can be summarized in a three-dimensional real matrix.

    Using five slices of this model, we can express the joint probability distribution for the angles, hidden variables and indices for glutamic acid (illustrated in Figure 1.2)

$$\begin{aligned}
p(\boldsymbol{\theta} = [\phi, \psi, & \chi_1, \chi_2, \chi_3], \boldsymbol{h}, \boldsymbol{b} = [\phi, \psi, \chi_{(1,\mathrm{E})}, \chi_{(2,\mathrm{E})}, \chi_{(3,\mathrm{E})}]) \\
&= p(b_1 = \phi)p(h_1 \mid b_1 = \phi)p(\theta_1 = \phi \mid h_1) \\
&\quad p(b_2 = \psi)p(h_2 \mid b_2 = \psi, h_1)p(\theta_2 = \psi \mid h_2) \\
&\quad p(b_3 = \chi_{(1,\mathrm{E})})p(h_3 \mid b_3 = \chi_{(1,\mathrm{E})}, h_2)p(\theta_2 = \chi_1 \mid h_3) \\
&\quad p(b_4 = \chi_{(2,\mathrm{E})})p(h_4 \mid b_4 = \chi_{(2,\mathrm{E})}, h_3)p(\theta_2 = \chi_2 \mid h_4) \\
&\quad p(b_5 = \chi_{(3,\mathrm{E})})p(h_5 \mid b_5 = \chi_{(3,\mathrm{E})}, h_4)p(\theta_2 = \chi_3 \mid h_5)
\end{aligned} \quad (1.13)$$

where slanted Greek letters are angle values and upright Greek letters are indices. For instance, setting the index $b_1 = \phi$ tells the model that the output of the first slices, θ_1, is the backbone angle ϕ. Similar setting the index $b_4 = \chi_{(2,\mathrm{E})}$, tells the model that the output of the forth slices, θ_4, is the χ_2 side chain angle in a glutamic acid, which is abbreviated E.

If we marginalize out the hidden variable, we can obtain the conditional distribution of the side chain angles, given the backbone angles using the expression

$$
\begin{aligned}
p(\boldsymbol{\theta}_{3,\dots} = \chi \mid \boldsymbol{\theta}_{1,2} &= [\phi, \psi], \boldsymbol{b}) \\
&= \frac{p(\boldsymbol{\theta} = [\phi, \psi, \chi] \mid \boldsymbol{b})}{p(\boldsymbol{\theta}_{1,2} = [\phi, \psi] \mid \boldsymbol{b})} \\
&= \frac{\sum_{\boldsymbol{h}} p(\boldsymbol{\theta} = [\phi, \psi, \chi] \mid \boldsymbol{h}) p(\boldsymbol{h} \mid \boldsymbol{b})}{\sum_{\boldsymbol{h}_{1,2}} p(\boldsymbol{\theta}_{1,2} = [\phi, \psi] \mid \boldsymbol{h}_{1,2}) p(\boldsymbol{h}_{1,2} \mid \boldsymbol{b})},
\end{aligned}
\tag{1.14}
$$

where the sum runs over all possible latent variable sequences, $\boldsymbol{\theta}_{3,\dots}$ are the appropriate number of side chain angles and the vector value of the index variable $\boldsymbol{b}$ is assumed to be selected accordingly. The sums in Equation (1.14) can efficiently be calculated using the forward algorithm, and efficient sampling can be done using the forward-backtrack algorithm [8]. Similar to TorusDBN, a maximum likelihood estimate for the parameters of the model can be obtained using (stochastic) EM [10, 45].

The BASILISK model can be used as a proposal distribution for the side chain angles in MCMC sampling of proteins. In such a setup, the TorusDBN model described in Section 1.5.2 can be used for proposing the backbone angles, and the BASILISK model can be used for proposing the side chain angles given the backbone angles using Equation (1.14).

In recent work by Navarro et al. [44] it was suggested modelling the conditional distribution of the side chain angles given the backbone angles using a multivariate Generalized von Mises (mGvM) based regression model. In this work, the first side chain angle of aspartate was modelled as a function of backbone angles, and the preliminary results were promising.

1.7 Discussion

In this chapter we have presented several models for the dihedral angles in proteins. They all work by introducing some simplifying assumptions, both in terms of functional form and dependency structure. This means that the models will not generate folded proteins. Instead, TorusDBN will generate samples that *locally* along the polypeptide chain will have a realistic structure, and BASILISK will generate samples that are *locally* compatible with the given backbone angles. As described earlier, this means that these models can be used as proposal distributions for MCMC sampling of proteins.

The models can also be used as a priori for protein structure determination from data. For instance Olsson et al. [46] showed that these models both improved the speed and accuracy in inferential structure determination from NMR data. Also, the models can be combined with global models using the so-called *reference ratio method* [16, 17, 21] to produce a full global probabilistic model of proteins, as shown by Valentin et al. [56]. The models are implemented in the open source simulation framework PHAISTOS [6].

Bibliography

[1] C. Andrieu, N. De Freitas, A. Doucet, and M. I. Jordan. An introduction to MCMC for machine learning. *Machine Learning*, 50(1-2):5–43, 2003.

[2] E. Batschelet. *Circular statistics in biology*. Academic Press Inc, 1981.

[3] Y. Bengio and P. Frasconi. An input output HMM architecture. In G. Tesauro, D. S. Touretzky, and T. K. Leen, editors, *Advances in Neural Information Processing Systems 7*, volume 7. MIT press, 1995.

[4] W. Boomsma, K. V. Mardia, C. C. Taylor, J. Ferkinghoff-Borg, A. Krogh, and T. Hamelryck. A generative, probabilistic model of local protein structure. *Proceedings of the National Academy of Sciences*, 105(26): 8932–8937, July 2008.

[5] W. Boomsma, J. Frellsen, and T. Hamelryck. Probabilistic models of local biomolecular structure and their applications. In T. Hamelryck, K. Mardia, and J. Ferkinghoff-Borg, editors, *Bayesian Methods in Structural Bioinformatics*, pages 233–254. Springer Berlin Heidelberg, Berlin, Heidelberg, 2012.

[6] W. Boomsma, J. Frellsen, T. Harder, S. Bottaro, K. E. Johansson, P. Tian, K. Stovgaard, C. Andreetta, S. Olsson, J. B. Valentin, L. D. Antonov, A. S. Christensen, M. Borg, J. H. Jensen, K. Lindorff-Larsen, J. Ferkinghoff-Borg, and T. Hamelryck. PHAISTOS: A framework for Markov chain Monte Carlo simulation and inference of protein structure. *Journal of Computational Chemistry*, 34(19):1697–1705, 2013.

[7] B. R. Brooks, R. E. Bruccoleri, B. D. Olafson, D. J. States, S. Swaminathan, and M. Karplus. CHARMM: A program for macromolecular energy, minimization, and dynamics calculations. *Journal of Computational Chemistry*, 4(2):187–217, 1983.

[8] S. L. Cawley and L. Pachter. HMM sampling and applications to gene finding and alternative splicing. *Bioinformatics*, 19(90002):36ii–41, 2003.

[9] T. Dean and K. Kanazawa. A model for reasoning about persistence and causation. *Computational Intelligence*, 5(2):142–150, 1989.

[10] A. P. Dempster, N. M. Laird, and D. Rubin. Maximum likelihood from incomplete data via the EM algorithm. *Journal of the Royal Statistical Society. Series B (Methodological)*, 39(1):1–38, 1977.

[11] R. L. Dunbrack and M. Karplus. Backbone-dependent rotamer library for proteins application to side-chain prediction. *Journal of Molecular Biology*, 230(2):543–574, 1993.

[12] R. Durbin, S. R. Eddy, A. Krogh, and G. Mitchison. *Biological Sequence Analysis: Probabilistic Models of Proteins and Nucleic Acids*. Cambridge University Press, Cambridge, 1998.

[13] J. Ferkinghoff-Borg. Monte Carlo methods for inference in high-dimensional systems. In T. Hamelryck, K. Mardia, and J. Ferkinghoff-Borg, editors, *Bayesian Methods in Structural Bioinformatics*, pages 49–93. Springer Berlin Heidelberg, Berlin, Heidelberg, 2012.

[14] J. Frellsen. *Probabilistic methods in macromolecular structure prediction*. PhD thesis, Department of Biology, University of Copenhagen, 2011.

[15] J. Frellsen, I. Moltke, M. Thiim, K. V. Mardia, J. Ferkinghoff-Borg, and T. Hamelryck. A probabilistic model of RNA conformational space. *PLoS Computational Biology*, 5(6):e1000406, June 2009.

[16] J. Frellsen, K. V. Mardia, M. Borg, J. Ferkinghoff-Borg, and T. Hamelryck. Towards a general probabilistic model of protein structure: The reference ratio method. In T. Hamelryck, K. Mardia, and J. Ferkinghoff-Borg, editors, *Bayesian Methods in Structural Bioinformatics*, pages 125–134, Berlin, Heidelberg, 2012. Springer Berlin Heidelberg.

[17] J. Frellsen, T. Hamelryck, and J. Ferkinghoff-Borg. Combining the multicanonical ensemble with generative probabilistic models of local biomolecular structure. In *Proceedings of the 59th World Statistics Congress of the International Statistical Institute*, pages 139–144, Hong Kong, 2014.

[18] J. Frellsen, O. Winther, Z. Ghahramani, and J. Ferkinghoff-Borg. Bayesian generalised ensemble Markov chain Monte Carlo. In A. Gretton and C. C. Robert, editors, *Proceedings of the 19th International Conference on Artificial Intelligence and Statistics*, volume 51 of *Proceedings of Machine Learning Research*, pages 408–416, Cadiz, Spain, 09–11 May 2016. PMLR.

[19] T. C. Gasser, R. W. Ogden, and G. A. Holzapfel. Hyperelastic modelling of arterial layers with distributed collagen fibre orientations. *Journal of The Royal Society Interface*, 3(6):15–35, 2006.

[20] T. Hamelryck. An overview of Bayesian inference and graphical models. In T. Hamelryck, K. Mardia, and J. Ferkinghoff-Borg, editors, *Bayesian Methods in Structural Bioinformatics*, pages 3–48. Springer Berlin Heidelberg, Berlin, Heidelberg, 2012.

[21] T. Hamelryck, M. Borg, M. Paluszewski, J. Paulsen, J. Frellsen, C. Andreetta, W. Boomsma, S. Bottaro, and J. Ferkinghoff-Borg. Potentials of mean force for protein structure prediction vindicated, formalized and generalized. *PloS One*, 5(11):e13714, 2010.

[22] T. Harder, W. Boomsma, M. Paluszewski, J. Frellsen, K. Johansson, and T. Hamelryck. Beyond rotamers: a generative, probabilistic model of side chains in proteins. *BMC Bioinformatics*, 11(1):306, 2010.

[23] W. K. Hastings. Monte Carlo sampling methods using Markov chains and their applications. *Biometrika*, 57(1):97–109, 1970.

[24] L. H. Holley and M. Karplus. Protein secondary structure prediction with a neural network. *Proceedings of the National Academy of Sciences*, 86(1):152–156, 1989.

[25] K. Y. Huang, G. A. Amodeo, L. Tong, and A. McDermott. The structure of human ubiquitin in 2-methyl-2,4-pentanediol: A new conformational switch. *Protein Science*, 20(3):630–639, 2011.

[26] Y. Iba. Extended ensemble Monte Carlo. *International Journal of Modern Physics C*, 12(5):623, 2001.

[27] A. Irbäck and S. Mohanty. PROFASI: a Monte Carlo simulation package for protein folding and aggregation. *Journal of Computational Chemistry*, 27(13):1548–1555, 2006.

[28] D. T. Jones. Successful ab initio prediction of the tertiary structure of NK-lysin using multiple sequences and recognized supersecondary structural motifs. *Proteins: Structure, Function, and Bioinformatics*, 29(S1): 185–191, 1997.

[29] T. A. Jones and S. Thirup. Using known substructures in protein model building and crystallography. *The EMBO Journal*, 5(4):819, 1986.

[30] W. L. Jorgensen and C. J. Swenson. Optimized intermolecular potential functions for amides and peptides. structure and properties of liquid amides. *Journal of the American Chemical Society*, 107(3):569–578, 1985.

[31] G. G. Krivov, M. V. Shapovalov, and R. L. Dunbrack. Improved prediction of protein side-chain conformations with SCWRL4. *Proteins: Structure, Function, and Bioinformatics*, 77(4):778–795, 2009.

[32] M. Le Van Quyen, L. E. Muller, B. Telenczuk, E. Halgren, S. Cash, N. G. Hatsopoulos, N. Dehghani, and A. Destexhe. High-frequency oscillations in human and monkey neocortex during the wakesleep cycle. *Proceedings of the National Academy of Sciences*, 113(33):9363–9368, 2016.

[33] C. Ley and T. Verdebout. *Modern Directional Statistics*. Chapman & Hall/CRC Press, Boca Raton, FL, 2017.

[34] S. C. Lovell, J. M. Word, J. S. Richardson, and D. C. Richardson. The penultimate rotamer library. *Proteins: Structure, Function, and Bioinformatics*, 40(3):389–408, 2000.

[35] S. C. Lovell, I. W. Davis, W. B. Arendall, P. I. de Bakker, J. M. Word, M. G. Prisant, J. S. Richardson, and D. C. Richardson. Structure validation by Cα geometry: ϕ, ψ and Cβ deviation. *Proteins: Structure, Function, and Bioinformatics*, 50(3):437–450, 2003.

[36] K. V. Mardia. Statistics of directional data. *Journal of the Royal Statistical Society. Series B (Methodological)*, 37(3):349–393, 1975.

[37] K. V. Mardia. Statistical approaches to three key challenges in protein structural bioinformatics. *Journal of the Royal Statistical Society: Series C (Applied Statistics)*, 62(3):487–514, 2013.

[38] K. V. Mardia and J. Frellsen. Statistics of bivariate von Mises distributions. In T. Hamelryck, K. Mardia, and J. Ferkinghoff-Borg, editors, *Bayesian Methods in Structural Bioinformatics*, pages 159–178. Springer Berlin Heidelberg, Berlin, Heidelberg, 2012.

[39] K. V. Mardia and P. E. Jupp. *Directional Statistics*. John Wiley & Sons, 2000.

[40] K. V. Mardia, C. C. Taylor, and G. K. Subramaniam. Protein bioinformatics and mixtures of bivariate von Mises distributions for angular data. *Biometrics*, 63(2):505–512, 2007.

[41] N. Metropolis, A. W. Rosenbluth, M. N. Rosenbluth, A. H. Teller, and E. Teller. Equation of state calculations by fast computing machines. *The Journal of Chemical Physics*, 21(6):1087–1092, 1953.

[42] K. P. Murphy. *Dynamic Bayesian Networks: Representation, Inference and Learning*. PhD thesis, University of California, Berkeley, 2002.

[43] I. Murray. *Advances in Markov chain Monte Carlo methods*. PhD thesis, University College London, Gatsby Computational Neuroscience Unit, 2007.

[44] A. K. Navarro, J. Frellsen, and R. E. Turner. The multivariate generalised von Mises distribution: Inference and applications. In *AAAI*, pages 2394–2400, 2017.

[45] S. F. Nielsen. The stochastic EM algorithm: estimation and asymptotic results. *Bernoulli*, 6(3):457–489, 2000.

[46] S. Olsson, W. Boomsma, J. Frellsen, S. Bottaro, T. Harder, J. Ferkinghoff-Borg, and T. Hamelryck. Generative probabilistic models extend the scope of inferential structure determination. *Journal of Magnetic Resonance*, 213(1):182 – 186, 2011.

[47] O. Ovaskainen, H. J. de Knegt, and M. del Mar Delgado. *Quantitative Ecology and Evolutionary Biology: Integrating Models with Data*. Oxford University Press, 2016.

[48] A. Pertsemlidis, J. Zelinka, J. W. Fondon, R. K. Henderson, and Z. Otwinowski. Bayesian statistical studies of the Ramachandran distribution. *Statistical Applications in Genetics and Molecular Biology*, 4 (1), 2005.

[49] J. W. Ponder and D. A. Case. Force fields for protein simulations. In *Protein Simulations*, volume 66 of *Advances in Protein Chemistry*, pages 27 – 85. Academic Press, 2003.

[50] G. N. Ramachandran, C. Ramakrishnan, and V. Sasisekharan. Stereochemistry of polypeptide chain configurations. *Journal of Molecular Biology*, 7(1):95–99, 1963.

[51] L. Saitta, A. Giordana, and A. Cornuéjols. *Phase Transitions in Machine Learning*. Cambridge University Press, 2011.

[52] M. V. Shapovalov and R. L. Dunbrack. A smoothed backbone-dependent rotamer library for proteins derived from adaptive kernel density estimates and regressions. *Structure*, 19(6):844–858, 2011.

[53] D. E. Shaw, P. Maragakis, K. Lindorff-Larsen, S. Piana, R. O. Dror, M. P. Eastwood, J. A. Bank, J. M. Jumper, J. K. Salmon, Y. Shan, and W. Wriggers. Atomic-level characterization of the structural dynamics of proteins. *Science*, 330(6002):341–346, 2010.

[54] K. T. Simons, C. Kooperberg, E. Huang, and D. Baker. Assembly of protein tertiary structures from fragments with similar local sequences using simulated annealing and Bayesian scoring functions. *Journal of Molecular Biology*, 268(1):209–225, 1997.

[55] H. Singh, V. Hnizdo, and E. Demchuk. Probabilistic model for two dependent circular variables. *Biometrika*, 89(3):719–723, 2002.

[56] J. B. Valentin, C. Andreetta, W. Boomsma, S. Bottaro, J. Ferkinghoff-Borg, J. Frellsen, K. V. Mardia, P. Tian, and T. Hamelryck. Formulation of probabilistic models of protein structure in atomic detail using the reference ratio method. *Proteins: Structure, Function, and Bioinformatics*, 82(2):288–299, 2014.

[57] S. Wang, J. Peng, J. Ma, and J. Xu. Protein secondary structure prediction using deep convolutional neural fields. *Scientific Reports*, 6, 2016.

[58] J. M. Word, S. C. Lovell, T. H. LaBean, H. C. Taylor, M. E. Zalis, B. K. Presley, J. S. Richardson, and D. C. Richardson. Visualizing and quantifying molecular goodness-of-fit: small-probe contact dots with explicit hydrogen atoms. *Journal of Molecular Biology*, 285(4):1711–1733, 1999.

[59] Y. Zhang. Progress and challenges in protein structure prediction. *Current Opinion in Structural Biology*, 18(3):342–348, 2008.

2

Statistics of Orientations of Symmetrical Objects

Richard Arnold

Victoria University of Wellington

Peter Jupp

University of St Andrews

CONTENTS

2.1   Ambiguous Rotations 25
     2.1.1   Symmetry Groups 31
     2.1.2   Symmetric Frames 32
2.2   From Symmetric Frames to Symmetric Arrays 33
2.3   Summary Statistics 35
2.4   Testing Uniformity 36
2.5   Distributions of Ambiguous Rotations 38
     2.5.1   A General Class of Distributions on $SO(3)/K$ 38
     2.5.2   Concentrated Distributions 39
2.6   Tests of Location 40
     2.6.1   One-Sample Tests 40
     2.6.2   Two-Sample Tests 41
2.7   Further Developments 41
2.8   Analysis of Examples 42
     2.8.1   Analysis of Example 1 42
     2.8.2   Analysis of Example 2 42
     2.8.3   Analysis of Example 3 42
     Acknowledgments 43
     Bibliography 43

2.1   Ambiguous Rotations

The main objects considered in directional statistics are directions (unit vectors), axes, rotations, frames, and subspaces. Techniques for analyzing such data can be found in, e.g., Fisher et al. (1987) for spheres and in Mardia and Jupp (2000) for more general spaces. Some modern methods for directional statistics are expounded in Ley and Verdebout (2017). Objects more complicated than those above arise naturally in various contexts, such as those in the following three examples.

Example 1 (Orientations of Diopside Crystals)

Diopside is a mineral that forms crystals in the monoclinic class. The orientation of such a crystal can be specified by a pair $(\mathbf{u}_0, \pm\mathbf{u}_1)$, where $\mathbf{u}_0$ is a unit vector and the axis $\pm\mathbf{u}_1$ is orthogonal to $\mathbf{u}_0$; see Figure 2.1. Figure 2.2 shows a sample of orientations $(\mathbf{u}_0, \pm\mathbf{u}_1)$ of 100 diopside crystals. The directions $\mathbf{u}_0$ are shown as triangles, and the axes $\pm\mathbf{u}_1$ as circles. The stereonets in Figure 2.3 show the $\mathbf{u}_0$ directions and $\pm\mathbf{u}_1$ axes given by the orientations of 34 diopside crystals from one region of a specimen and 37 diopside crystals from another region. □

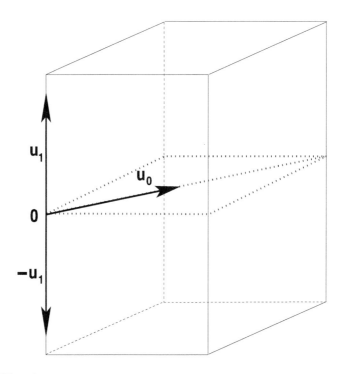

FIGURE 2.1
Orientation of a monoclinic crystal. $\mathbf{u}_0$ and $\mathbf{u}_1$ are orthogonal unit vectors.

Example 2 (Orientations of Ilmenite Crystals)

Ilmenite crystals are trigonal with three-fold rotational symmetry about some axis, and so the orientation of such a crystal can be described by a set of

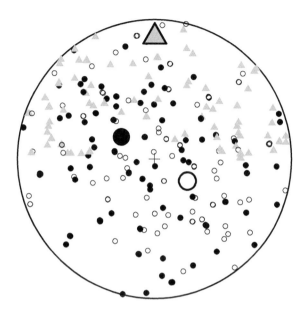

FIGURE 2.2

Orientations of diopside crystals. Stereonet of $\mathbf{u}_0$ vectors, shown as triangles, and $\pm\mathbf{u}_1$ axes, shown as circles, given by orientations of 100 diopside crystals. The disc is a stereographic projection of these vectors and axes, showing the whole of the sphere, so that each axis appears twice with filled circles denoting the lower ends of axes and open circles the upper ends. The sample mean is shown in large symbols.

unit vectors $(\mathbf{u}_1, \mathbf{u}_2, \mathbf{u}_3)$, normal to the rectangular faces of the crystal. Thus $\mathbf{u}_1, \mathbf{u}_2, \mathbf{u}_3$ are coplanar and $\mathbf{u}_i^{\mathsf{T}}\mathbf{u}_j = -1/2$ for $i \neq j$. Because of the symmetry, $(\mathbf{u}_1, \mathbf{u}_2, \mathbf{u}_3)$ cannot be distinguished from $(\mathbf{u}_2, \mathbf{u}_3, \mathbf{u}_1)$ or $(\mathbf{u}_3, \mathbf{u}_1, \mathbf{u}_2)$; see Figure 2.4. Figure 2.5 shows a sample of vectors $(\mathbf{u}_1, \mathbf{u}_2)$ representing orientations $(\mathbf{u}_1, \mathbf{u}_2, \mathbf{u}_3)$ of 42 ilmenite crystals chosen randomly from a larger data set. $\qquad\qquad\square$

Example 3 (Focal Mechanisms of Earthquakes)

When an earthquake occurs, tectonic stress overwhelms the strength of rock and a rupture occurs on a fault plane: the block on one side of the fault slides against the block on the other side. Taking one block as the reference, the normal, $\mathbf{n}$, to that block, and the slip, $\mathbf{u}$, of the other block against it are a pair of 3-vectors describing the focal mechanism of the earthquake (Stein & Wysession, 2003, p. 217); see Figure 2.6. This pair of vectors has two ambiguities: (i) freedom in the choice of reference block means that $(\mathbf{n}, \mathbf{u})$

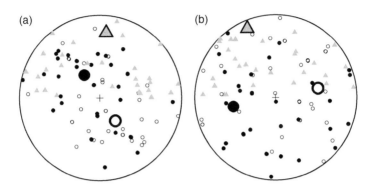

FIGURE 2.3

Orientations of diopside crystals. Stereonets of $\mathbf{u}_0$ vectors, shown as triangles, and $\pm\mathbf{u}_1$ axes, shown as circles, given by orientations of two samples of diopside crystals. Each disc is a stereographic projection of these axes and vectors, showing the whole of the sphere, so that each axis appears twice. (a): 34 orientations from one region of a specimen. (b): 37 orientations from another region. The sample means are shown as large symbols.

cannot be distinguished from $(-\mathbf{n}, -\mathbf{u})$, (ii) unless the fault plane is visible, the only information on $\mathbf{n}$ and $\mathbf{u}$ is provided by seismometer traces, a major limitation of which is that $(\mathbf{n}, \mathbf{u})$ cannot be distinguished from $(\mathbf{u}, \mathbf{n})$. These ambiguities can be accommodated by transforming the normal and slip vectors into the compressional, P, and tensional, T, axes $\pm\mathbf{p} = \pm(\mathbf{n} - \mathbf{u})/\sqrt{2}$ and $\pm\mathbf{t} = \pm(\mathbf{n} + \mathbf{u})/\sqrt{2}$. The null axis, or A axis, is defined as $\pm\boldsymbol{\alpha} = \pm\mathbf{p} \times \mathbf{t}$.

Figure 2.7 displays on stereonets the T, A, and P axes for three clusters of earthquakes. These diagrams are stereographic projections with the downward vertical at the center, North at the top, East to the right, and show the orientations of the three axes in each case. Each symbol indicates an intersection of the corresponding axis with a small unit sphere enclosing the fault plane. Only the lower half of the sphere is plotted, so that each axis appears only once on the diagram.

The stereonet in Figure 2.7(a) shows the focal mechanisms of 50 earthquakes that took place between two very large events on September 4, 2010 and February 22, 2011 near Christchurch, New Zealand. The P axes are mostly close to the horizontal plane, with a South-East to North-West alignment. However, there are several observations where the P axis is closer to vertical. The T axes are mostly close to the horizontal, and as a result of the horizontal P and T axes, the A axes are all nearly vertical. Geophysicists are interested in whether the earthquake of February 22, 2011 changed the pattern of earth-

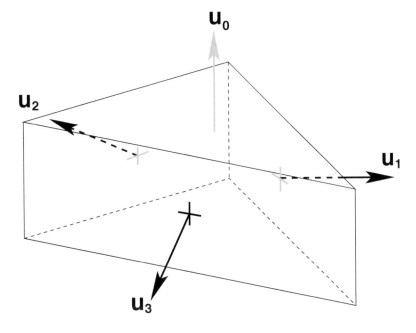

FIGURE 2.4
Idealized trigonal crystal with unit vectors $\mathbf{u}_1, \mathbf{u}_2, \mathbf{u}_3$ normal to the rectangular faces. $\mathbf{u}_0 = (2/\sqrt{3})\,\mathbf{u}_1 \times \mathbf{u}_2$.

quakes, and the stereonet in Figure 2.7(b) plots 50 earthquakes in the subsequent 7-month period. It shows a similar concentrated distribution of P and T axes. A more diverse group of 32 focal mechanisms comes from the north of the South Island of New Zealand, and is plotted in the stereonet in Figure 2.7(c). These focal mechanisms are much more scattered across the stereonet than the events near Christchurch. □

An important feature common to Examples 1–3 is that, in each example, each observation can be obtained by rotation of a standard (idealized) observation. In Example 1 the pair $(\mathbf{u}_0, \pm\mathbf{u}_1)$ can be obtained by rotating the pair $(\mathbf{e}_1, \pm\mathbf{e}_2)$ by the rotation matrix $(\mathbf{u}_0, \mathbf{u}_1, \mathbf{u}_0 \times \mathbf{u}_1)$, where $\mathbf{e}_1, \mathbf{e}_2, \mathbf{e}_3$ is the standard orthonormal basis of $\mathbb{R}^3$. In Example 2 the triple $(\mathbf{u}_1, \mathbf{u}_2, \mathbf{u}_3)$ of coplanar vectors (suitably ordered) can be obtained by rotating the triple $(\mathbf{e}_1, -(1/2)\mathbf{e}_1 + (\sqrt{3}/2)\mathbf{e}_2, -(1/2)\mathbf{e}_1 - (\sqrt{3}/2)\mathbf{e}_2)$ by a suitable rotation matrix. In Example 3 the triple $(\pm\mathbf{p}, \pm\mathbf{t}, \pm\boldsymbol{\alpha})$ of orthogonal axes can be obtained by rotating the triple $(\pm\mathbf{e}_1, \pm\mathbf{e}_2, \pm\mathbf{e}_3)$ by the rotation matrix $(\mathbf{p}, \mathbf{t}, \boldsymbol{\alpha})$. In each example, the rotation is not unique but subject to some ambiguity (depending on the symmetry of the example) arising from the choice of unit vectors to represent axes. We shall consider statistical methods for handing observations that are general *ambiguous rotations*, which are rotations of $\mathbb{R}^3$ for which it is not possible to distinguish a rotation $\mathbf{X}$ from $\mathbf{XR}$ for any rotation $\mathbf{R}$ in

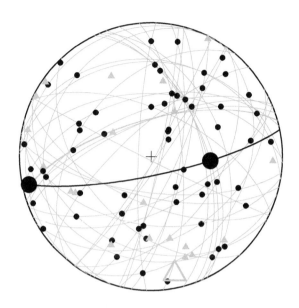

FIGURE 2.5
Orientations $(\mathbf{u}_1, \mathbf{u}_2, \mathbf{u}_3)$ of 42 ilmenite crystals. For clarity, for each crystal only the near side of the stereogram is shown. The gray great circles are the intersections of the planes containing $\mathbf{u}_1, \mathbf{u}_2$, and $\mathbf{u}_3$ with the near side of the sphere, and the filled circles are the locations of any of these vectors that occur on the near side of the sphere. The gray triangles are those vectors $\mathbf{u}_0 = (2/\sqrt{3})\,\mathbf{u}_1 \times \mathbf{u}_2$ that lie on the near side of the sphere. The sample mean is shown using large symbols. For this sample the mean direction of the $\mathbf{u}_0$ lies on the far side of the sphere and is plotted as a large triangle.

some given finite subgroup K of the rotation group, $SO(3)$. From the mathematical point of view, the sample space is the quotient $SO(3)/K$ of $SO(3)$ by K. Ambiguous rotations in $SO(3)/K$ occur generally as orientations of rigid objects with symmetry group K, each orientation being obtained by rotation of some standard orientation. For $\mathbf{X}$ in $SO(3)$ we shall denote the equivalence class $\{\mathbf{XR} : \mathbf{R} \in K\}$ of $\mathbf{X}$ in $SO(3)/K$ by $[\mathbf{X}]$. The spaces $SO(3)/K$ arise in many scientific contexts: several groups K of low order occur as the symmetry groups of crystals; the icosahedral group is the symmetry group of some carborane molecules (Jemmis, 1982), of most closed-shell viruses (Harrison, 2013), of the natural quasicrystal, icosahedrite (Bindi et al., 2011), and of the blue phases of some liquid crystals (Seideman, 1990, Section 6.1.2). A detailed description of methods of analyzing data on ambiguous rotations is given by Arnold et al. (2018).

An important class of ambiguous rotations consists of the *orthogonal axial frames*, which consist of sets $(\pm\mathbf{u}_1, \pm\mathbf{u}_2, \pm\mathbf{u}_3)$ of orthogonal axes in $\mathbb{R}^3$. More

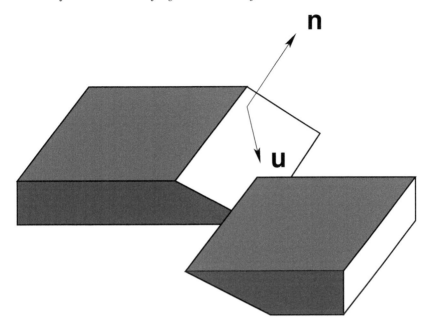

FIGURE 2.6
Schematic diagram of a slide along a fault. The unit vector, $\mathbf{n}$, is normal to
the two blocks, and the unit vector, $\mathbf{u}$, gives the direction of the slip.

generally, we can consider orthogonal axial r-frames $(\pm\mathbf{u}_1, \ldots, \pm\mathbf{u}_r)$ of orthogonal axes in $\mathbb{R}^p$ with $1 \leq r \leq p$. Orthogonal axial frames arise in seismology
(as in Example 3), as axes of orthorhombic crystals, and as principal axes of
variance matrices. The case $r = p = 3$ corresponds to $K = D_2$ in the general
setting of $SO(3)/K$ described below. Methods for handling orthogonal axial
frames with general values of r and p have been developed in some detail in
Arnold and Jupp (2013). In particular, there are constructions analogous to
those described below for $SO(3)/K$. The bottom line of each of Tables 2.1–2.4
provides values appropriate for orthogonal axial r-frames in $\mathbb{R}^3$.

2.1.1 Symmetry Groups

The finite subgroups of $SO(3)$ are known also as the point groups of the first
kind. The classification result for these groups, given, e.g., in Miller (1972,
Section 2.4), states that any such group is isomorphic to one of the following:
the cyclic groups, C_r, for $r = 1, 2, \ldots$, the dihedral groups, D_r, for $r = 2, 3, \ldots$,
the tetrahedral group, T, the octahedral group, O, and the icosahedral group,
Y. These groups are listed in Table 2.1, together with the frames of vectors
that will be used to represent elements of the sample spaces $SO(3)/K$. The
group C_1 has one element, the identity, $\mathbf{I}_3$.

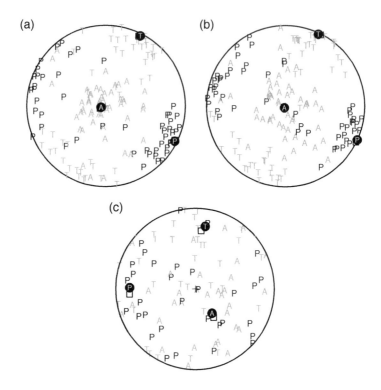

FIGURE 2.7

Focal mechanisms of earthquakes. Stereonets of T, A, and P axes given by focal mechanisms of three clusters of earthquakes in New Zealand. Each disc is a stereographic projection of the directions of these axes, but showing only the lower half of the sphere, so that each axis appears only once. The center of the diagram is the downward vertical direction, and the directions on the circumference are all horizontal. North is at the top, and East to the right. (a): 50 focal mechanisms close to Christchurch in the period September 4, 2010–February 21, 2011. (b): 50 focal mechanisms near Christchurch in the period February 23,–October 2, 2011. (c): 32 focal mechanisms in northern South Island in the period 2004–2010. The sample means are shown as labelled filled circles. The open squares (obscured in panels (a) and (b)) denote approximate sample means obtained using a method based on Gram–Schmidt orthonormalization and described in Arnold and Jupp (2013, Section 2.2).

2.1.2   Symmetric Frames

Instead of considering ambiguous rotations in $SO(3)/K$ as equivalence classes of rotations, it is more useful to consider them as K-*frames*, i.e., equivalence classes of frames (sets of vectors or axes in $\mathbb{R}^3$) obtained from one another by rotations in K. For Examples 1, 2, and 3 the C_2-frames, C_3-frames, and

D_2-frames are those determined by $(\mathbf{u}_0, \pm\mathbf{u}_1)$, $(\mathbf{u}_1, \mathbf{u}_2, \mathbf{u}_3)$, and $(\pm\mathbf{p}, \pm\mathbf{t}, \pm\boldsymbol{\alpha})$, respectively.

For $K = C_r$ with $r \geq 3$ or $K = D_r$ with $r \geq 3$, it is convenient to take the vectors of the frame to be unit normals to the sides of a regular r-gon; for $K = C_2$ we take a unit vector and an axis orthogonal to it; for $K = D_2$ we take a pair of orthogonal axes; for $K = T$, O, or Y, it is convenient to take the vectors to be unit normals to the sides of a regular tetrahedron, cube, or dodecahedron, respectively. Permutation of the vectors of the frame by the action of K leads to ambiguity. This ambiguity is removed by passing to the corresponding K-frame, i.e., the equivalence class of the frame under such permutations. The K-frames will be denoted by square brackets, e.g., for $K = C_r$, $[\mathbf{u}_1, \ldots, \mathbf{u}_r]$ denotes the K-frame arising from $(\mathbf{u}_1, \ldots, \mathbf{u}_r)$. By a *symmetric frame*, we shall mean a K-frame for some K. The frames that we consider are listed in Table 2.1, together with an indication of the ambiguities.

TABLE 2.1
Symmetry groups and frames.

Group, K		Name	$|K|$	Frame	
C_1		trivial	1	$(\mathbf{u}_1, \mathbf{u}_2, \mathbf{u}_3)$	$\mathbf{u}_1, \mathbf{u}_2, \mathbf{u}_3$ orthonormal, $\mathbf{u}_3 = \mathbf{u}_1 \times \mathbf{u}_2$
C_2		cyclic	2	$(\mathbf{u}_0, \pm\mathbf{u}_1)$	$\mathbf{u}_0, \mathbf{u}_1$ orthonormal
C_r	$(r \geq 3)$	cyclic	r	$(\mathbf{u}_1, \ldots, \mathbf{u}_r)$	$\mathbf{u}_1, \ldots, \mathbf{u}_r$ coplanar, known up to cyclic order, $\mathbf{u}_i^{\mathsf{T}}\mathbf{u}_{i-1} = \cos(2\pi/r)$ for $i = 2, \ldots, r$
D_2		dihedral	4	$(\pm\mathbf{u}_1, \pm\mathbf{u}_2)$	orthogonal axes
D_r	$(r \geq 3)$	dihedral	$2r$	$(\mathbf{u}_1, \ldots, \mathbf{u}_r)$	$\mathbf{u}_1, \ldots, \mathbf{u}_r$ coplanar, known up to cyclic order and reversal, $\mathbf{u}_i^{\mathsf{T}}\mathbf{u}_{i-1} = \cos(2\pi/r)$ for $i = 2, \ldots, r$
$T = A_4$		tetrahedral	12	$\{\mathbf{u}_1, \ldots, \mathbf{u}_4\}$	$\mathbf{u}_i^{\mathsf{T}}\mathbf{u}_j = -1/3$ for $i \neq j$
$O = \Sigma_4$		octahedral = cubic	24	$\{\pm\mathbf{u}_1, \pm\mathbf{u}_2, \pm\mathbf{u}_3\}$	orthogonal axes
$Y = A_5$		icosahedral = dodecahedral	60	$\{\pm\mathbf{u}_1, \ldots, \pm\mathbf{u}_6\}$	$|\mathbf{u}_i^{\mathsf{T}}\mathbf{u}_j| = 5^{-1/2}$ for $i \neq j$
$\mathbb{Z}_2^r$			2^r	$(\pm\mathbf{u}_1, \ldots, \pm\mathbf{u}_r)$	orthogonal axes in $\mathbb{R}^p$

The $\mathbf{u}_i$ are unit vectors. $|K|$ is the number of elements in K. Bottom row refers to orthogonal axial r-frames in $\mathbb{R}^p$.

Special cases of the symmetries listed in Table 2.1 arise in the 7 crystal systems: triclinic, monoclinic, trigonal, tetragonal, orthorhombic, hexagonal, and cubic with symmetry groups C_1, C_2, C_3, C_4, D_2, D_6, and O, respectively.

2.2 From Symmetric Frames to Symmetric Arrays

In order to carry out statistics on $SO(3)/K$, we shall transform points in $SO(3)/K$ into vectors in an inner-product space E on which $SO(3)$ acts, using a function $\mathbf{t}: SO(3)/K \to E$. (This is an example of the embedding approach to directional statistics; see, e.g., Mardia and Jupp (2000, Section 10.8).) We require that (i) $\mathbf{t}$ is one-to-one, (ii) $\langle \mathbf{t}([\mathbf{VU}]), \mathbf{t}([\mathbf{VW}])\rangle = \langle \mathbf{t}([\mathbf{U}]), \mathbf{t}([\mathbf{W}])\rangle$ for $\mathbf{V}$ in $SO(3)$ and $[\mathbf{U}], [\mathbf{W}]$ in $SO(3)/K$, (iii) if $[\mathbf{U}]$ is uniformly distributed on $SO(3)/K$ then $E\{\mathbf{t}([\mathbf{U}])\} = \mathbf{0}$. Since E is an inner-product space, we can use our Euclidean intuition. We shall choose $\mathbf{t}$ to be a particularly simple embedding, $\mathbf{t}_K$, of $SO(3)/K$ into an appropriate space of symmetric arrays. These $\mathbf{t}_K$ are given in Table 2.2 and are based on symmetric r-way arrays $\otimes^r \mathbf{u}_i$, which can be thought of as "rth powers" of vectors. For vectors $\mathbf{v}$ with $\mathbf{v} = (v_1, v_2, v_3)^{\mathsf{T}}$, the r-way array $\otimes^r \mathbf{v}$ has $(j_1, \ldots, j_r)$-th entry

$$(\otimes^r \mathbf{v})_{j_1, \ldots, j_r} = \prod_{k=1}^{r} v_{j_k}, \quad 1 \le j_1, \ldots, j_r \le 3, \qquad (2.1)$$

e.g., $\otimes^2 \mathbf{v} = \mathbf{v}\mathbf{v}^T$. Some of the $\mathbf{t}_K$ involve also the arrays $\text{symm}(\otimes^{r/2}\mathbf{I}_3)$ for r even. These are defined by

$$\left\{ \text{symm}(\otimes^{r/2}\mathbf{I}_3) \right\}_{j_1, \ldots, j_r} = \frac{1}{r!} \sum_{\sigma \in \Sigma_r} \prod_{i=1}^{r/2} \delta_{\sigma(j_{2i-1})\sigma(j_{2i})}, \qquad (2.2)$$

where Σ_r is the group of permutations of $\{1, \ldots, r\}$ and $\delta_{ij} = 1$ for $i = j$, $\delta_{ij} = 0$ otherwise. Thus, e.g., $\text{symm}(\otimes^2\mathbf{I}_3)_{j_1 j_2 j_3 j_4} = (1/3)\,(\delta_{j_1 j_2}\delta_{j_3 j_4} + \delta_{j_1 j_3}\delta_{j_2 j_4} + \delta_{j_1 j_4}\delta_{j_2 j_3})$.

For many of the tests and other objects that we discuss below it is not necessary to construct the representations $\mathbf{t}_K([\mathbf{U}])$ explicitly; it is sufficient to compute inner products $\langle \mathbf{t}_K([\mathbf{U}]), \mathbf{t}_K([\mathbf{V}])\rangle$ as required, where $\langle \cdot, \cdot \rangle$ is the standard inner product on the space of symmetric arrays.

TABLE 2.2
Some embeddings $\mathbf{t}_K : SO(3)/K \to E$.

Group, K	$\mathbf{t}_K$
C_1	$\mathbf{t}_{C_1}(\mathbf{u}_1, \mathbf{u}_2, \mathbf{u}_3) = (\mathbf{u}_1, \mathbf{u}_2, \mathbf{u}_3)$
C_2	$\mathbf{t}_{C_2}(\mathbf{u}_0, \pm\mathbf{u}_1) = (\mathbf{u}_0, \mathbf{u}_1\mathbf{u}_1^{\mathsf{T}} - (1/3)\mathbf{I}_3)$
$C_r \quad (r \ge 3)$	
r odd	$\mathbf{t}_{C_r}([\mathbf{u}_1, \ldots, \mathbf{u}_r]) = \left(\mathbf{u}_0, \sum_{i=1}^{r} \otimes^r \mathbf{u}_i\right)$
r even	$\mathbf{t}_{C_r}([\mathbf{u}_1, \ldots, \mathbf{u}_r]) = \left(\mathbf{u}_0, \sum_{i=1}^{r} \otimes^r \mathbf{u}_i - r/(r+1)\,\text{symm}(\otimes^{r/2}\mathbf{I}_3)\right)$
D_2	$\mathbf{t}_{D_2}(\pm\mathbf{u}_1, \pm\mathbf{u}_2) = (\mathbf{u}_1\mathbf{u}_1^{\mathsf{T}} - (1/3)\mathbf{I}_3, \mathbf{u}_2\mathbf{u}_2^{\mathsf{T}} - (1/3)\mathbf{I}_3, \mathbf{u}_3\mathbf{u}_3^{\mathsf{T}} - (1/3)\mathbf{I}_3)$
$D_r \quad (r \ge 3)$	
r odd	$\mathbf{t}_{D_r}([\mathbf{u}_1, \ldots, \mathbf{u}_r]) = \sum_{i=1}^{r} \otimes^r \mathbf{u}_i$
r even	$\mathbf{t}_{D_r}([\mathbf{u}_1, \ldots, \mathbf{u}_r]) = \sum_{i=1}^{r} \otimes^r \mathbf{u}_i - r/(r+1)\,\text{symm}(\otimes^{r/2}\mathbf{I}_3)$
T	$\mathbf{t}_T(\{\mathbf{u}_1, \mathbf{u}_2, \mathbf{u}_3, \mathbf{u}_4\}) = \otimes^3\mathbf{u}_1 + \otimes^3\mathbf{u}_2 + \otimes^3\mathbf{u}_3 + \otimes^3\mathbf{u}_4$
O	$\mathbf{t}_O(\{\pm\mathbf{u}_1, \pm\mathbf{u}_2, \pm\mathbf{u}_3\}) = \otimes^4\mathbf{u}_1 + \otimes^4\mathbf{u}_2 + \otimes^4\mathbf{u}_3 - (3/5)\,\text{symm}(\otimes^2\mathbf{I}_3)$
Y	$\mathbf{t}_Y(\{\pm\mathbf{u}_1, \ldots, \pm\mathbf{u}_6\}) = \sum_{i=1}^{6} \otimes^{10}\mathbf{u}_i - (6/11)\,\text{symm}(\otimes^5\mathbf{I}_3)$
$\mathbb{Z}_2^r$	$\mathbf{t}(\pm\mathbf{u}_1, \ldots, \pm\mathbf{u}_r) = (\mathbf{u}_1\mathbf{u}_1^{\mathsf{T}} - (1/p)\mathbf{I}_p, \ldots, \mathbf{u}_r\mathbf{u}_r^{\mathsf{T}} - (1/p)\mathbf{I}_p)$

For C_r with $r \ge 3$, $\mathbf{u}_0 = \{\sin(2\pi/r)\}^{-1}\,\mathbf{u}_1 \times \mathbf{u}_2$. For D_2, $\mathbf{u}_3 = \pm\mathbf{u}_1 \times \mathbf{u}_2$. The symmetric arrays $\otimes^r \mathbf{u}_i$ and $\text{symm}(\otimes^{r/2}\mathbf{I}_3)$ are defined in (2.1) and (2.2), respectively. Bottom row refers to orthogonal axial r-frames in $\mathbb{R}^p$.

Define ρ^2 by

$$\rho^2 = \|\mathbf{t}([\mathbf{U}])\|^2, \tag{2.3}$$

which has the same value for all $\mathbf{U}$ in $SO(3)$. Then $\mathbf{t}$ embeds $SO(3)/K$ in the sphere of radius ρ with center the origin in the vector space E.

TABLE 2.3
Inner products of transforms of symmetric frames.

Group, K	Inner product
C_1	$\langle \mathbf{t}_{C_1}(\mathbf{u}_1, \mathbf{u}_2, \mathbf{u}_3), \mathbf{t}_{C_1}(\mathbf{v}_1, \mathbf{v}_2, \mathbf{v}_3) \rangle = \mathbf{u}_1^\mathsf{T}\mathbf{v}_1 + \mathbf{u}_2^\mathsf{T}\mathbf{v}_2 + \mathbf{u}_3^\mathsf{T}\mathbf{v}_3$
C_2	$\langle \mathbf{t}_{C_2}(\mathbf{u}_0, \pm\mathbf{u}_1), \mathbf{t}_{C_2}(\mathbf{v}_0, \pm\mathbf{v}_1) \rangle = \mathbf{u}_0^\mathsf{T}\mathbf{v}_0 + (\mathbf{u}_1^\mathsf{T}\mathbf{v}_1)^2 - 1/3$
$C_r \quad (r \geq 3)$	
r odd	$\langle \mathbf{t}_{C_r}([\mathbf{u}_1, \ldots, \mathbf{u}_r]), \mathbf{t}_{C_r}([\mathbf{v}_1, \ldots, \mathbf{v}_r]) \rangle = \mathbf{u}_0^\mathsf{T}\mathbf{v}_0 + \sum_{i=1}^{r}\sum_{j=1}^{r}(\mathbf{u}_i^\mathsf{T}\mathbf{v}_j)^r$
r even	$\langle \mathbf{t}_{C_r}([\mathbf{u}_1, \ldots, \mathbf{u}_r]), \mathbf{t}_{C_r}([\mathbf{v}_1, \ldots, \mathbf{v}_r]) \rangle = \mathbf{u}_0^\mathsf{T}\mathbf{v}_0 + \sum_{i=1}^{r}\sum_{j=1}^{r}(\mathbf{u}_i^\mathsf{T}\mathbf{v}_j)^r - r^2/(r+1)$
D_2	$\langle \mathbf{t}_{D_2}(\pm\mathbf{u}_1, \pm\mathbf{u}_2), \mathbf{t}_{D_2}(\pm\mathbf{v}_1, \pm\mathbf{v}_2) \rangle = (\mathbf{u}_1^\mathsf{T}\mathbf{v}_1)^2 + (\mathbf{u}_2^\mathsf{T}\mathbf{v}_2)^2 + (\mathbf{u}_3^\mathsf{T}\mathbf{v}_3)^2 - 1$
$D_r \quad (r \geq 3)$	
r odd	$\langle \mathbf{t}_{D_r}([\mathbf{u}_1, \ldots, \mathbf{u}_r]), \mathbf{t}_{D_r}([\mathbf{v}_1, \ldots, \mathbf{v}_r]) \rangle = \sum_{i=1}^{r}\sum_{j=1}^{r}(\mathbf{u}_i^\mathsf{T}\mathbf{v}_j)^r$
r even	$\langle \mathbf{t}_{D_r}([\mathbf{u}_1, \ldots, \mathbf{u}_r]), \mathbf{t}_{D_r}([\mathbf{v}_1, \ldots, \mathbf{v}_r]) \rangle = \sum_{i=1}^{r}\sum_{j=1}^{r}(\mathbf{u}_i^\mathsf{T}\mathbf{v}_j)^r - r^2/(r+1)$
T	$\langle \mathbf{t}_T(\{\mathbf{u}_1, \mathbf{u}_2, \mathbf{u}_3, \mathbf{u}_4\}), \mathbf{t}_T(\{\mathbf{v}_1, \mathbf{v}_2, \mathbf{v}_3, \mathbf{v}_4\}) \rangle = \sum_{i=1}^{4}\sum_{j=1}^{4}(\mathbf{u}_i^\mathsf{T}\mathbf{v}_j)^3$
O	$\langle \mathbf{t}_O(\{\pm\mathbf{u}_1, \pm\mathbf{u}_2, \pm\mathbf{u}_3\}), \mathbf{t}_O(\{\pm\mathbf{v}_1, \pm\mathbf{v}_2, \pm\mathbf{v}_3\}) \rangle = \sum_{i=1}^{3}\sum_{j=1}^{3}(\mathbf{u}_i^\mathsf{T}\mathbf{v}_j)^4 - 9/5$
Y	$\langle \mathbf{t}_Y(\{\pm\mathbf{u}_1, \ldots, \pm\mathbf{u}_6\}), \mathbf{t}_Y(\{\pm\mathbf{v}_1, \ldots, \pm\mathbf{v}_6\}) \rangle = \sum_{i=1}^{6}\sum_{j=1}^{6}(\mathbf{u}_i^\mathsf{T}\mathbf{v}_j)^{10} - 36/11$
$\mathbb{Z}_2^r$	$\langle \mathbf{t}(\pm\mathbf{u}_1, \ldots, \pm\mathbf{u}_r), \mathbf{t}(\pm\mathbf{v}_1, \ldots, \pm\mathbf{v}_r) \rangle = \sum_{i=1}^{r}(\mathbf{u}_i \mathbf{v}_i^\mathsf{T})^2 - 1$

For C_r with $r \geq 3$, $\mathbf{u}_0 = \{\sin(2\pi/r)\}^{-1}\mathbf{u}_1 \times \mathbf{u}_2$. For D_2, $\mathbf{u}_3 = \pm\mathbf{u}_1 \times \mathbf{u}_2$.
Bottom row refers to orthogonal axial r-frames in $\mathbb{R}^p$.

2.3 Summary Statistics

Observations $[\mathbf{U}_1], \ldots, [\mathbf{U}_n]$ in $SO(3)/K$ can usefully be summarized by the sample mean $\bar{\mathbf{t}}$ of their images by $\mathbf{t}$, i.e., by $\bar{\mathbf{t}} = n^{-1}\sum_{k=1}^{n}\mathbf{t}([\mathbf{U}_k])$. The *sample mean* $[\bar{\mathbf{U}}]$ is defined as the $[\mathbf{U}]$ in $SO(3)/K$ that maximizes $\langle \mathbf{t}([\mathbf{U}]), \bar{\mathbf{t}} \rangle$. If $[\mathbf{U}_1], \ldots, [\mathbf{U}_n]$ are generated by a continuous distribution then $[\bar{\mathbf{U}}]$ is unique with probability 1. If $K = C_2$ then the sample mean of $[\mathbf{U}_1], \ldots, [\mathbf{U}_n]$ is $\bar{\mathbf{t}} = \left(n^{-1}\sum_{k=1}^{n}\mathbf{u}_{k0}, n^{-1}\sum_{k=1}^{n}\mathbf{u}_{k1}\mathbf{u}_{k1}^\mathsf{T} - (1/3)\mathbf{I}_3\right)$, where $[\mathbf{U}_k]$ is represented by the C_2-frame $(\mathbf{u}_{k0}, \pm\mathbf{u}_{k1})$ for $k = 1, \ldots, n$. If $K = C_3$ then the sample mean, $\bar{\mathbf{t}}$, of $[\mathbf{U}_1], \ldots, [\mathbf{U}_n]$ can be expressed as $\bar{\mathbf{t}} = [\bar{\mathbf{u}}_1, \bar{\mathbf{u}}_2, \bar{\mathbf{u}}_3]$ for some vectors $\bar{\mathbf{u}}_1, \bar{\mathbf{u}}_2, \bar{\mathbf{u}}_3$ in $\mathbb{R}^3$. It is determined by $\bar{\mathbf{u}}_0^\mathsf{T}\mathbf{v}_0 + \sum_{i=1}^{3}\sum_{j=1}^{3}(\bar{\mathbf{u}}_i^\mathsf{T}\mathbf{v}_j)^3 = n^{-1}\sum_{k=1}^{n}\left\{\mathbf{u}_{k0}^\mathsf{T}\mathbf{v}_0 + \sum_{i=1}^{3}\sum_{j=1}^{3}(\mathbf{u}_{ki}^\mathsf{T}\mathbf{v}_j)^3\right\}$, where $[\mathbf{u}_{1k}, \mathbf{u}_{2k}, \mathbf{u}_{3k}]$ is obtained by rotating some standard K-frame by $\mathbf{U}_k$ in $SO(3)$ for $k = 1, \ldots, n$. Here $[\mathbf{v}_1, \mathbf{v}_2, \mathbf{v}_3]$ is an arbitrary C_3-frame and $\mathbf{v}_0 = (2/\sqrt{3})\mathbf{v}_1 \times \mathbf{v}_2$. If $K = D_2$ then the sample mean of $[\mathbf{U}_1], \ldots, [\mathbf{U}_n]$ is $\bar{\mathbf{T}} = n^{-1}\sum_{i=1}^{n}\mathbf{t}([\mathbf{U}_i]) = (\bar{\mathbf{T}}_1, \bar{\mathbf{T}}_2, \bar{\mathbf{T}}_3)$, where $\bar{\mathbf{T}}_j = n^{-1}\sum_{i=1}^{n}\left(\mathbf{u}_{ij}\mathbf{u}_{ij}^\mathsf{T} - (1/3)\mathbf{I}_3\right)$ for $j = 1, 2, 3$ and $[\mathbf{U}_i] = (\pm\mathbf{u}_{i1}, \pm\mathbf{u}_{i2}, +\mathbf{u}_{i3})$. For general K, there does not appear to be an explicit expression for $[\bar{\mathbf{U}}]$. In the cases $K = C_r$ (with $r \geq 2$) or D_r (with $r \geq 2$) there are the following explicit recipes for an approximation to $[\bar{\mathbf{U}}]$. For $r = 2$, let $(\mathbf{u}_{10}, \pm\mathbf{u}_{11}), \ldots, (\mathbf{u}_{n0}, \pm\mathbf{u}_{n1})$ or $(\pm\mathbf{u}_{11}, \pm\mathbf{u}_{12}), \ldots, (\mathbf{u}_{n1}, \pm\mathbf{u}_{n2})$ be symmetric frames representing n elements of $SO(3)/C_2$ or $SO(3)/D_2$, respectively. For $r \geq 3$, let $[\mathbf{u}_{11}, \ldots, \mathbf{u}_{1r}], \ldots, [\mathbf{u}_{n1}, \ldots, \mathbf{u}_{nr}]$ be symmetric frames representing n elements of $SO(3)/C_r$ or $SO(3)/D_r$. The first step is to estimate the K-invariant axis. For $K = D_2$, define $\pm\mathbf{u}_{i0}$ by $\pm\mathbf{u}_{i0} = \pm\mathbf{u}_{i1} \times \mathbf{u}_{i2}$ for $i = 1, \ldots, n$. For $r \geq 3$, define $\mathbf{u}_{i0}$ by $\mathbf{u}_{i0} = \{\sin(2\pi/r)\}^{-1} \mathbf{u}_{i1} \times \mathbf{u}_{i2}$, $i = 1, \ldots, n$. In the case $K = C_r$ with $r \geq 2$, define $\tilde{\mathbf{u}}_0$ to be the mean direction of the unit vectors $\mathbf{u}_{10}, \ldots, \mathbf{u}_{n0}$, i.e., the unit vector that is a positive multiple of $\sum_{i=1}^n \mathbf{u}_{i0}$. In the case $K = D_r$ with $r \geq 2$, define the axis $\pm\tilde{\mathbf{u}}_0$ to be the mean of the axes $\pm\mathbf{u}_{10}, \ldots, \pm\mathbf{u}_{n0}$, i.e., the dominant principal axis of $\sum_{i=1}^n \mathbf{u}_{i0}\mathbf{u}_{i0}^{\mathsf{T}}$. Choose orthonormal vectors $\mathbf{e}_1$ and $\mathbf{e}_2$ that are normal to $\tilde{\mathbf{u}}_0$. For $i = 1, \ldots, n$, define $\theta_{i1}, \ldots, \theta_{ir}$ by $\|(I_3 - \tilde{\mathbf{u}}_0\tilde{\mathbf{u}}_0^{\mathsf{T}})\mathbf{u}_{ij}\|^{-1}(I_3 - \tilde{\mathbf{u}}_0\tilde{\mathbf{u}}_0^{\mathsf{T}})\mathbf{u}_{ij} = \cos(\theta_{ij})\,\mathbf{e}_1 + \sin(\theta_{ij})\,\mathbf{e}_2$ for $j = 1, \ldots, r$. Define $\hat{\theta}$ as the solution in $[0, 2\pi/r)$ of $r\hat{\theta}$ is the mean direction of $r\theta_{11}, \ldots, r\theta_{1r}, \ldots, r\theta_{n1}, \ldots, r\theta_{nr}$ and put $\tilde{\mathbf{u}}_j = \cos(\hat{\theta} + j2\pi/r)\,\mathbf{e}_1 + \sin(\hat{\theta} + j2\pi/r)\,\mathbf{e}_2$ for $j = 1, \ldots, r$. Then $(\tilde{\mathbf{u}}_0, \pm\tilde{\mathbf{u}}_1)$ (for $K = C_2$ and $r = 2$) or $(\pm\tilde{\mathbf{u}}_1, \pm\tilde{\mathbf{u}}_2)$ (for $K = D_2$ and $r = 2$) or $[\tilde{\mathbf{u}}_1, \ldots, \tilde{\mathbf{u}}_r]$ (for $r \geq 3$) is an approximation to the sample mean. If the directions $\mathbf{u}_{10}, \ldots, \mathbf{u}_{n0}$ (for $SO(3)/C_r$) or the axes $\pm\mathbf{u}_{10}, \ldots, \pm\mathbf{u}_{n0}$ (for $SO(3)/D_r$) are concentrated then with high probability the above approximate sample means are close to the exact sample means. For $K = D_2$, the above approximation to $[\bar{\mathbf{U}}]$ is not the same as that given in Section 2.2 of Arnold and Jupp (2013).

The *dispersion* is defined as

$$d = \rho^2 - \|\bar{\mathbf{t}}\|^2, \tag{2.4}$$

analogous to the quantity $1 - \bar{R}^2$ used for spherical data; see Mardia and Jupp (2000, p. 164). The dispersion satisfies the inequalities $0 \leq d \leq \rho^2$, where ρ^2 is defined in (2.3). Since $\mathbf{t}$ is one-to-one, $d = 0$ if and only if $[\mathbf{U}_1] = \cdots = [\mathbf{U}_n]$. Transformation of $[\mathbf{U}_1], \ldots, [\mathbf{U}_n]$ to $[\mathbf{V}\mathbf{U}_1], \ldots, [\mathbf{V}\mathbf{U}_n]$ with $\mathbf{V}$ in $SO(3)$ leaves d unchanged. If $K = C_1$ then $d = 3 - \text{trace}\,(\bar{\mathbf{R}}^2)$, where $\bar{\mathbf{R}} = (\bar{\mathbf{X}}^T\bar{\mathbf{X}})^{1/2}$ with $\bar{\mathbf{X}} = \bar{\mathbf{t}}$, the sample mean of $\mathbf{X} = (\mathbf{u}_1, \mathbf{u}_2, \mathbf{u}_3)$, as in Mardia and Jupp (2000, p. 290). If $K = C_2$ then $d = 2 - n^{-2}\sum_{k=1}^n\sum_{\ell=1}^n \{\mathbf{u}_{k0}^{\mathsf{T}}\mathbf{u}_{\ell0} + (\mathbf{u}_{k1}^{\mathsf{T}}\mathbf{u}_{\ell1})^2\}$, where $[\mathbf{U}_k]$ is represented by the C_2-frame $(\mathbf{u}_{k0}, \pm\mathbf{u}_{k1})$ for $k = 1, \ldots, n$. If $K = C_3$ then $d = 13/4 - n^{-1}\sum_{k=1}^n \left\{\mathbf{u}_{k0}^{\mathsf{T}}\bar{\mathbf{u}}_0 + \sum_{i=1}^3\sum_{j=1}^3 (\mathbf{u}_{ki}^{\mathsf{T}}\bar{\mathbf{u}}_j)^3\right\}$. If $K = D_2$ then $d = 2 - \sum_{j=1}^r \text{tr}\,(\bar{\mathbf{T}}_j^2)$.

2.4 Testing Uniformity

The most basic hypothesis about a distribution on $SO(3)/K$ is that it is uniform, i.e., invariant under left multiplication by $SO(3)$, in which $\mathbf{V}$ in

$SO(3)$ maps $[\mathbf{U}]$ to $[\mathbf{VU}]$. Since $\mathrm{E}\{\mathbf{t}([\mathbf{U}])\} = \mathbf{0}$ for $\mathbf{U}$ uniformly distributed on $SO(3)/K$, it is intuitively reasonable to reject uniformity if $\bar{\mathbf{t}}$ is far from $\mathbf{0}$, i.e., if $n\|\bar{\mathbf{t}}\|^2$ is large. For all sample sizes significance can be assessed using simulation from the uniform distribution on $SO(3)/K$. For large samples, the following asymptotic result can be used. A proof is given in the Appendix to Arnold et al. (2018).

Proposition 1

Given a random sample on $SO(3)/K$, define S by

$$S = (\nu/\rho^2)n\|\bar{\mathbf{t}}_K\|^2 = n\nu(1 - d/\rho^2), \tag{2.5}$$

where ρ^2 and d are given by (2.3) and (2.4), respectively, and ν is the dimension of E.

(i) For $K = C_1, D_r$ with $r \geq 2$, T, O or Y, under uniformity, the asymptotic distribution of S is $S \sim \chi_\nu^2$, as $n \to \infty$.

(ii) For $K = C_2$,
$$S = (1/3)S_R + (2/15)S_B,$$

where $S_R = 3n\bar{R}^2$ is the Rayleigh statistic for uniformity of $\mathbf{u}_0$ and $S_B = (15/2)n\{\mathrm{tr}(\bar{\mathbf{T}}^2) - (1/3)\}$ is the Bingham statistic for uniformity of $\pm\mathbf{u}_1$, $\bar{R}$ being the mean resultant length of $\mathbf{u}_0$ and $\bar{\mathbf{T}}$ being the sample scatter matrix of $\pm\mathbf{u}_1$. Under uniformity, S_R and S_B are asymptotically independent with asymptotic distributions χ_3^2 and χ_5^2, respectively.

(iii) For $K = C_r$ with $r \geq 3$,

$$(\rho_C^2/\nu_C)S = (1/3)S_R + (\rho_D^2/\nu_D)S_D,$$

where the subscripts C and D refer respectively to C_r-frames and the corresponding D_r-frames obtained by replacing the directed normal to the plane of a C_r-frame by the undirected normal, and $S_R = 3n\bar{R}^2$ is the Rayleigh statistic for uniformity of $\mathbf{u}_0$. Under uniformity, S_R and S_D are asymptotically independent with asymptotic distributions χ_3^2 and $\chi_{\nu_D}^2$, respectively.

Values of ρ^2 and ν are given in Table 2.4. In the case $K = C_1$ (i.e., no ambiguity), S is the Rayleigh statistic (Mardia & Jupp, 2000, p. 287) for testing uniformity on $SO(3)$. In the case $K = D_2$, S is the statistic given in Arnold and Jupp (2013, Section 3) for testing uniformity on $O(3)/\mathbb{Z}_2^3$.

Remark 1

The embeddings $\mathbf{t}$ provide not only tests of uniformity of ambiguous rotations, as described in this section, but also permutational multi-sample tests, tests of symmetry, tests of independence, and of goodness-of-fit, using the general machinery of Wellner (1979), Jupp and Spurr (1983), and (1985), and Jupp (2005). Two-sample tests are considered in Section 2.6.2.

TABLE 2.4
Values of squared radius, ρ^2, and dimension, ν

Group	ρ^2	ν
C_1	3	9
C_2	5/3	8
$C_r \quad (r \geq 3)$		
r odd	$1 + 2^{1-r}r^2$	$(r+2)(r+1)/2 + 3$
r even	$1 + r^2 2^{1-r}\left\{1 + 2^{-1}\binom{r}{r/2}\right\} - r^2/(r+1)$	$(r+2)(r+1)/2 + 3$
D_2	2	10
$D_r \quad (r \geq 3)$		
r odd	$2^{1-r}r^2$	$(r+2)(r+1)/2$
r even	$r^2 2^{1-r}\left\{1 + 2^{-1}\binom{r}{r/2}\right\} - r^2/(r+1)$	$(r+2)(r+1)/2$
T	32/9	10
O	6/5	15
Y	18816/6875	66
$\mathbb{Z}_2^r$	r	$r(p-1)(p+2)/2$

Bottom row refers to orthogonal axial r-frames in $\mathbb{R}^p$.

2.5 Distributions of Ambiguous Rotations

2.5.1 A General Class of Distributions on $SO(3)/K$

The scalar product $\langle \mathbf{t}([\mathbf{U}]), \mathbf{t}([\mathbf{V}]) \rangle$ provides a measure of closeness between $[\mathbf{U}]$ and $[\mathbf{V}]$, and so an appealing class of distributions on $SO(3)/K$ consists of those with densities of the form

$$f([\mathbf{U}]; [\mathbf{M}], \kappa) = g\left(\langle \mathbf{t}([\mathbf{U}]), \mathbf{t}([\mathbf{M}])\rangle; \kappa\right), \qquad (2.6)$$

where $g(\cdot; \kappa)$ is a suitable known function and $[\mathbf{M}] \in SO(3)/K$. The parameter $[\mathbf{M}]$ measures location and κ measures concentration. If $g(\cdot; \kappa)$ is a strictly increasing function, as in (2.7) or León et al. (2006) with $\kappa > 0$, then the mode is $[\mathbf{M}]$.

In the case $K = C_1$, the densities (2.6) depend on $\mathbf{U}$ only through $\text{trace}(\mathbf{U}\mathbf{M}^{\mathsf{T}})$ and the axes and the rotation angles of the random rotations are independent, with the axes being uniformly distributed. These distributions were introduced by Bingham et al. (2009) under the name of *uniform axis-random spin distributions* and by Hielscher et al. (2010) under the name of *radially symmetric distributions*.

For $K \neq C_1$, elements of $SO(3)/K$ do not have well-defined axes and, in general, the distributions on $SO(3)$ with densities $\tilde{f}$ of the form $\tilde{f}(\mathbf{U}) = f([\mathbf{U}]; [\mathbf{M}], \kappa)$ do not have uniformly distributed axes.

Taking $g(x; \kappa)$ proportional to $e^{\kappa x}$ in (2.6) gives the densities of the exponential form

$$f([\mathbf{U}]; [\mathbf{M}], \kappa) = c(\kappa)^{-1} \exp\{\kappa \langle \mathbf{t}([\mathbf{U}]), \mathbf{t}([\mathbf{M}]) \rangle\}. \qquad (2.7)$$

For $\kappa > 0$, the mode is $[\mathbf{M}]$ and the maximum likelihood estimate of $[\mathbf{M}]$ is the sample mean. The family (2.7) is a curved exponential family and is a subfamily of the crystallographic exponential family introduced by Boogaart (2002, Section 3.2). For $K = C_1$, (2.7) is the density of the matrix Fisher distribution with parameter matrix $\kappa \mathbf{M}$ and $c(\kappa) = {_0}F_1(3/2, (\kappa^2/4)\mathbf{I}_3)$ (Mardia & Jupp, 2000, Section 13.2.3), where ${_0}F_1$ is the hypergeometric function of matrix argument defined in (A.28) of Mardia and Jupp (2000). For $K = D_2$, (2.7) is the density of the equal concentration frame Watson distributions considered by Arnold and Jupp (2013, Section 6.1). Taking $g(x; \kappa)$ proportional to $(1 + x)^\kappa$ in (2.6) gives the densities of the form

$$f([\mathbf{U}]; [\mathbf{M}], \kappa) = c(\kappa)^{-1} \{1 + \langle \mathbf{t}([\mathbf{U}]), \mathbf{t}([\mathbf{M}]) \rangle\}^\kappa. \qquad (2.8)$$

For $K = C_1$, these densities are those of the de la Vallée Poussin distributions introduced by Schaeben (1997), and, under the name of Cayley distributions, by León et al. (2006). Taking $g(x; \kappa) = 1 + \kappa x$ with $0 \leq \kappa \leq \rho^{-2}$ in (2.6) gives the densities

$$f([\mathbf{U}]; [\mathbf{M}], \kappa) = 1 + \kappa \langle \mathbf{t}([\mathbf{U}]), \mathbf{t}([\mathbf{M}]) \rangle \qquad (2.9)$$

of the frame cardioid distributions, which are discussed briefly in Section 5.1 of Arnold et al. (2018).

For orthogonal axial frames (i.e., $K = D_2$), the families of distributions with densities (2.7) can be extended to a hierarchy of curved exponential models that are submodels of the frame Bingham family, which have densities of the form

$$f([\mathbf{U}]; \mathbf{A}_1, \mathbf{A}_2, \mathbf{A}_3) = c(\mathbf{A}_1, \mathbf{A}_2, \mathbf{A}_3)^{-1} \exp\left(\sum_{j=1}^{3} \mathbf{u}_j^{\mathsf{T}} \mathbf{A}_j \mathbf{u}_j\right), \qquad (2.10)$$

where $[\mathbf{U}] = (\pm \mathbf{u}_1, \pm \mathbf{u}_2, \pm \mathbf{u}_3)$, $\mathbf{A}_1, \mathbf{A}_2, \mathbf{A}_3$ are symmetric 3×3 matrices, which may be assumed to have trace zero, and $c(\mathbf{A}_1, \mathbf{A}_2, \mathbf{A}_3)$ is the normalizing constant. There are straightforward extensions to orthogonal axial r-frames in $\mathbb{R}^p$; see Arnold and Jupp (2013, Table 1).

2.5.2 Concentrated Distributions

A standard coordinate system on $SO(3)$ is given by the inverse of a restriction of the exponential map $\mathbf{S} \mapsto \sum_{k=0}^{\infty}(k!)^{-1}\mathbf{S}^k$ from the space of skew-symmetric 3×3 matrices to $SO(3)$. This can be modified to provide coordinate systems on $SO(3)/K$. Let $[\mathbf{M}]$ be an element of $SO(3)/K$. There are neighborhoods

$\mathcal{N}_{[\mathbf{M}]}$ of $[\mathbf{M}]$ in $SO(3)/K$ and $\mathcal{V}$ of $\mathbf{0}$ in $\mathbb{R}^3$ such that each $[\mathbf{U}]$ in $\mathcal{N}_{[\mathbf{M}]}$ can be written uniquely as $[\mathbf{U}] = [\mathbf{M}\exp\{\mathbf{A}(\mathbf{v})\}]$, where

$$\mathbf{A}(\mathbf{v}) = \begin{pmatrix} 0 & -v_3 & v_2 \\ v_3 & 0 & -v_1 \\ -v_2 & v_1 & 0 \end{pmatrix}$$

with $\mathbf{v} = (v_1, v_2, v_3)^{\mathsf{T}}$ in $\mathcal{V}$. Define $p_{[\mathbf{M}]}$ from $\mathcal{N}_{[\mathbf{M}]}$ to $\mathcal{V}$ by $p_{[\mathbf{M}]}([\mathbf{U}]) = \mathbf{v}$, where $[\mathbf{U}] = [\mathbf{M}\exp\{\mathbf{A}(\mathbf{v})\}]$. Then $p_{[\mathbf{M}]}$ is a coordinate system on $\mathcal{N}_{[\mathbf{M}]}$. Second-order Taylor expansion about $\mathbf{0}$ of $[\mathbf{U}]$ as a function of $\mathbf{v}$, together with some computer algebra, gives the high-concentration asymptotic distribution of $[\mathbf{U}]$.

Proposition 2

  For $[\mathbf{U}]$ near $[\mathbf{M}]$ in $SO(3)/K$ put $[\mathbf{U}] = [\mathbf{M}\exp\{\mathbf{A}(\mathbf{v})\}]$ for $\mathbf{v}$ near $\mathbf{0}$ in $\mathbb{R}^3$. If $[\mathbf{U}]$ has density (2.7) with $\mathbf{t} = \mathbf{t}_K$ as in Table 2.2 then the asymptotic distribution of $\kappa^{1/2}\mathbf{v}$ as $\kappa \to \infty$ is trivariate normal with mean $\mathbf{0}$ and variance $\boldsymbol{\Sigma}$, where $\boldsymbol{\Sigma}$ is given in Table 2.5. Similarly, if $[\mathbf{U}]$ has density (2.8) with $\mathbf{t} = \mathbf{t}_K$ then the asymptotic distribution of $(\kappa/2)^{1/2}\mathbf{v}$ is this trivariate normal distribution.

TABLE 2.5

High-concentration asymptotic variance, $\boldsymbol{\Sigma}$, of $\kappa^{1/2}\mathbf{v}$.

Group		$\boldsymbol{\Sigma}$
C_1		$(1/2)\mathbf{I}_3$
C_2		$\mathrm{diag}(1/2, 1/4, 1/6)$
C_r	$(r \geq 3)$	$\mathrm{diag}\left[(1 + rA_r)^{-1}, (1 + rA_r)^{-1}, \{2rA_r - r(r-1)A_{r-1}\}^{-1}\right]$
D_2		$(1/4)\mathbf{I}_3$
D_r	$(r \geq 3)$	$\mathrm{diag}\left[(rA_r)^{-1}, (rA_r)^{-1}, \{rA_r - r(r-1)A_{r-1}\}^{-1}\right]$
T		$0.070\mathbf{I}_3$
O		$(1/8)\mathbf{I}_3$
Y		$0.026\mathbf{I}_3$

For C_r and D_r, v_3 is the component of $\mathbf{v}$ normal to the plane of $\mathbf{u}_1, \ldots, \mathbf{u}_r$ and $A_r = \sum_{k=1}^{r} \cos(k2\pi/r)^r$. The coefficients of $\mathbf{I}_3$ in $\boldsymbol{\Sigma}$ for $K = T, Y$ are rounded to 3 decimal places.

2.6   Tests of Location

2.6.1   One-Sample Tests

Let $[\mathbf{M}]$ be an element of $SO(3)/K$ which is some measure of location of a distribution on $SO(3)/K$. There are various tests of the null hypothesis that $[\mathbf{M}] = [\mathbf{M}_0]$, where $[\mathbf{M}_0]$ is some given element of $SO(3)/K$. The case $K = D_2$ was considered in Arnold and Jupp (2013, Section 8).

For $K = C_2$, $[\mathbf{M}_0]$ can be represented by a C_2-frame $(_0, \pm_1)$. If the standard C_2-frame is taken as $(\mathbf{e}_0, \pm\mathbf{e}_1)$, where $(\mathbf{e}_0, \mathbf{e}_1, \mathbf{e}_2)$ is some specified orthonormal basis of $\mathbb{R}^3$, then one possibility for $\mathbf{M}_0$ is $\mathbf{M}_0 = (_0, _1, _0 \times_1)$ and C_2 can be identified with $\{\mathbf{I}, \mathbf{M}_0\mathbf{R}_\pi\mathbf{M}_0^\mathsf{T}\}$, where $\mathbf{R}_\pi = \mathrm{diag}(1, -1, -1)$.

One way of formalizing the idea that a distribution of random elements $[\mathbf{X}]$ of $SO(3)/K$ is located around $[\mathbf{M}_0]$ is as the hypothesis that the distribution is symmetrical about $[\mathbf{M}_0]$, in the sense that the distribution of $[\mathbf{M}_0\mathbf{R}\mathbf{M}_0^\mathsf{T}\mathbf{X}]$ is the same as that of $[\mathbf{X}]$, for any $\mathbf{R}$ in K. For a sample $[\mathbf{U}_1], \ldots, [\mathbf{U}_n]$ summarized by the sample mean $\bar{\mathbf{t}}$ of $\mathbf{t}$, an appealing measure of the squared distance between the sample and $[\mathbf{M}_0]$ is $\|\bar{\mathbf{t}} - \mathbf{t}([\mathbf{M}_0])\|^2$. It is appropriate to reject the null hypothesis for large values of $\|\bar{\mathbf{t}} - \mathbf{t}([\mathbf{M}_0])\|^2$. If the distribution of $[\mathbf{U}]$ is symmetrical about $[\mathbf{M}_0]$ then significance can be assessed by comparing the observed value of $\|\bar{\mathbf{t}} - \mathbf{t}([\mathbf{M}_0])\|^2$ with its randomization distribution, given by N pseudo-samples, in the tth of which $[\mathbf{U}_1], \ldots, [\mathbf{U}_n]$ is replaced by $[\mathbf{M}_0\mathbf{R}_t\mathbf{M}_0^\mathsf{T}\mathbf{U}_1], \ldots, [\mathbf{M}_0\mathbf{R}_t\mathbf{M}_0^\mathsf{T}\mathbf{U}_n]$, for $t = 1, \ldots, N$, where $\mathbf{R}_1, \ldots, \mathbf{R}_N$ are independent and distributed uniformly on K.

If $[\mathbf{U}_1], \ldots, [\mathbf{U}_n]$ is a sample from a concentrated distribution with density (2.7) and mode $[\mathbf{M}]$ then it is sensible to test $H_0 : [\mathbf{M}] = [\mathbf{M}_0]$ by applying Hotelling's 1-sample T^2 test to $p_{[\mathbf{M}_0]}([\mathbf{U}_1]), \ldots, p_{[\mathbf{M}_0]}([\mathbf{U}_n])$, where $p_{[\mathbf{M}_0]}$ is a suitable projection onto the tangent space to $SO(3)/K$ at $[\mathbf{M}_0]$.

2.6.2   Two-Sample Tests

Suppose that two independent random samples $[\mathbf{U}_1], \ldots, [\mathbf{U}_n]$ and $[\mathbf{V}_1], \ldots, [\mathbf{V}_m]$ on $SO(3)/K$ are summarized by the sample means $\bar{\mathbf{t}}_1$ and $\bar{\mathbf{t}}_2$ of $\mathbf{t}([\mathbf{U}_1]), \ldots, \mathbf{t}([\mathbf{U}_n])$ and $\mathbf{t}([\mathbf{V}_1]), \ldots, \mathbf{t}([\mathbf{V}_m])$. Then the squared distance between the two samples can be measured by $\|\bar{\mathbf{t}}_1 - \bar{\mathbf{t}}_2\|^2$. It is appropriate to reject the null hypothesis that the parent populations are the same if $\|\bar{\mathbf{t}}_1 - \bar{\mathbf{t}}_2\|^2$ is large. Significance can be assessed by comparing the observed value of $\|\bar{\mathbf{t}}_1 - \bar{\mathbf{t}}_2\|^2$ with its randomization distribution, obtained by sampling from the potential values corresponding to the partitions of the combined sample into samples of sizes n and m.

Suppose that $[\mathbf{U}_1], \ldots, [\mathbf{U}_n]$ and $[\mathbf{V}_1], \ldots, [\mathbf{V}_m]$ are samples from concentrated distributions with density (2.7) on $SO(3)/K$. Let $[\widehat{\mathbf{M}}]$ be the maximum likelihood estimate of the mode $[\mathbf{M}]$ under the null hypothesis that the parent populations are the same. Then the null hypothesis can be tested by applying Hotelling's 2-sample T^2 test to $p_{[\widehat{\mathbf{M}}]}([\mathbf{U}_1]), \ldots, p_{[\widehat{\mathbf{M}}]}([\mathbf{U}_n])$ and $p_{[\widehat{\mathbf{M}}]}([\mathbf{V}_1]), \ldots, p_{[\widehat{\mathbf{M}}]}([\mathbf{V}_m])$, where $p_{[\widehat{\mathbf{M}}]}$ is a suitable projection onto the tangent space to $SO(3)/K$ at $[\widehat{\mathbf{M}}]$.

2.7   Further Developments

Some regression models and tests of independence are considered in Section 7 of Arnold et al. (2018).

An important way in which orthogonal axial frames arise is as sets of principal axes of variance matrices. Although sets of principal axes are often considered as sets of unit vectors, the ambiguity of signs means that it is more appropriate to consider them as orthogonal axial frames. It is shown in Arnold and Jupp (2013, Theorem 2) that for large samples or concentrated distributions or diffuse multivariate normal distributions, the distributions of orthogonal axial frames derived from sample variance matrices are asymptotically of the form (2.10) with the parameter matrices commuting. For concentrated distributions, these distributions are asymptotically those derived by Anderson (1963).

2.8   Analysis of Examples

2.8.1   Analysis of Example 1

Application of the randomization test of uniformity based on S of (2.5) to the diopside data depicted in Figure 2.2 yields a p-value less than 0.001, leading to decisive rejection of uniformity.

Application of the two-sample permutation test above to the two samples shown in Figure 2.3 yields a p-value of 0.089 for equality of the populations of the orientations in the two regions, so the hypothesis of equality is not rejected.

2.8.2   Analysis of Example 2

The p-value of the randomization test of uniformity based on S of (2.5) applied to the ilmenite data depicted in Figure 2.5 is 0.873, so that uniformity is not rejected. (Comparison of S_R+S_D, where S_R and S_D are defined in Proposition 1(iii), with the large-sample asymptotic χ^2_{13} distribution gives the p-value as 0.948.)

2.8.3   Analysis of Example 3

The sample means of the three clusters in Figure 2.7 are shown as filled and labelled circles on the stereonets. The open squares denote approximate sample means obtained using a method based on Gram–Schmidt orthonormalization and described in Arnold and Jupp (2013, Section 2.2). In the concentrated earthquake clusters from Christchurch the approximate sample means are very close to the sample means, and are almost indistinguishable from them in the stereonets. In the diffuse data set from northern South Island the approximate sample means differ only slightly from the sample means.

The dispersions of the three samples in Figure 2.7 are $d = 1.237$, 1.276 and 1.902. The dispersions of the two Christchurch samples are similar to one another, and the sample from northern South Island has higher dispersion.

For each of the Christchurch clusters shown in Figure 2.7, the test based on S rejects uniformity decisively with p-values less than 10^{-30}. In accordance with the visual message of Figure 2.7, the northern South Island cluster shows no evidence of lack of uniformity, as the test for uniformity on $O(3)/\mathbb{Z}_2^3$ has p-value 0.107.

The two-sample permutation test of §1.6.2 yields a p-value of 0.890 for equality of the populations of the alignments of the two Christchurch clusters but < 0.001 for equality of the first Christchurch cluster and the cluster from northern South Island.

Acknowledgments

We are grateful to David Mainprice for providing the crystallographic data and to John Townend and Steven Sherburn for supplying the New Zealand earthquake data.

Bibliography

[1] Anderson, T. W. 1963. Asymptotic theory for principal components analysis. *Ann. Math. Statist.* 34: 122–148.

[2] Arnold, R. and Jupp, P. E. 2013. Statistics of orthogonal axial frames. *Biometrika* 100: 571–586.

[3] Arnold, R., Jupp, P. E. and Schaeben, H. 2018. Statistics of ambiguous rotations. *J. Multivariate Anal.* 165: 73–85.

[4] Bindi, L., Steinhardt, P., Yao, N. and Lu, P. J. 2011. Icosahedrite, $Al_6Cu_{24}Fe_{13}$, the first natural quasicrystal. *Amer. Mineralogist* 96: 928–31.

[5] Bingham, M. A., Nordman, D. J. and Vardeman, S. B. 2009. Modeling and inference for measured crystal orientations and a tractable class of symmetric distributions for rotations in three dimensions. *J. Amer. Statist. Assoc.* 104: 1385–97.

[6] van den Boogaart, K. G. 2002. *Statistics for Individual Crystallographic Orientation Measurements.* Aachen: Shaker Verlag.

[7] Fisher, N. I., Lewis, T. and Embleton, B. J. J. 1987. *Statistical Analysis of Spherical Data*. Cambridge: Cambridge Univ. Press.

[8] Harrison, S. C. 2013. Principles of Virus Structure. In: *Fields Virology*, 6th ed., ed. D. M. Knipe, P. M. Howley, J. L. Cohen et al. 52–86. Philadelphia: Wolters Kluwer Health/Lippincott Williams and Wilkins.

[9] Hielscher, R., Schaeben, H. and Siemes, H. 2010. Orientation distribution within a single hematite crystal. *Math. Geosciences* 42: 359–375.

[10] Jemmis, E. D. 1982. Overlap control and stability of polyhedral molecules. *closo*-carboranes. *J. Am. Chem. Soc.* 104: 7017–7020.

[11] Jupp, P. E. 2005. Sobolev tests of goodness of fit of distributions on compact Riemannian manifolds. *Ann. Statist.* 33: 2957–2966.

[12] Jupp, P. E. and Spurr, B. D. 1983. Sobolev tests for symmetry of directional data. *Ann. Statist.* 11: 1225–1231.

[13] Jupp, P. E. and Spurr, B. D. 1985. Sobolev tests for independence of directions, *Ann. Statist.* 3: 1140–1155.

[14] León, C., Massé, J.-C. and Rivest, L.-P. 2006. A statistical model for random rotations. *J. Multivariate Anal.* 97: 412–430.

[15] Ley, C. and T. Verdebout, T. 2017. *Modern Directional Statistics*. Chapman & Hall/CRC.

[16] Mardia, K. V. and Jupp, P. E. 2000. *Directional Statistics*. Chichester: Wiley.

[17] Miller, W. 1972. *Symmetry Groups and their Applications*. New York: Academic Press.

[18] Schaeben, H. 1997. A simple standard orientation density function: The hyperspherical de la Vallée Poussin kernel. *Phys. Stat. Sol. (b)* 200: 367–376.

[19] Seideman, T. 1990. The liquid-crystalline blue phases. *Rep. Prog. Phys.* 53: 659–705.

[20] Stein, S. and Wysession, M. 2003. *An Introduction to Seismology, Earthquakes and Earth Structure*. Malden, MA: Blackwell.

[21] Wellner, J. A. 1979. Permutation tests for directional data. *Ann. Statist.* 7: 929–43.

3

Correlated Cylindrical Data

Francesco Lagona

University of Roma Tre

CONTENTS

3.1 Correlated Cylindrical Data 47
3.2 Cylindrical Hidden Markov Models 49
    3.2.1 The Abe–Ley Density 49
    3.2.2 Modeling a Cylindrical Time Series 49
    3.2.3 Modelling a Cylindrical Spatial Series 51
3.3 Identification of Sea Regimes 52
3.4 Segmentation of Current Fields 54
3.5 Outline 55
    Acknowledgments 57
    Bibliography 57

Cylindrical data are pairs of angles and intensities that are simultaneously observed on the same statistical unit. The name *cylindrical* is motivated by the special domain of these data because the pair of an angle and an intensity can be described as a point on a cylinder. Correlated cylindrical data arise in environmental studies when cylindrical measurements are repeatedly taken along time or across space, respectively taking the form of a cylindrical time series or a cylindrical spatial series. Popular examples of cylindrical time series include temporal sequences of wind directions and pollutant concentrations [7], wind directions and speeds [18], wildfire sizes and orientations [8], wave directions and heights [4, 17, 24]. Examples of cylindrical spatial series include satellite images of wind directions and speeds [19], wave directions and heights that are generated by deterministic wave models [22, 23], speeds and directions of marine currents recorded by a network of high-frequency radars [17], as well as telemetry data of animal movement [9].

The analysis of correlated cylindrical data is complicated by the special topology of the support on which the measurements are taken (the cylinder), and to the difficulties in modeling the cross-correlations between angular and linear measurements across space and along time. Additional complications arise from the skewness and the multimodality of the marginal distribution of the data. Indeed, intensities are typically negatively skewed and directional data are rarely symmetric; multimodality may arise as well as the data are often observed under heterogeneous conditions that vary in time or space.

This chapter focuses on a specific approach to the analysis of correlated cylindrical data which has been recently proposed in the literature [15, 16, 21]. It is based on the specification of hidden Markov models (HMMs) that parsimoniously account for the specific features of cylindrical temporal and spatial series. Taking an HMM approach, more precisely, the data distribution is approximated by a mixture of cylindrical densities, whose parameters vary across time or space according to a latent (hidden) multinomial Markov process. This approach extends traditional HMMs for cylindrical data, where intensities and directions are conditionally independent given the latent state generated by the multinomial process [8, 17]. Conditional independence facilitates both the specification and the estimation of a bivariate non-normal HMM and can be motivated by borrowing arguments from the latent-class literature [14]. It is however a restrictive assumption in many empirical applications and may not properly accommodate the complex shape of cylindrical distributions. In a Bayesian setting Mastrantonio et al. [18] recently explored the HMM to jointly model cylindrical time series, by assuming a projected normal distribution to account for correlations between variables. This specification relaxes the conditional independence assumption between circular and linear variables. However, as stated by the authors, major drawbacks of this approach include difficulties in parameter interpretation, because different sets of parameters can give really similar distributional shapes.

Such issues can be addressed by exploiting parametric cylindrical densities, that is bivariate densities with a cylindrical support, known up to a number of intuitively appealing parameters. In addition, mixtures of cylindrical densities provide a flexible distributional extension to allow for multimodal cylindrical data. Assuming that the mixture parameters vary according to the evolution of a latent process across time or space represents a further extension to capture unobserved temporal or spatial heterogeneity and to allow for temporal or spatial correlation.

The choice of a suitable cylindrical density and a suitable multinomial process depends on the settings of the motivating data and the research goals. In practice, this choice is also a compromise between model flexibility and computational complexity. The models illustrated in this chapter are based on the Abe–Ley density [1], a five-parameter bivariate density on the cylinder. It parsimoniously accommodates correlated and skew cylindrical data by means of parameters that can be easily interpreted in terms of traditional concepts such as location, shape, scale, skewness, and concentration. Motivated by two cases studies, two HMMs are discussed. The first HMM is a time-varying mixture of Abe–Ley densities, whose parameters evolve according to a latent homogeneous Markov chain. It is exploited to model a cylindrical time series. The second HMM, exploited for the analysis of a cylindrical spatial series, is a space-varying mixture of Abe–Ley densities, whose parameters evolve according to a Potts model, that is a one-parameter homogeneous Markov random field.

The rest of the chapter is organized as follows. Two case studies of correlated cylindrical data are presented in Section 3.1. The cylindrical HMMs in time and space are introduced in Section 3.2, and applied to the case studies in Sections 3.3 and 3.4, respectively. Finally, Section 3.5 provides an outline of the chapter.

3.1 Correlated Cylindrical Data

Environmental agencies routinely collect correlated cylindrical data. Figure 3.1 summarizes two typical examples that often arise in marine research.

The first example (Figure 3.1, top) is the scatter plot of semihourly wave directions and heights, recorded in the period 2/15/2010 – 3/16/2010 by the buoy of Ancona, located in the Adriatic Sea at about 30 km from the coast. These data are collected to identify sea regimes during the study period, i.e., typical distributions that the data take under specific conditions. In wintertime, relevant events in the Adriatic Sea are typically generated by the southeastern Sirocco wind and the northern Bora wind. Bora and Sirocco episodes are usually associated with high waves. The Adriatic sea is however mostly characterized by waves of moderate height. As a result, the distribution of wave heights is negatively skewed. Less perceivable is the skewness of the distribution of wave direction. It can be explained by recalling the orography of the Adriatic Sea, as the waves tend to travel from north-northwest and southeasterly along the major axis of the basin, where they can travel freely, without being obstructed by physical obstacles, such as coastlines.

Although some distinct patterns appear in this scatter plot, their interpretation is difficult due to the weak correlation between the measurements, the skewness and the multimodality of the data. Multimodality, weak correlation, and skewness are often held responsible for the inaccuracy of numerical wave models in the Adriatic Sea and they are traditionally explained as the result of the complex orography of this basin, combined with time-varying weather conditions. This motivates the development of special methods in order to segment these data into a small number of latent classes, conditionally on which the distribution of the data takes a shape that is easier to interpret than the shape taken by the marginal distribution.

The second example (Figure 3.1, bottom) is obtained by the spatial pattern of surface current vectors which indicate the direction and the speed of the current across a regular grid of points having a horizontal resolution of 2.5 km × 2.5 km. These surface measurements were obtained in the Northern Adriatic Sea, on the 2008/05/22 at 18:00, by a network of high-frequency (HF) radars installed along the coast of the area of interest. HF radars measure

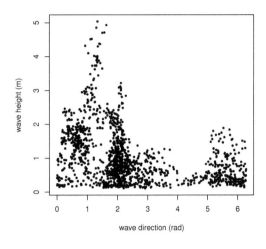

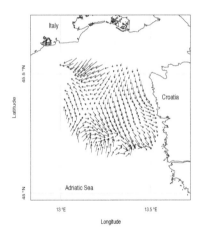

FIGURE 3.1

Top: Wave directions and heights, observed at the Ancona buoy of the Adriatic basin in wintertime; angles indicate the directions from which wave travels (North = 0 rad). Bottom: the spatial distribution of sea currents in the Northern Adriatic Sea (arrow length is proportional to current speed).

surface currents by detecting the Doppler shift of an electromagnetic wave transmitted at a certain frequency. The salient features of this spatial data pattern can be interpreted by recalling that the northern Adriatic Sea often shows a mixing between currents driven by the Bora wind (a typical northern wind of this area) and cyclonic circulation currents. The distribution of current speed is multimodal because most of the currents in the study area flow at moderate rates. The skewness of the data distribution is due to the occurrence of acceleration events during Bora episodes.

The large amount of data heterogeneity displayed by the picture suggests that interpreting the data as a mixture of two regimes (Bora and cyclonic circulation) may however be too restrictive, motivating the estimation of an optimal segmentation of the study area.

3.2 Cylindrical Hidden Markov Models

The proposed cylindrical HMMs integrate Abe–Ley densities with latent Markov processes that evolve along time or across space. In this section, the Abe–Ley density is first introduced as a convenient cylindrical density. Second, it is combined with a latent Markov chain to describe an HMM for cylindrical time series. This approach is finally extended to the spatial setting, by integrating Abe–Ley densities with a latent Potts model, a one-parameter multinomial Markov field.

3.2.1 The Abe–Ley Density

A cylindrical sample is a pair $\mathbf{z} = (x, y)$, where $x \in [0, 2\pi)$ is a point in the circle and y is a point on the positive semi-line $[0, +\infty)$. The Abe–Ley density is defined on the cylinder $[0, 2\pi) \times [0, +\infty)$ and takes the form

$$f(x, y) = \frac{\alpha \beta^\alpha}{2\pi \cosh(\kappa)} \left(1 + \lambda \sin(x - \mu)\right) y^{\alpha-1} \exp\left(-(\beta y)^\alpha (1 - \tanh(\kappa) \cos(x - \mu))\right)$$

(3.1)

The properties of this density are extensively described by Abe and Ley [1]. We summarize here the most important advantages of this density. First, the meaning of the parameters is intuitively appealing. Precisely, $\alpha > 0$ is a (linear) shape parameter, $\beta > 0$ is a (linear) scale parameter, $\mu \in [0, 2\pi)$ is a (circular) location parameter, $\kappa > 0$ is a (circular) concentration parameter and, finally, $\lambda \in [-1, 1]$ is a (circular) skewness parameter. Second, the normalizing constant $2\pi \cosh(\kappa)$ is numerically tractable, and this facilitates maximum likelihood estimation. Third, under this model, the univariate marginal and conditional distributions are available in closed form, and this simplifies simulation routines.

3.2.2 Modeling a Cylindrical Time Series

A cylindrical time series is a bivariate sequence with linear and circular components, say $\mathbf{z} = (\mathbf{z}_t, t = 0, \ldots, T)$, $\mathbf{z}_t = (x_t, y_t)$, $x_t \in [0, 2\pi)$ and $y_t \in [0, +\infty)$. Taking an HMM approach, we assume that the distribution of the data is driven by the evolution of an unobserved Markov chain with K states, which represent (time-varying) latent classes. A Markov chain is essentially a multinomial process in discrete time. Accordingly, we introduce a sequence $\boldsymbol{\xi} = (\boldsymbol{\xi}_t, t = 0, \ldots, T)$ of multinomial variables $\boldsymbol{\xi}_t = (\xi_{t1} \ldots \xi_{tK})$ with one trial and K classes, whose binary components represent class membership at time t. The joint distribution $p(\boldsymbol{\xi})$ of the chain is fully known up to K initial probabilities $\pi_k = P(\xi_{0k} = 1), k = 1, \ldots, K, \sum_k p_k = 1$, and K^2 transition probabilities $\pi_{hk} = P(\xi_{tk} = 1 | \xi_{t-1,h} = 1), h, k = 1, \ldots, K, \sum_k \pi_{hk} = 1$. Formally, we assume that

$$p(\boldsymbol{\xi}) = \prod_{k=1}^{K} \pi_k^{\xi_{0k}} \prod_{t=1}^{T} \prod_{h=1}^{K} \prod_{k=1}^{K} \pi_{hk}^{\xi_{t-1,h}\xi_{tk}}. \qquad (3.2)$$

The specification of the cylindrical HMM is completed by assuming that the observations are conditionally independent, given a realization of the Markov chain. As a result, the conditional distribution of the observed process, given the latent process, takes the form of a product density, say

$$f(\mathbf{z}|\boldsymbol{\xi}) = \prod_{t=0}^{T} \prod_{k=1}^{K} f_k(\mathbf{z}_t)^{\xi_{tk}}, \qquad (3.3)$$

where $f_k(\mathbf{z}), k = 1, \ldots, K$ are K Abe–Ley densities, known up to a vector of parameters $\boldsymbol{\theta}_k$. The observed process specifically depends on $5 \times K$ parameters. The joint density of the observed data and the unobserved class memberships is therefore given by

$$f(\mathbf{z}|\boldsymbol{\xi})p(\boldsymbol{\xi}).$$

By integrating this distribution with respect to the segmentation $\boldsymbol{\xi}$, we obtain the marginal distribution of the observed data, known up to the parameters $(\boldsymbol{\pi}, \boldsymbol{\theta}_1 \ldots \boldsymbol{\theta}_K)$. Under this setting, the maximum likelihood estimates, $\hat{\boldsymbol{\pi}}$ and $\hat{\boldsymbol{\theta}}$, of the parameters can be obtained by maximizing the likelihood function

$$L(\boldsymbol{\pi}, \boldsymbol{\theta}; \mathbf{z}) = \sum_{\boldsymbol{\xi}} p(\boldsymbol{\xi}; \boldsymbol{\pi}) f(\mathbf{z}|\boldsymbol{\xi}; \boldsymbol{\theta}). \qquad (3.4)$$

A specific algorithm for maximizing this likelihood function is provided by Lagona et al. [16] who also give details on simulation routines to obtain bootstrap standard errors of the estimates.

An interesting limiting case of the proposed cylindrical HMM is obtained when $\pi_{hk} = \pi_k$, for each $h, k = 1 \ldots K$. In this case, the transition probabilities of the chain do not depend on the previous state and the Markov process collapses into a sequence of independent multinomial random variables. Under this setting, the model reduces to a latent class model for cylindrical data. Assuming transition probabilities that do not depend on the previous state can be useful in settings that involve occasional environmental factors, occurring independently during the observation period [13]. It is however unreasonable in studies where the outcome is influenced by weather conditions that are naturally correlated in time, such as in studies of wave dynamics.

3.2.3 Modelling a Cylindrical Spatial Series

A cylindrical spatial series can be represented as a bivariate vector of angles x_i and intensities y_i, observed at a lattice of n observation points, say $\mathbf{z} = (\mathbf{z}_i, i = 1, \dots, n)$, where $\mathbf{z}_i = (x_i, y_i)$, $x_i \in [0, 2\pi)$, and $y_i \in [0, +\infty)$. In this spatial setting, a hidden Markov model can be specified by introducing a (latent) multinomial process $\boldsymbol{\xi}$ that evolves across space and assuming that the cylindrical observations are conditionally independent given the pattern generated by this process, say

$$f(\mathbf{z}|\boldsymbol{\xi}) = \prod_{i=1}^{n} \prod_{k=1}^{K} f_k(\mathbf{z}_i)^{\xi_{ik}}, \tag{3.5}$$

where $f_k(\mathbf{z}), k = 1, \dots, K$ are K Abe–Ley densities, defined on the cylinder $[0, 2\pi) \times [0, +\infty)$ and known up to K parameters $\boldsymbol{\theta}_k, k = 1 \dots K$. The observed process specifically depends on $5 \times K$ parameters.

There are several ways to specify the joint distribution of spatial multinomial processes. The Potts model is a special multinomial process in discrete space which depends on a single interaction parameter ρ and on a neighborhood structure among the observation sites [6]. A neighborhood structure on $S = \{1 \dots i \dots n\}$, say $E \subset S^2$, is a symmetric and nonreflexive binary relationship, such that $(i, j) \in E \Rightarrow (j, i) \in E$ (symmetry) and $(i, i) \notin E$ (non-reflexivity). According to the structure E, each spatial index i is associated with a neighborhood $N(i) = \{j \in S : (i, j) \in E\}$ of adjacent observations points. The structure E can be interpreted as the set of the edges of an undirected graph with vertices in S and it is often conveniently specified by means of a $n \times n$ symmetric adjacency matrix $\mathbf{C}$, whose generic entry c_{ij} is equal to 1 if $(i, j) \in E$ and 0 otherwise (diagonal entries c_{ii} are all equal to zero, due to the nonreflexivity of E). Under the Potts model, the joint distribution of the segmentation $\boldsymbol{\xi} = (\boldsymbol{\xi}_1 \dots \boldsymbol{\xi}_n)$ is given by

$$p(\boldsymbol{\xi}) = \frac{\exp(\frac{\rho}{2} \sum_{i=1}^{n} \sum_{j:c_{ij}=1} \boldsymbol{\xi}_i^{\mathsf{T}} \boldsymbol{\xi}_j)}{W(\rho)} \tag{3.6}$$

where $W(\rho)$ is the normalizing constant.

When $\rho = 0$, the Potts model reduces to a sequence of independent multinomial distributions. Otherwise, the multinomial components of the process are spatially dependent. In particular, under the Potts model, the univariate conditional distributions

$$p(\xi_{ik} = 1 \mid \boldsymbol{\xi}_1 \dots \boldsymbol{\xi}_{i-1}, \boldsymbol{\xi}_{i+1}, \dots \boldsymbol{\xi}_n) = \frac{\exp(\rho \xi_{ik} \sum_{j:c_{ij}=1} \xi_{jk})}{\sum_{k=1}^{K} \exp(\rho \xi_{ik} \sum_{j:c_{ij}=1} \xi_{jk})} \tag{3.7}$$

depend only on the observations in the neighborhood (spatial Markov property). As a result, the Potts model is a special one-parameter Markov random field.

Under this setting, the joint distribution of the observed data and the unobserved class memberships is given by $f(\mathbf{z}|\boldsymbol{\xi})p(\boldsymbol{\xi})$. By integrating this distribution with respect to the spatial segmentation $\boldsymbol{\xi}$, we obtain the marginal distribution of the spatial cylindrical series, known up to the parameters $(\rho, \boldsymbol{\theta}_1 \dots \boldsymbol{\theta}_K)$. The maximum likelihood estimates of these parameters can be obtained by maximizing the likelihood function

$$L(\rho, \boldsymbol{\theta}; \mathbf{z}) = \sum_{\boldsymbol{\xi}} p(\boldsymbol{\xi}; \rho) f(\mathbf{z}|\boldsymbol{\xi}; \boldsymbol{\theta}). \tag{3.8}$$

A specific algorithm for maximizing this likelihood function is provided by Lagona and Picone [21] who also give details on simulation routines to obtain bootstrap standard errors of the estimates.

Because of the Markov property of the Potts model, the proposed model can be referred to as a cylindrical hidden Markov field (HMF). Hidden MRFs are popular models in spatial statistics, since the seminal paper by Besag [2]. They can be seen as an extension of hidden Markov models, exploited in time series analysis, to the spatial setting. In the proposed framework, the segmentation labels generated by the Potts model can be interpreted as latent classes, which cluster observation sites according to label-specific Abe–Ley cylindrical distributions.

3.3   Identification of Sea Regimes

The cylindrical HMM introduced in Section 3.2.2 was estimated from the time series of wave directions and heights illustrated in Section 3.1, by varying the number K of components from 2 to 4. In this application, the BIC statistic suggested a model with $K = 3$ components. Figure 3.2 overlaps the inferred state-specific densities on the data points, filled with gray levels according to the posterior membership probabilities $P(\xi_{tk} = 1 \mid \hat{\boldsymbol{\theta}}, \hat{\boldsymbol{\pi}}, \mathbf{z})$, under each state (black indicates $\hat{\pi}_{tk} = 1$).

The first component of the model is associated with high waves coming from the North. These waves are generated by northern Bora jets that blow along the major axis of the Adriatic basin. These waves are highly concentrated around one modal direction. Under a Bora episode, most of the wind energy is transferred to the sea surface and, as a result, most of the data with the highest waves in the sample are clustered within this regime. Interestingly, the distribution of wave directions is negatively skewed, due to the orographic conditions of the basin: only the waves that travel along the major axis of the basis are allowed to reach elevated heights.

The second component of the model is associated with periods of calm sea. Under this regime, moderate waves are uniformly distributed around the circle of directions. As a result, the dependence between height and direction

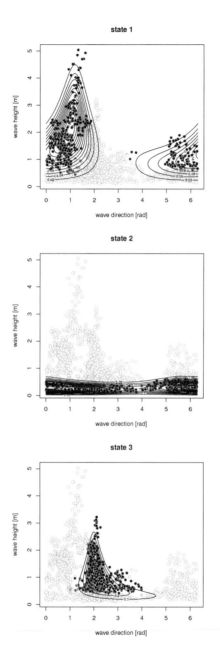

FIGURE 3.2
State-specific distribution of the data, according to a cylindrical hidden Markov model with three states; points are filled with gray levels according to the estimated posterior membership probabilities (black indicates a probability equal to 1).

is weak and the shape of the joint distribution of wave height and direction is essentially symmetric.

The third component is associated with Sirocco episodes. In studies of the Adriatic Sea, detection of the Sirocco regime is very important because it exposes Venice to the famous flooding tides when occurring in combination with luni-solar astronomical forces. Compared to the first regime, wave heights and directions appear more synchronized. In this second regime, waves travel southeasterly along the major axis of the basin, driven by winds that blow from a similar directional angle. The positive skewness parameter is the result of the orography effect. While this effect was negative for the northern waves in the first component of the model, it is positive under the third component, which clusters waves that travel southeasterly.

3.4   Segmentation of Current Fields

The surface current field illustrated in Section 3.1 was segmented by a number of cylindrical hidden Markov random fields (Section 3.2.3), by varying the number K of latent classes from 2 to 5. The BIC statistic suggested a model with $K = 4$ components. Figures 3.3 and 3.4 overlap the class-specific cylindrical densities on the data points, filled with gray levels according to the posterior class-membership probabilities $P(\xi_{ik} = 1 \mid \mathbf{z}; \hat{\rho}, \hat{\boldsymbol{\theta}})$ (black indicates that the posterior probability is equal to 1), under each class. These figures further include the spatial pattern of the sites that are clustered within each latent class, according to the estimated class-membership probabilities (a black arrow indicates $P(\xi_{ik} = 1 \mid \mathbf{z}; \hat{\rho}, \hat{\boldsymbol{\theta}}) = 1$).

Class 1 (Figure 3.3: top) is associated with northern Bora jets that blow northeasterly. Most of the currents that flow at the fastest rates in the sample are clustered within this regime. The high value of the concentration parameter further indicates that speed and directions are highly correlated and that currents are highly concentrated around one modal direction.

Class 2 (Figure 3.3: bottom) is instead associated with cyclonic circulation flows. Most of the moderate currents that travel along the Istrian coast are clustered under this regime. Interestingly, the distribution of the speeds within this class is positively skewed, due to the orographic conditions of the basin: only the currents that travel along the major axis of the Adriatic basin are allowed to reach elevated speed.

Classes 3 and 4 (Figure 3.4) can be instead interpreted as transition states between the two opposite regimes identified by the first to classes. These classes, indeed, cluster moderate currents which are uniformly distributed around the circle of directions. As a result, the dependence between speed and direction is weak and the shape of the joint distribution of speeds and directions is essentially symmetric.

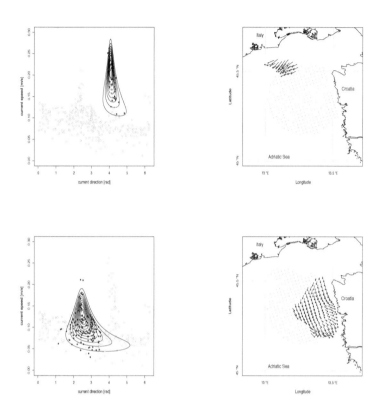

FIGURE 3.3
State-specific (top: state 1 – bottom: state 2) joint distributions (left) and spatial distributions (right) of the data, according to a cylindrical hidden Markov random field model with four states; points and arrows are colored with gray levels according to the estimated posterior membership probabilities (black indicates a probability equal to 1);

3.5 Outline

Hidden Markov models provide a flexible framework to analyze spatial and time series of cylindrical data. This approach allows to segment cylindrical time series and to cluster cylindrical spatial data according to a finite number of latent classes, associated with specific Abe–Ley densities that describe the distribution of the data under each class. These models depend on a number of parameters that increase linearly (in the case of a spatial HMRF) and

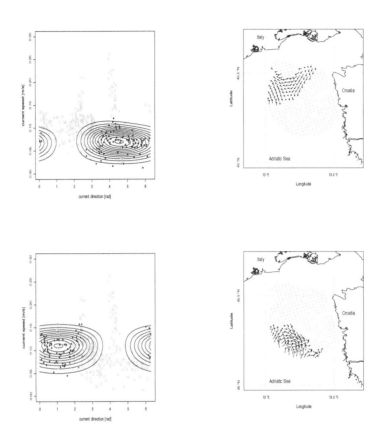

FIGURE 3.4
State-specific ((top: state 3 - bottom: state 4)) joint distributions (left) and spatial distributions (right) of the data, according to a cylindrical hidden Markov random field model with 4 states; points and arrows are colored with grey levels according to the estimated posterior membership probabilities (black indicates a probability equal to 1).

more than linearly (in the case of a temporal HMM) with the number of latent classes. Large sample sizes are in general required to attain a reasonable precision of the estimates. In this chapter, the sample sizes of the presented case studies are large enough to allow the numerical maximization of the likelihood of a temporal HMM with three states and a spatial HMRF with four states, respectively. When parameter estimation is feasible and numerically stable, parametric bootstrap methods offer as a viable strategy to estimate the precision of the estimates (through confidence intervals that are based on the empirical distribution of the bootstrap estimates). The precision of the estimates, however, does not only depend on the number of parameters and the sample size, but also on the amount of correlation of the data. Correlation introduces redundancy in the information that the data provide about the parameters. As a result, in the case of strongly correlated data, the sample size that is required to get a given precision might be much larger than the size that is required for a sample of almost independent data.

The proposed approach was motivated by segmentation issues that arise in marine studies. In marine research, the identification of environmental regimes that meaningfully segment wave and current processes is crucial in studies of the drift of floating objects and oil spills [11], in the design of off-shore structures [5], and in studies of sediment transport [12] and coastal erosion [20]. Nevertheless, HMMs can be easily adapted to a wide range of real-world data, including, for example, ecological studies of animal behavior, where direction and speed of movements are recorded across space or time, as well as environmental studies that involve time sequence or spatial patterns of wind speeds and directions.

From a methodological viewpoint, the model offers a number of advantages. It flexibly accommodates spatial correlation, linear-circular correlation, skewness and multimodality by means of parameters that can be easily interpreted in terms of traditional concepts such as location, shape, scale, skewness, and concentration. In the considered case studies of marine research, this approach offered a parsimonious description of wave dynamics and surface current patterns.

Acknowledgments

Francesco Lagona is supported by the 2015 PRIN supported project "Environmental processes and human activities: capturing their interactions via statistical methods," funded by the Italian Ministry of Education, University and Scientific Research.

Bibliography

[1] Toshihiro Abe and Christophe Ley. A tractable, parsimonious and flexible model for cylindrical data, with applications. *Econometrics and Statistics*, 4(Supplement C):91 – 104, 2017.

[2] Julian Besag. On statistical analysis of dirty pictures. *Journal of the Royal Statistical Society B*, 48:259–302, 1986.

[3] J. Bulla, F. Lagona, A. Maruotti, and M. Picone. A multivariate hidden markov model for the identification of sea regimes from incomplete skewed and circular time series. *Journal of Agricultural, Biological, and Environmental Statistics*, 17(4):544–567, 2012.

[4] Jan Bulla, Francesco Lagona, Antonello Maruotti, and Marco Picone. Environmental conditions in semi-enclosed basins: A dynamic latent class approach for mixed-type multivariate variables. *Journal de la Société Française de Statistique*, 156:114–137, 2015.

[5] OM Faltinsen. *Sea Loads on Ships and Offshore Structures*. Cambridge University Press, 1990.

[6] C Gaetan and X Guyon. *Spatial Statistics and Modelling*. Springer, 2010.

[7] E. García-Portugués, R.M. Crujeiras, and W. González-Manteiga. Exploring wind direction and so2 concentration by circular-linear density estimation. *Stochastic Environmental Research and Risk Assessment*, 27(5):1055–1067, 2013.

[8] Eduardo García-Portugués, Ana M.G. Barros, Rosa M. Crujeiras, Wenceslao González-Manteiga, and José Pereira. A test for directional-linear independence, with applications to wildfire orientation and size. *Stochastic Environmental Research and Risk Assessment*, 28:1261–1275, 2014.

[9] Ephraim M. Hanks, Mevin B. Hooten, and Mat W. Alldredge. Continuous-time discrete-space models for animal movement. *The Annals of Applied Statistics*, 9:145–165, 2015.

[10] Hajo Holzmann, Axel Munk, Max Suster, and Walter Zucchini. Hidden markov models for circular and linear-circular time series. *Environmental and Ecological Statistics*, 13:325–347, 2006.

[11] G Huang, A Wing-Keung Law, and Z Huang. Wave-induced drift of small floating objects in regular waves. *Ocean Engineering*, 38:712–718, 2011.

[12] K-R Jin and Z-G Ji. Case study: Modeling of sediment transport and wind-wave impact in Lake Okeechobee. *Journal of Hydraulic Engineering*, 130:1055–1067, 2004.

[13] Francesco Lagona, Dmitri Jdanov, and Maria Shkolnikova. Latent time-varying factors in longitudinal analysis: a linear mixed hidden markov model for heart rates. *Statistics in Medicine*, 33(23):4116–4134, 2014.

[14] Francesco Lagona and Marco Picone. Model-based clustering of multivariate skew data with circular components and missing values. *Journal of Applied Statistics*, 39(5):927–945, 2012.

[15] Francesco Lagona and Marco Picone. Model-based segmentation of spatial cylindrical data. *Journal of Statistical Computation and Simulation*, 86(13):2598–2610, 2016.

[16] Francesco Lagona, Marco Picone, and Antonello Maruotti. A hidden markov model for the analysis of cylindrical time series. *Environmetrics*, 26(8):534–544, 2015.

[17] Francesco Lagona, Marco Picone, Antonello Maruotti, and Simone Cosoli. A hidden markov approach to the analysis of space-time environmental data with linear and circular components. *Stochastic Environmental Research and Risk Assessment*, 29(2):397–409, 2015.

[18] Gianluca Mastrantonio, Antonello Maruotti, and Giovanna Jona-Lasinio. Bayesian hidden markov modelling using circular-linear general projected normal distribution. *Environmetrics*, 26(2):145–158, 2015.

[19] Danny Modlin, Montserrat Fuentes, and Brian Reich. Circular conditional autoregressive modeling of vector fields. *Environmetrics*, 23(1):46–53, 2012.

[20] A Pleskachevsky, DP Eppel, and H Kapitza. Interaction of waves, currents and tides, and wave-energy impact on the beach area of sylt island. *Ocean Dynamics*, 59:451–461, 2009.

[21] Monia Ranalli, Francesco Lagona, Marco Picone, and Enrico Zambianchi. Segmentation of sea current fields by cylindrical hidden markov models: a composite likelihood approach. *Journal of the Royal Statistical Society: Series C (Applied Statistics)*, forthcoming.

[22] F. Wang, A.E. Gelfand, and G. Jona-Lasinio. Joint spatio-temporal analysis of a linear and a directional variable: space-time modeling of wave heights and wave directions in the Adriatic Sea. *Statistica Sinica*, 25:25–39, 2015.

[23] Fangpo Wang and Alan E. Gelfand. Modeling space and space-time directional data using projected gaussian processes. *Journal of the American Statistical Association*, 109(508):1565–1580, 2014.

[24] Fangpo Wang, Alan E. Gelfand, and G. Jona Lasinio. Joint spatio-temporal analysis of a linear and a directional variable: modeling of wave heights and wave directions in the Adriatic Sea. *Statistica Sinica*, 25:25–39, 2015.

4

Toroidal Diffusions and Protein Structure Evolution

Eduardo García-Portugués,[1] **Michael Golden,**[2] **Michael Sørensen**[3]

[1] *Department of Statistics, Carlos III University of Madrid (Spain)*
[2] *Department of Statistics, University of Oxford (UK)*
[3] *Department of Mathematical Sciences, University of Copenhagen (Denmark)*
[4] *Department of Mathematics, University of Leeds (UK)*

Kanti V. Mardia,[2,4] **Thomas Hamelryck,**[5,6] **Jotun Hein**[2]

[5] *Department of Biology, University of Copenhagen (Denmark)*
[6] *Department of Computer Science, University of Copenhagen (Denmark)*

CONTENTS

4.1	Introduction	62
	4.1.1 Protein Structure	62
	4.1.2 Protein Evolution	64
	4.1.3 Toward a Generative Model of Protein Evolution	66
4.2	Toroidal Diffusions	67
	4.2.1 Toroidal Ornstein–Uhlenbeck Analogues	70
	4.2.2 Estimation for Toroidal Diffusions	72
	4.2.3 Empirical Performance	75
4.3	ETDBN: An Evolutionary Model for Protein Pairs	77
	4.3.1 Hidden Markov Model Structure	77
	4.3.2 Site-Classes: Constant Evolution and Jump Events	80
	4.3.3 Model Training	81
	4.3.4 Benchmarks	83
4.4	Case Study: Detection of a Novel Evolutionary Motif	87
4.5	Conclusions	90
	Acknowledgments	91
	Bibliography	91

4.1   Introduction

Toroidal diffusions, this is, continuous-time Markovian processes on the torus, are useful statistical tools for modelling the evolution of a protein's backbone throughout its dihedral angles representation. This chapter reviews a class of time-reversible ergodic diffusions, which can be regarded as the toroidal analogues of the celebrated Ornstein–Uhlenbeck process, and presents their application to the construction of an evolutionary model for pairs of related proteins that aims to provide new insights into the relationship between protein sequence and structure evolution.

The chapter is organized as follows. The rest of this section provides a brief background on protein structure and protein evolution, while it outlines the fundamentals of ETDBN (standing for Evolutionary Torus Dynamic Bayesian Network), a probabilistic model for protein evolution. Section 4.2 studies toroidal diffusions, the main methodological innovation behind ETDBN. The important challenges that arise in the estimation of the diffusion parameters require the consideration of tractable approximate likelihoods and, among the several approaches, the one yielding a specific approximation to the transition density of the wrapped normal process is shown to give the best empirical performance on average. ETDBN is described in detail in Section 4.3: its structure as a hidden Markov model featuring a wrapped normal process and two continuous-time Markov chains, its training from real data, and its empirical performance in several benchmarks. A distinctive feature of ETDBN is that it allows for both "smooth" and "catastrophic" conformational changes on the protein structure evolution by combining two evolutionary regimes within each hidden node. In addition, ETDBN provides new insights into the relationship between sequence and structure evolution through the analysis of hidden states. These two points are thoroughly illustrated in the case study given in Section 4.4.

4.1.1   Protein Structure

Proteins are large and complex biomolecules that are vital to all forms of life – from virus to human [6, 7]. Their main functions include defense against infections, catalyzing chemical reactions, transfer of signals between cells, transport of other molecules such as oxygen, and providing structure and support for cells, tissues, and organs. Chemically, proteins are simply linear polymers of amino acids, which for most proteins fall into 20 different types.

The amino acid sequence of a protein is encoded in the DNA of its matching gene, and is easy to obtain experimentally. However, most proteins also adopt a specific three-dimensional shape, which is the result of a folding process in which the linear polymer folds into a compact shape. This process is driven by the so-called hydrophobic effect – fatty amino acids repel water and become buried in the so-called hydrophobic core at the center of the protein [6]. The three-dimensional shape of a protein is often crucial for its function. Unfortunately, unlike a protein's amino acid sequence, a protein's three-dimensional structure is hard to obtain, requiring expensive and elaborate experimental techniques such as X-ray crystallography or nuclear magnetic resonance. Therefore, computational techniques to predict the three-dimensional structure of proteins are in high demand and an active area of research [7]. Currently, most structure prediction methods are essentially heuristic in character, but sophisticated probabilistic models of protein structure and sequence are increasingly being developed, applied, and accepted [3, 27, 34].

From a geometrical point of view, a protein's shape can be fully specified by a set of bond lengths between atom pairs, bond angles between three connected atoms, and *dihedral angles* between four connected atoms. The ϕ (specified by C^{i-1}-N^i-$C\alpha^i$-C^i atoms) and ψ (N^i-$C\alpha^i$-C^i-N^{i+1}) dihedral angles describe the part that is common to all amino acids and they form the most important degrees of freedom of the protein's structure, together with the ω dihedral angle which describes the configuration of the planar peptide bond between consecutive amino acids. The ω angle is unusual in the sense that it adopts values very close to either $0°$ or $180°$, rather than a continuous range of values as is the case for ϕ and ψ. As the great majority of ω angles are close to $0°$, this angle is often not a crucial degree of freedom. The (ϕ, ψ) dihedral angles essentially describe the overall curve of the linear part of the protein polymer, namely the protein's backbone structure. In addition, each amino acid also has a variable side-chain, which can contain up to four dihedral angles. The dihedral angles of the side-chains (gray boxes in Figure 4.1) can be fairly well predicted given the (ϕ, ψ) angles, and are thus a less critical degree of freedom.

A benefit of the dihedral representation is that it bypasses the need for *structural alignment*, which is the process of rotating and translating the three-dimensional Cartesian coordinates of the protein structures to closely align their atoms. Protein structure models on Cartesian coordinates require a structural alignment of proteins to compare them [18], a potential source of uncertainty. In addition, the dihedral representation introduces a simple distance between two conformations of the same protein: the average of the *angular distances* (see [13] for further details) between pairs of dihedral angles at each amino acid. This distance is a computationally faster alternative to the Root Mean Squared Deviation (RMSD) between the structurally aligned three-dimensional atomic coordinates of two proteins.

The structure of a protein can be further used to label local regions of a protein by their *secondary structure*. Secondary structure is a coarse-grained description of protein structure where each amino acid residue in a protein is assigned a label associated to the main structural motif which they belong to. Hence, knowledge of the secondary structure notably constrains the conformational possibilities of the dihedral angles. The two most important secondary structure classes are α-helices and β-sheets (see the cartoon depictions

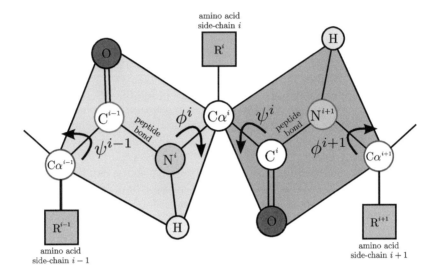

FIGURE 4.1
Small section of the protein backbone showing the ϕ and ψ dihedral angles between the planar atomic configurations.

of helices and flat arrows, respectively, in Figure 4.2). ETDBN parametrizes a protein as a discrete sequence of amino acids ($\mathbf{A}$), a continuous sequence of dihedral angles ($\mathbf{X}$), and a discrete sequence of secondary structure labels ($\mathbf{S}$). More precisely, a protein comprised of n amino acids is encoded mathematically as $\mathbf{P} := (\mathbf{A}, \mathbf{X}, \mathbf{S}) \equiv (\mathbf{P}^1, \ldots, \mathbf{P}^n)$, where $\mathbf{P}^i := (A^i, \mathbf{X}^i, S^i)$, $\mathbf{A} := (A^1, \ldots, A^n)$, $\mathbf{X} := (\mathbf{X}^1, \ldots, \mathbf{X}^n)$, $\mathbf{S} := (S^1, \ldots, S^n)$, and $\mathbf{X}^i := \langle \phi^i, \psi^i \rangle$, $i = 1, \ldots, n$. This notation is used extensively in Section 4.3.

4.1.2   Protein Evolution

Two or more proteins are termed *homologous* if they share a common ancestor. A descendant protein is assumed to diverge from ancestral proteins via a process of mutation. Multiple homologous proteins that have diverged from a common ancestor will have dependencies (similarities) in their sequence and structure. These dependencies can be represented by a phylogenetic tree (see Figure 4.2). One way in which these evolutionary dependencies manifest themselves is in the degree of amino acid sequence similarity shared among the homologous proteins. Their strength is assumed to be a result of two major factors: the time since the common ancestor and the rate of evolution.

   Failing to account for evolutionary dependencies can lead to misleading inferences [9]. For example, strong signals of structural conservation may be wrongly attributed to the selective maintenance of structural features due to their supposed functional importance. In reality, these structural similarities may simply be due to the close evolutionary relatedness of the proteins being analyzed. On the other hand, accounting for evolutionary dependencies allows information from homologous proteins to be incorporated in a principled manner. This can lead to more accurate inferences, such as the prediction of a protein structure from a homologous protein sequence and structure, known as *homology modelling* [1]. While stochastic models do not yet out-perform standard homology modelling approaches in terms of predictive accuracy, they provide a statistical foundation that allows for evolutionary parameters and their associated uncertainties to be estimated in a rigorous manner.

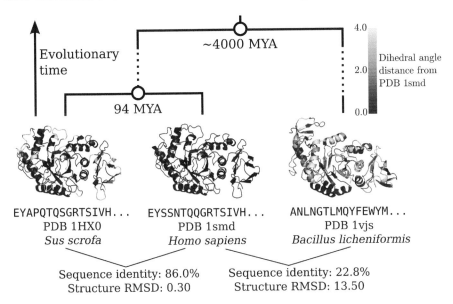

FIGURE 4.2
A phylogenetic tree relating three homologous amylase proteins, comparing their sequence identity and structural divergence. Overlaid on the cartoon representations of PDB 1hx0 (left) and PDB 1vjs (right) are the dihedral angle distances compared to PDB 1smd (center) at each aligned amino acid position. The degree of structural divergence (reported as "Structure RMSD") is measured using the RMSD between the aligned atomic coordinates of the corresponding PDB (Protein Data Bank; [21]) files. The wild boar (*Sus scrofa*) and human (*Homo sapiens*) amylases share an ancestor 94 Million Years Ago (MYA). The third amalyse from *Bacillus licheniformis* shares a common ancestor with the other two species approximately 4000 MYA. The common (unobserved) ancestors are represented by white nodes. Notice that the dependencies are weaker the further back in time two proteins share an ancestor.

A common property of the stochastic processes used to model molecular evolution is *time-reversibility*. While biological processes are not expected to obey time-reversibility, it is nevertheless considered a reasonable assumption that provides computational advantages. For example, in a pairwise evolutionary model, such as ETDBN, time-reversibility together with the *pulley principle* [8] allows the phylogenetic tree to be re-rooted on an observed protein (where the root is assumed to be drawn from the stationary distribution of the stochastic process), avoiding a costly integration over an unobserved ancestral protein (see Figure 4.3 for an illustration).

Two homologous proteins can not only differ in amino acid identity due to mutations that have occurred since their common ancestor, but also by the insertion and deletion of amino acids. Insertions and deletions are collectively referred to as *indels*. Accounting for this in practice involves a *sequence alignment* of both proteins. A sequence alignment is constructed by attempting to identify positions that have evolved via mutation alone (termed *homologous* positions) and positions that have incurred indels. Gap characters (typically denoted by a '−') are inserted in either sequence at indel positions such that both sequences with newly inserted gaps have the same length, thus forming a sequence alignment. Each position in this sequence alignment is referred to as an *aligned site*. Section 4.3 makes extensive use of this terminology.

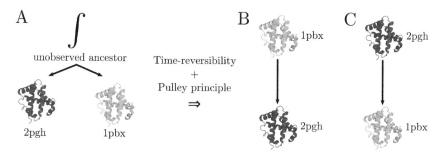

FIGURE 4.3
Illustration of the pulley principle in a pairwise phylogeny. To calculate the likelihood of two homologous proteins it is usually necessary to integrate out the sequence and structure of the unobserved ancestor at the root of the tree (A). However, under a time-reversible stochastic process, the pulley principle (i.e., assuming that the protein at the root of the tree is drawn from the stationary distribution of the process) allows any of the two observed proteins to be set as the root (B or C) without altering the likelihood. This has the advantage of avoiding a costly integration.

4.1.3   Toward a Generative Model of Protein Evolution

A key step in modelling protein structure evolution is selecting a structural representation and an adequate stochastic process. The first investigations of protein structure evolution represented protein structure using the three-dimensional Cartesian coordinates of protein backbone atoms and employed diffusions processes to model the relationship between structural distance and sequence similarity [14, 15]. More recent publications [5, 18] likewise used the three-dimensional Cartesian coordinates of amino acid $C\alpha$ atoms to represent protein structure together with Ornstein–Uhlenbeck (OU) diffusions to construct Bayesian probabilistic models of protein structure evolution. These models emphasize estimation of evolutionary parameters such as the evolutionary time between species, tree topologies and sequence alignment, and attempt to fully account for sources of uncertainty. For the sake of computational tractability, the aforementioned approaches treat the Cartesian coordinates as evolving independently. From a generative perspective, the Gaussian-like and independence assumptions on the evolution of the $C\alpha$ atoms will lead to evolved proteins with $C\alpha$ atoms that are unnaturally dispersed in space.

Rather than using a Cartesian coordinate representation, ETDBN [13] uses a dihedral angle representation of protein structures motivated by the non-evolutionary TorusDBN model [3, 4] (see also [26] in the present book). TorusDBN represents a single protein structure as a sequence of dihedral angle pairs, which are modelled using bivariate von Mises distributions [25, 26]. The evolution, rather than the distribution, of dihedral angles in ETDBN is modelled using a novel diffusion process [11] aimed to provide a more realistic tool for capturing the evolution of the underlying protein structure manifold. This diffusive process is coupled with two continuous-time Markov chains that model the evolution of amino acids and secondary structures labels. An additional coupling is introduced such that an amino acid change can lead to a *jump* in dihedral angles and a change in diffusion process, allowing the model to capture changes in amino acid that are directionally coupled with changes in dihedral angle or secondary structure. As in [5] and [18], the indel evolutionary process is also modelled to account for sequence alignment uncertainty by summing over all possible histories of insertions and deletions using a birth-death process as a prior [32].

For computational expediency, in the development of ETDBN it was key that the toroidal diffusion used to model the evolution of dihedral angles was time-reversible and allowed for a tractable likelihood approximation for arbitrary times. Additionally, it was desirable to efficiently sample dihedral angles under the diffusion to perform inference. The next section outlines in detail the development of a diffusion meeting the above criteria, which was guided by the goal of finding a toroidal OU-like process. The treatment in detail of ETDBN is therefore postponed until Section 4.3.

4.2 Toroidal Diffusions

The three-dimensional backbone of a protein comprised by n amino acids can be described as a sequence of $n-2$ pairs of dihedral angles $\{(\phi, \psi)_i\}_{i=1}^{n-2}$ (the first and last pairs are often disregarded due to missing ϕ and ψ angles, respectively). Therefore, a statistical tool for modelling the evolution of a protein's backbone using its true degrees of freedom is a continuous-time stochastic process on the torus $\mathbb{T}^p = [-\pi, \pi) \times \overset{p}{\cdots} \times [-\pi, \pi)$ (with $-\pi$ and π identified), with $p = n - 2$ or $p = 2$, depending on whether the backbone is modelled as a whole or piecewisely as a combination of pairs of dihedral angles, respectively.

One of the first continuous-time processes on the circle ($p = 1$) was proposed by [22] as the solution to the Stochastic Differential Equation (SDE)

$$\mathrm{d}X_t = \alpha \sin(\mu - X_t)\mathrm{d}t + \sigma\mathrm{d}W_t, \tag{4.1}$$

where $\{W_t\}$ is a Wiener process, $\alpha > 0$ is the drift strength, $\mu \in [-\pi, \pi)$ is the circular mean, and $\sigma > 0$ is the diffusion coefficient. This process, termed as the von Mises process, is attracted to μ, with a drift approximately linear in the neighborhood of μ. The process is ergodic and its stationary distribution (sdi) is a $\mathrm{vM}\left(\mu, \frac{2\alpha}{\sigma^2}\right)$, the von Mises (vM) distribution with mean μ and concentration $\frac{2\alpha}{\sigma^2}$, usually regarded as a circular analogue of the Gaussian distribution. The similarities of (4.1) with the celebrated OU process

$$\mathrm{d}X_t = \alpha(\mu - X_t)\mathrm{d}t + \sigma\mathrm{d}W_t, \tag{4.2}$$

whose sdi is a $\mathcal{N}\left(\mu, \frac{\sigma^2}{2\alpha}\right)$, supported Kent's [22] claim about the vM process being "the circular analogue of the OU process on the line."

Despite the similarities of (4.1) and (4.2), only the latter presents a closed-form analytical expression for its transition probability density (tpd), thus making its maximum likelihood inference fully tractable. The unavailability of the tpd is usually the case for the majority of *diffusions*, the continuous-time Markovian processes solving SDEs. In a general setting, the tpd of the p-dimensional Euclidean diffusion

$$\mathrm{d}\mathbf{X}_t = b(\mathbf{X}_t)\mathrm{d}t + \sigma(\mathbf{X}_t)\mathrm{d}\mathbf{W}_t, \tag{4.3}$$

where $b : \mathbb{R}^p \to \mathbb{R}^p$ is the drift function, $\sigma : \mathbb{R}^p \to \mathbb{R}^{p \times p}$ is the diffusion coefficient, and $\mathbf{W}_t = (W_{t,1}, \ldots, W_{t,p})'$ is a vector of p independent standard Wiener processes ($'$ denotes transposition), is denoted as $p_t(\cdot \,|\, \mathbf{x}_s)$. It represents the density function of the conditional distribution of $\mathbf{X}_{t+s}$ given $\mathbf{X}_s = \mathbf{x}_s$. The tpd is only given implicitly as the solution to the Fokker–Planck equation, this is, the Partial Differential Equation (PDE)

$$\frac{\partial}{\partial t}p_t(\mathbf{x}\,|\,\mathbf{x}_s) = -\sum_{i=1}^{p}\frac{\partial}{\partial x_i}(b_i(\mathbf{x})p_t(\mathbf{x}\,|\,\mathbf{x}_s))$$
$$+ \frac{1}{2}\sum_{i,j=1}^{p}\frac{\partial^2}{\partial x_i \partial x_j}(V_{ij}(\mathbf{x})p_t(\mathbf{x}\,|\,\mathbf{x}_s)), \tag{4.4}$$

with $\mathbf{x}, \mathbf{x}_s \in \mathbb{R}^p$, $V(\cdot) := \sigma(\cdot)\sigma(\cdot)'$, and initial condition $p_0(\mathbf{x}\,|\,\mathbf{x}_s) = \delta(\mathbf{x} - \mathbf{x}_s)$ ($\delta(\cdot)$ represents Dirac's delta). This PDE has no explicit solution except for very few particular choices of b (e.g., linear) and V (e.g., constant).

Defining diffusive processes $\{\mathbf{\Theta}_t\}$ whose state space is $\mathbb{T}^p$, such as (4.1), requires certain caution for achieving proper transitions of the process through the identified points $-\pi$ and π. A useful construction consists in regarding $\{\Theta_t\}$ as a Euclidean process $\{\mathbf{X}_t\}$ that is *wrapped* into its principal angles by the *wrapping operator* $\mathrm{cmod}\,(\cdot) := ((\cdot + \pi) \mod 2\pi) - \pi$.

Definition 1 (Toroidal diffusion) *The stochastic process* $\{\Theta_t\} \subset \mathbb{T}^p$ *is said to be a* toroidal diffusion *if it arises as the wrapping* $\Theta_t = \mathrm{cmod}\,(\mathbf{X}_t)$ *of the diffusion* (4.3), *and if b and σ are 2π-periodical:*

$$b(\mathbf{x} + 2\mathbf{k}\pi) = b(\mathbf{x}), \quad \sigma(\mathbf{x} + 2\mathbf{k}\pi) = \sigma(\mathbf{x}), \quad \forall \mathbf{k} \in \mathbb{Z}^p, \forall \mathbf{x} \in \mathbb{R}^p.$$

The toroidal diffusion coming from the wrapping of (4.3) *is denoted as* $\mathrm{d}\Theta_t = b(\Theta_t)\mathrm{d}t + \sigma(\Theta_t)\mathrm{d}\mathbf{W}_t$.

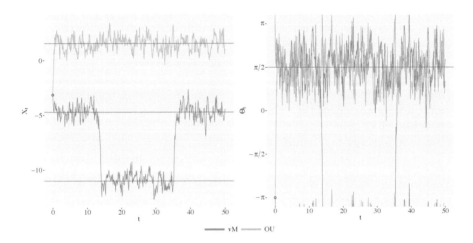

FIGURE 4.4
Trajectories of the vM and OU processes on the circle for $t \in [0, 50]$, $\alpha = 5$, $\mu = \frac{\pi}{2}$, $\sigma = 2$, and starting point $x_0 = -\pi$. The trajectories in the left correspond to the unwrapped processes $\{X_t\}$, whereas the ones in the right are their wrapped versions $\{\Theta_t\}$. The vM process is not ergodic in $\mathbb{R}$ but its wrapping is ergodic in $[-\pi, \pi)$. The OU and WOU processes are ergodic in $\mathbb{R}$ and $[-\pi, \pi)$, respectively. However, the WOU process is *not* Markovian and is not a toroidal diffusion. Note that only the vM process is able to travel in both directions toward μ.

With the previous definition and notation, for a given t, $\mathbf{X}_t = \Theta_t + 2\mathrm{wind}(\mathbf{X}_t)\pi$, where $\mathrm{wind}(\mathbf{X}_t) := \lfloor \frac{\mathbf{X}_t + \pi}{2\pi} \rfloor \in \mathbb{Z}^p$ is the *winding number* of $\mathbf{X}_t$. The fact that b and σ are required to be periodic implies that $\{\mathbf{X}_t\}$ is a non-ergodic process in $\mathbb{R}^p$ and that $\{\Theta_t\}$ is a Markovian process. Otherwise $\Theta_{t_2} \mid (\Theta_{t_1}, \Theta_{t_0})$, with $t_2 > t_1 > t_0$, would not depend only on Θ_{t_1} but also on $\mathrm{wind}(\mathbf{X}_{t_1})$ in $\mathbf{X}_{t_1} = \Theta_{t_1} + 2\mathrm{wind}(\mathbf{X}_{t_1})\pi$. The construction of the vM diffusion and the Wrapped OU (WOU) process on the circle is illustrated in Figure 4.4.

In the rest of this section we illustrate how to construct OU-like toroidal diffusions and focus on the particular bivariate diffusion we employ in ETDBN, whose properties and approximate inference are analyzed.

4.2.1   Toroidal Ornstein–Uhlenbeck Analogues

Let f be a probability density function (pdf) over $\mathbb{R}^p$. The so-called *Langevin diffusions* are the family of diffusions (4.3) such that the $i = 1, \ldots, p$ entries of the drift are given by

$$b_i(\mathbf{x}) = \frac{1}{2} \sum_{j=1}^p V_{ij}(\mathbf{x}) \frac{\partial}{\partial x_j} \log f(\mathbf{x})$$

$$+ \det V(\mathbf{x})^{\frac{1}{2}} \sum_{j=1}^p \frac{\partial}{\partial x_j} \left(V_{ij}(\mathbf{x}) \det V(\mathbf{x})^{-\frac{1}{2}} \right). \tag{4.5}$$

If $V(\mathbf{x}) = \sigma(\mathbf{x})\sigma(\mathbf{x})' = \Sigma$ with Σ a covariance matrix, i.e., if the diffusion coefficient is constant, then the Langevin diffusions are of the form

$$\mathrm{d}\mathbf{X}_t = \frac{1}{2} \Sigma \nabla \log f(\mathbf{X}_t) \mathrm{d}t + \Sigma^{\frac{1}{2}} \mathrm{d}\mathbf{W}_t. \tag{4.6}$$

Under mild regularity conditions on f and σ, these diffusions are ergodic with stationary density f. The construction of Langevin toroidal diffusions is achieved by wrappings of Langevin diffusions, imposing now that f is a toroidal density: $\int_{\mathbb{T}^p} f(\boldsymbol{\theta}) \mathrm{d}\boldsymbol{\theta} = 1$ and $f(\boldsymbol{\theta} + 2\mathbf{k}\pi) = f(\boldsymbol{\theta}) \; \forall \boldsymbol{\theta} \in \mathbb{T}^p, \mathbf{k} \in \mathbb{Z}^p$. The following result guarantees that the stationary density of such toroidal diffusion is indeed f, with the addition of a highly convenient characterization.

Proposition 1 ([11]) *Assume $\{\Theta_t\}$ is obtained from the wrapping of a Langevin diffusion $\{\mathbf{X}_t\}$ with drift (4.5) given by a strictly positive toroidal density f. Assume that the second derivatives of both f and the entries of V are locally uniformly Hölder continuous, and that V is 2π-periodical. Then $\{\Theta_t\}$ is the unique toroidal time-reversible diffusion that is ergodic with stationary density f and prescribed V.*

The above result roots on Kent's [23] characterization of ergodic time-reversible diffusions on manifolds and is particularly useful for constructing OU toroidal analogues. To that aim, consider first the multivariate OU process

$$\mathrm{d}\mathbf{X}_t = \mathbf{A}(\boldsymbol{\mu} - \mathbf{X}_t)\mathrm{d}t + \Sigma^{\frac{1}{2}} \mathrm{d}\mathbf{W}_t, \tag{4.7}$$

with $\boldsymbol{\mu} \in \mathbb{R}^p$, Σ a covariance matrix, and $\mathbf{A}$ such that $\mathbf{A}^{-1}\Sigma$ is a covariance matrix. This process has sdi equal to $\mathcal{N}(\boldsymbol{\mu}, \frac{1}{2}\mathbf{A}^{-1}\Sigma)$ and, in virtue of [23]'s characterization, (4.7) is the unique time-reversible diffusion with Gaussian sdi and constant diffusion coefficient. Therefore, analogues of the OU process in $\mathbb{T}^p$ follow by wrapping Langevin diffusions for toroidal pdfs that are Gaussian analogues. One of them is the vM due to important Gaussian-like characterizations (see Section 2.2.4 of [18]). Nevertheless, the Wrapped Normal (WN) exhibits also important similarities with the Gaussian (*ibid*, Section 2.2.6) and, contrary to the vM, it appears in Gaussian-related limit laws (see

Section 4.3.2 of [24]). For this reason, as well as for its better tractability, the focus on obtaining an OU-like toroidal process is on the Langevin diffusion associated to a WN density, referred to below as the *WN process*.

The pdf of a WN in T^p, $\mathrm{WN}(\boldsymbol{\mu}, \boldsymbol{\Sigma})$, is given by $f_{\mathrm{WN}}(\boldsymbol{\theta}; \boldsymbol{\mu}, \boldsymbol{\Sigma}) := \sum_{\mathbf{k} \in \mathbb{Z}^p}$ $\phi_{\boldsymbol{\Sigma}}(\boldsymbol{\theta} - \boldsymbol{\mu} + 2\mathbf{k}\pi)$, with $\boldsymbol{\mu} \in \mathrm{T}^p$, $\boldsymbol{\Sigma}$ a covariance matrix, and $\phi_{\boldsymbol{\Sigma}}$ the pdf of a $\mathcal{N}(\mathbf{0}, \boldsymbol{\Sigma})$. Its interpretation is simple: the series recovers the probability mass spread outside T^p, in a periodic fashion, such that f_{WN} becomes a density in T^p. Set $\mathrm{WN}\left(\boldsymbol{\mu}, \frac{1}{2}\mathbf{A}^{-1}\boldsymbol{\Sigma}\right)$ as the sdi to compare with (4.7). Then the WN process follows from wrapping (4.6) for the pdf associated to the previous sdi:

$$d\boldsymbol{\Theta}_t = \mathbf{A}\left(\boldsymbol{\mu} - \boldsymbol{\Theta}_t - \sum_{\mathbf{k} \in \mathbb{Z}^p} 2\mathbf{k}\pi w_{\mathbf{k}}(\boldsymbol{\Theta}_t)\right)dt + \boldsymbol{\Sigma}^{\frac{1}{2}}d\mathbf{W}_t, \qquad (4.8)$$

$$w_{\mathbf{k}}(\boldsymbol{\theta}) := \frac{\phi_{\frac{1}{2}\mathbf{A}^{-1}\boldsymbol{\Sigma}}(\boldsymbol{\theta} - \boldsymbol{\mu} + 2\mathbf{k}\pi)}{\sum_{\in \mathbb{Z}^p} \phi_{\frac{1}{2}\mathbf{A}^{-1}\boldsymbol{\Sigma}}(\boldsymbol{\theta} - \boldsymbol{\mu} + 2\pi)}.$$

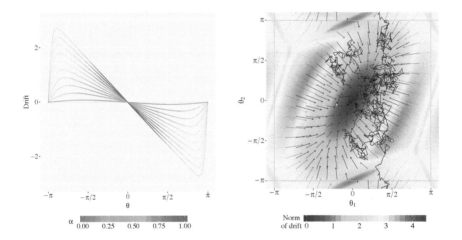

FIGURE 4.5
Drifts for the WN process for $p = 1, 2$, with shading proportional to the stationary density. Left: drifts in $p = 1$ for $\alpha = i/10$, $i = 1, \ldots, 10$, $\mu = 0$, and $\sigma = 1$. Note the periodicity of the drift, its linearity around μ, and how the maxima drifts shift toward antipodality as α increases. Right: drift vector field in $p = 2$ for $\mathbf{A} = (1.5, -0.5; -0.5, 1)$, $\boldsymbol{\mu} = (0, 0)$, and $\boldsymbol{\Sigma} = \sigma^2\mathbf{I}$ with $\sigma = 1.5$. The color gradient represents the intensity of the drift, measured as the norm of the arrows. Note that the stronger drifts are located at the regions with lowest density. A trajectory starting at $\left(\frac{\pi}{2}, -\frac{\pi}{2}\right)$ (round facet) is displayed evolving for $t \in [0, 5]$ toward its final point (triangular facet).

Illustrative drifts of the WN process are shown in Figure 4.5. For $p = 1$, $\mathbf{A} = \alpha$ and $\boldsymbol{\Sigma} = \sigma^2$. In this case the WN drift is a smoothed binding of lines with slope close to $-\alpha$ that go through $(\mu, 0)$ and such that they are bended to pass through $(\mu \pm \pi, 0)$. Hence, the drift behaves almost linearly in a neighborhood of μ (equilibrium point, stable) and rapidly decays to pass across $\mu \pm \pi$ (equilibrium point, unstable). The drift maxima vary from $\mu \pm \pi$ (if $\frac{\sigma^2}{2\alpha} \to 0$, the sdi is degenerate at μ) to $\mu \pm \frac{\pi}{2}$ (if $\frac{\sigma^2}{2\alpha} \to \infty$, the sdi is uniform and the drift is null). When $p = 2$, the vector field of the drift has a characteristic tessellated structure formed by hexagonal-like tiles *anchored* at the points $\mu + \mathbf{k}_0\pi$, $\mathbf{k}_0 \in \{-1, 0, 1\}^p \setminus \{\mathbf{0}\}$, where the drift is null. $\boldsymbol{\Sigma}$ alters the tessellation structure non-trivially by modifying $\{w_\mathbf{k}(\boldsymbol{\theta}) : \mathbf{k} \in \mathbb{Z}^p\}$. When $\boldsymbol{\Sigma} = \sigma^2\mathbf{I}$, the larger (respectively, smaller) σ, the more spread (concentrated) the distribution $\{w_\mathbf{k}(\boldsymbol{\theta}) : \mathbf{k} \in \mathbb{Z}^p\}$ is, resulting in flat (peaked) drifts with smooth (rough) transitions in the limits defining the tessellation.

Finally, note that it is easy to parametrize the 2×2 drift matrices $\mathbf{A}$ such that $\mathbf{A}^{-1}\boldsymbol{\Sigma}$ is a covariance matrix. For $\boldsymbol{\Sigma} = (\sigma_1^2, 0; 0, \sigma_2^2)$, these are $\mathbf{A} = (\alpha_1, \frac{\sigma_1}{\sigma_2}\alpha_3; \frac{\sigma_2}{\sigma_1}\alpha_3, \alpha_2)$, with $\alpha_1, \alpha_2 > 0$ and $\alpha_3^2 < \alpha_1\alpha_2$. In particular, the dependence between components is modelled by α_3, as is evident from the stationary covariance matrix: $\frac{1}{2}\mathbf{A}^{-1}\boldsymbol{\Sigma} = \frac{1}{2(\alpha_1\alpha_2 - \alpha_3^2)} (\alpha_2, \sigma_1^2; -\alpha_3\sigma_1\sigma_2; -\alpha_3\sigma_1\sigma_2, \alpha_1\sigma_2^2)$.

## 4.2.2	Estimation for Toroidal Diffusions

The Maximum Likelihood Estimator (MLE) of the parameter $\boldsymbol{\lambda}$ of

$$d\boldsymbol{\Theta}_t = b(\boldsymbol{\Theta}_t; \boldsymbol{\lambda})dt + \sigma(\boldsymbol{\Theta}_t; \boldsymbol{\lambda})d\mathbf{W}_t, \tag{4.9}$$

when the sample is a discretized trajectory $\{\boldsymbol{\Theta}_{\Delta i}\}_{i=0}^N$ in the time interval $[0, T]$, $T = N\Delta$, is given by $\hat{\boldsymbol{\lambda}}_{\mathrm{MLE}} := \arg\max_{\boldsymbol{\lambda} \in \Lambda} l(\boldsymbol{\lambda}; \{\boldsymbol{\Theta}_{\Delta i}\}_{i=0}^N)$, where, using the Markovianity of (4.9), the log-likelihood is given by

$$l\left(\boldsymbol{\lambda}; \{\boldsymbol{\Theta}_{\Delta i}\}_{i=0}^N\right) = \log p(\boldsymbol{\Theta}_0; \boldsymbol{\lambda}) + \sum_{i=1}^N \log p_\Delta(\boldsymbol{\Theta}_{\Delta i} \mid \boldsymbol{\Theta}_{\Delta(i-1)}; \boldsymbol{\lambda}). \tag{4.10}$$

Here $p_\Delta(\cdot \mid \cdot; \boldsymbol{\lambda})$ is the tpd of (4.9) and the first term in (4.10) is set to the sdi of (4.9) if the process is assumed to start in the stationary regime. The MLE can rarely be readily obtained, as usually no explicit expression for the tpd exists. In the following two estimation strategies that rely on an *approximate likelihood function*, where the unknown tpd is replaced by an approximation, are studied. For the sake of brevity, we suppress $\boldsymbol{\lambda}$ in the notation.

We note also that the tpd can be computed by solving numerically the PDE (4.4). This is a computationally expensive task, too demanding for computing the MLE for $p > 1$, but useful for obtaining insightful visualizations of the tpd (see Figure 4.6) and for constructing the accuracy benchmark employed in Section 4.2.3 to test the more computationally expedient approximate likelihoods. We refer to [11] for the details of how to solve this PDE for $p = 1, 2$.

Adapted Pseudo-Likelihoods

The above PDE solution is too costly for obtaining the MLE. The Euler pseudo-tpd is a cheap computational alternative. The Euler scheme arises as the first order discretization of the process, where the drift and the diffusion coefficient are approximated constantly between $\boldsymbol{\Theta}_{\Delta(i-1)}$ and $\boldsymbol{\Theta}_{\Delta(i)}$:

$$\boldsymbol{\Theta}_{\Delta i} = \mathrm{cmod}\left(\boldsymbol{\Theta}_{\Delta(i-1)} + b(\boldsymbol{\Theta}_{\Delta(i-1)})\Delta + \sqrt{\Delta}\sigma(\boldsymbol{\Theta}_{\Delta(i-1)})\mathbf{Z}^i\right), \qquad (4.11)$$

where $\mathbf{Z}^i \sim \mathcal{N}(\mathbf{0},\mathbf{I})$, $i = 1,\dots,N$, and $\mathrm{cmod}\,(\cdot) = ((\cdot + \pi) \mod 2\pi) - \pi$. This yields the Euler pseudo-tpd

$$p_\Delta^{\mathrm{E}}(\boldsymbol{\theta}\,|\,\boldsymbol{\varphi}) = f_{\mathrm{WN}}\left(\boldsymbol{\theta};\boldsymbol{\varphi} + b(\boldsymbol{\varphi})\Delta, V(\boldsymbol{\varphi})\Delta\right), \qquad \boldsymbol{\theta},\,\boldsymbol{\varphi} \in \mathrm{T}^p.$$

Sampling trajectories of (4.9) with separation time Δ_s can be done by using (4.11) iteratively for $\Delta = \Delta_s/M$, $M > 1$, and then thinning conveniently the sampled trajectory.

An improvement on the Euler scheme is the Shoji–Ozaki [31] scheme. It employs a linear approximation for the drift, $b(\mathbf{X}_t) \approx b(\mathbf{X}_s) + \mathbf{J}_s(\mathbf{X}_t - \mathbf{X}_s)$ for $t \in [s, s+\Delta)$ ($\mathbf{J}_s = J(\mathbf{X}_s)$ denotes the Jacobian of b at $\mathbf{X}_s$), and approximates the diffusion coefficient constantly. This results in a linear SDE that can be solved explicitly. Wrapping this solution provides the Shoji–Ozaki pseudo-tpd:

$$p_\Delta^{\mathrm{SO}}(\boldsymbol{\theta}\,|\,\boldsymbol{\varphi}) = f_{\mathrm{WN}}\left(\boldsymbol{\theta};E_\Delta(\boldsymbol{\varphi}), V_\Delta(\boldsymbol{\varphi})\right), \qquad \boldsymbol{\theta},\,\boldsymbol{\varphi} \in \mathrm{T}^p,$$

where, assuming that $V(\boldsymbol{\varphi})^{-1}J(\boldsymbol{\varphi})$ is symmetric (the case for Langevin diffusions), $E_\Delta(\boldsymbol{\varphi}) = \boldsymbol{\varphi} + J(\boldsymbol{\varphi})^{-1}(\exp\{J(\boldsymbol{\varphi})\Delta\} - \mathbf{I})b(\boldsymbol{\varphi})$ and $V_\Delta(\boldsymbol{\varphi}) = \frac{1}{2}J(\boldsymbol{\varphi})^{-1}(\exp\{2J(\boldsymbol{\varphi})\Delta\} - \mathbf{I})V(\boldsymbol{\varphi})$. If the real parts of the eigenvalues of $J(\boldsymbol{\varphi})$ are negative, then $p_\Delta^{\mathrm{SO}}(\boldsymbol{\theta}\,|\,\boldsymbol{\varphi}) \xrightarrow[\Delta\to\infty]{} f_{\mathrm{WN}}\left(\boldsymbol{\theta};\boldsymbol{\varphi} - J(\boldsymbol{\varphi})b(\boldsymbol{\varphi}), -\frac{1}{2}J(\boldsymbol{\varphi})^{-1}V(\boldsymbol{\varphi})\right)$ (see top row of Figure 4.6). Otherwise, the pseudo-tpd degenerates into a uniform density when $\Delta \to \infty$, as Euler's tpd always does (see Figure 4.6). The Euler and Shoji–Ozaki pseudo-likelihoods follow by replacing the tpd by the pseudo-tpds in (4.10).

Transition Density Approximation for the WN Process

We consider now a specific analytic approximation for the tpd of the WN process aimed to work equally well irrespectively of Δ (a vital aspect for ETDBN where there is little control over Δ) and to cope with its potential multimodality (the number of potential modes is $2p$). The approximation relies on the connection of the WN process with (4.7), whose tpd is $p_t(\cdot\,|\,\mathbf{x}_s;\mathbf{A},\boldsymbol{\mu},\boldsymbol{\Sigma}) = \phi_{\boldsymbol{\Gamma}_t}(\cdot - \boldsymbol{\mu}_t)$, where $\boldsymbol{\mu}_t := \boldsymbol{\mu} + e^{-t\mathbf{A}}(\mathbf{x}_s - \boldsymbol{\mu})$ and $\boldsymbol{\Gamma}_t := \int_0^t e^{-s\mathbf{A}}\boldsymbol{\Sigma}e^{-s\mathbf{A}'}\mathrm{d}s$. We denote by WOU, standing for wrapped multivariate OU process, to the wrapping of the process (4.7). Assuming that $\mathbf{X}_s \sim \mathcal{N}(\boldsymbol{\mu}, \frac{1}{2}\mathbf{A}^{-1}\boldsymbol{\Sigma})$, the conditional density of the WOU process follows from the tpd of (4.7) as

$$p_t^{\mathrm{WOU}}(\boldsymbol{\theta}\,|\,\boldsymbol{\theta}_s;\mathbf{A},\boldsymbol{\mu},\boldsymbol{\Sigma}) := \sum_{\mathbf{m}\in\mathbb{Z}^p} f_{\mathrm{WN}}(\boldsymbol{\theta};\boldsymbol{\mu}_t^{\mathbf{m}},\boldsymbol{\Gamma}_t)w_{\mathbf{m}}(\boldsymbol{\theta}_s) \qquad (4.12)$$

where $\boldsymbol{\mu}_t^{\mathbf{m}} := \boldsymbol{\mu} + e^{-t\mathbf{A}}(\boldsymbol{\theta}_s - \boldsymbol{\mu} + 2\mathbf{m}\pi)$. The conditional density (4.12) is the wrapping of the tpd of (4.7) plus a weighting by the sdi of the winding numbers of $\mathbf{X}_s$, resembling the structure of the WN drift: a weighting of linear drifts by the winding number sdi to achieve periodicity. Albeit (4.12) is *not* the tpd of the WN process, both behave similarly in several key situations, as shown in the next result. Note that sampling from (4.12) is immediate and requires no intermediate steps as in (4.11): (i) simulate $\mathbf{M}$ from the discrete distribution $\mathbb{P}\{\mathbf{M}=\mathbf{m}\}=w_{\mathbf{m}}(\boldsymbol{\theta}_s)$, $\mathbf{m}\in\mathbb{Z}^p$; ($ii$) sample from a $\mathcal{N}(\boldsymbol{\mu}_t^{\mathbf{M}},\boldsymbol{\Gamma}_t)$ and wrap the output by cmod $(\cdot)$.

Corollary 1 ([11]) For almost all $\boldsymbol{\theta},\boldsymbol{\theta}_s\in\mathbb{T}^p$ and for all $t>0$, p_t^{WOU} approximates p_t, the true tpd of the WN process, in the following ways (the dependence on the parameters is omitted):

1. *Point mass:* $\lim_{t\to 0}p_t(\boldsymbol{\theta}\,|\,\boldsymbol{\theta}_s)=\lim_{t\to 0}p_t^{\mathrm{WOU}}(\boldsymbol{\theta}\,|\,\boldsymbol{\theta}_s)=\delta(\boldsymbol{\theta}-\boldsymbol{\theta}_s)$.

2. *Sdi-correct:* $\lim_{t\to\infty}p_t(\boldsymbol{\theta}\,|\,\boldsymbol{\theta}_s)\quad=\quad\lim_{t\to\infty}p_t^{\mathrm{WOU}}(\boldsymbol{\theta}\,|\,\boldsymbol{\theta}_s)\quad=$ $f_{\mathrm{WN}}(\boldsymbol{\theta};\boldsymbol{\mu},\frac{1}{2}\mathbf{A}^{-1}\boldsymbol{\Sigma})$.

3. *Time-reversibility:* As p_t does, p_t^{WOU} satisfies

$$p_t^{\mathrm{WOU}}(\boldsymbol{\theta}\,|\,\boldsymbol{\theta}_s)f_{\mathrm{WN}}(\boldsymbol{\theta}_s;\boldsymbol{\mu},\tfrac{1}{2}\mathbf{A}^{-1}\boldsymbol{\Sigma})=p_t^{\mathrm{WOU}}(\boldsymbol{\theta}_s\,|\,\boldsymbol{\theta})f_{\mathrm{WN}}(\boldsymbol{\theta};\boldsymbol{\mu},\tfrac{1}{2}\mathbf{A}^{-1}\boldsymbol{\Sigma}).$$

4. *High-concentration:* if $\mathbf{A}^{-1}\boldsymbol{\Sigma}\to\mathbf{0}$ and $\boldsymbol{\Sigma}$ bounded, $p_t(\boldsymbol{\theta}\,|\,\boldsymbol{\theta}_s)\approx$ $p_t^{\mathrm{WOU}}(\boldsymbol{\theta}\,|\,\boldsymbol{\theta}_s)$.

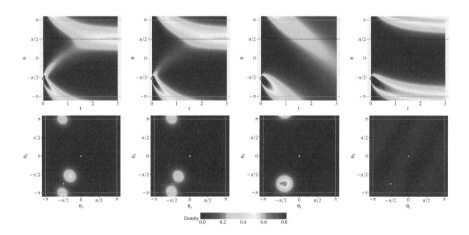

FIGURE 4.6
From left to right columns: numerical solution of the PDE, WOU tpd approximation, Euler pseudo-tpd, and Shoji–Ozaki pseudo-tpd. Top row: approximations for $p_t(\cdot\,|\,\theta_0)$ with $t\in[0.01,3]$, for the one-dimensional WN process with $\alpha=1$, $\mu=\frac{\pi}{2}$ (horizontal line), $\sigma=1$, and $\theta_0=-\frac{\pi}{2}-0.1$ (round facet). Bottom row: approximations for $p_t(\cdot\,|\,\boldsymbol{\theta}_0)$ at $t=0.25$, for the two-dimensional WN process with $\mathbf{A}=(1,0.5;0.5,1)$, $\boldsymbol{\mu}=(0,0)$ (triangular facet), $\boldsymbol{\Sigma}=\mathbf{I}$, and $\boldsymbol{\theta}_0=(-\frac{\pi}{2},-\frac{3\pi}{4})$ (round facet). Note the *explosion* in the Shoji–Ozaki tpd, a consequence of negative drift eigenvalues at $\boldsymbol{\theta}_0$.

The tractability of (4.12) degenerates quickly with the dimension, but it can be readily computed for $p = 2$ by a series of computational tricks. Specifically, $e^{-t\mathbf{A}}$ can be obtained in virtue of Corollary 2.4 of [2]: for any 2×2 matrix $\mathbf{A}$, $e^{t\mathbf{A}} = a(t)\mathbf{I} + b(t)\mathbf{A}$ with $a(t) := e^{s(\mathbf{A})t}\left(\cosh(q(\mathbf{A})t) - s(\mathbf{A})\frac{\sinh(q(\mathbf{A})t)}{q(\mathbf{A})}\right)$, $b(t) := e^{s(\mathbf{A})t}\frac{\sinh(q(\mathbf{A})t)}{q(\mathbf{A})}$, $s(\mathbf{A}) := \frac{\text{tr}[\mathbf{A}]}{2}$, and $q(\mathbf{A}) := \sqrt{|\det(\mathbf{A} - s\mathbf{I})|}$. The fact that $\mathbf{A}^{-1}\mathbf{\Sigma}$ is symmetric plus the previous formula gives

$$\mathbf{\Gamma}_t = \frac{1}{2}\mathbf{A}^{-1}(\mathbf{I} - \exp\{-2\mathbf{A}t\})\mathbf{\Sigma} = s(t)\frac{1}{2}\mathbf{A}^{-1}\mathbf{\Sigma} + i(t)\mathbf{\Sigma},$$

with $s(t) = 1 - a(-2t)$ and $i(t) = -\frac{1}{2}b(-2t)$. This expression gives a neat interpolation between the infinitesimal and stationary covariance matrices, specifically convenient for computing the tpd for several t's.

4.2.3 Empirical Performance

The goodness-of-fit of the tpd approximations has a direct influence on the resulting approximate MLEs for the WN process. To measure the closeness of the approximation, we consider the Kullback–Leibler (KL) divergence of the approximation $p_t^{\text{A}}(\cdot \,|\, \boldsymbol{\theta}_s)$ ($\text{A} = \text{E}, \text{SO}, \text{WOU}$) to $p_t(\cdot \,|\, \boldsymbol{\theta}_s)$ by *weighting* by the sdi $\text{WN}\left(\boldsymbol{\mu}, \frac{1}{2}\mathbf{A}^{-1}\mathbf{\Sigma}\right)$ the contributions of each initial point $\boldsymbol{\theta}_s$ to the divergence. Since the PDE solution is obtained for an initial condition of the form $\text{WN}(\boldsymbol{\theta}_s, \sigma_0^2\mathbf{I})$, we consider the same initial condition for the approximations to remove the bias in the comparison: $p_{t,\sigma_0^2}^{\text{A}}(\boldsymbol{\theta} \,|\, \boldsymbol{\theta}_s) := \int_{\mathbb{T}^p} p_t^{\text{A}}(\boldsymbol{\theta} \,|\, \boldsymbol{\varphi})f_{\text{WN}}(\boldsymbol{\varphi}; \boldsymbol{\theta}_s, \sigma_0^2)\mathrm{d}\boldsymbol{\varphi}$. The divergence measure we consider is then

$$\text{D}_{t,\sigma_0^2}^{\text{A}} := \int_{\mathbb{T}^p}\int_{\mathbb{T}^p} p_{t,\sigma_0^2}^{\text{PDE}}(\boldsymbol{\theta} \,|\, \boldsymbol{\theta}_s) \log\left(\frac{p_{t,\sigma_0^2}^{\text{PDE}}(\boldsymbol{\theta} \,|\, \boldsymbol{\theta}_s)}{p_{t,\sigma_0^2}^{\text{A}}(\boldsymbol{\theta} \,|\, \boldsymbol{\theta}_s)}\right) f_{\text{WN}}\left(\boldsymbol{\theta}_s; \boldsymbol{\mu}, \tfrac{1}{2}\mathbf{A}^{-1}\mathbf{\Sigma}\right)\mathrm{d}\boldsymbol{\theta}\mathrm{d}\boldsymbol{\theta}_s.$$

Figure 4.7 shows the $\text{D}_{t,\sigma_0^2}^{\text{A}}$ curves in log-scale for the WN process with $p = 2$, under three illustrative drift strengths and diffusivities. As it can be seen, WOU outperforms the other approximations under all scenarios and times. Besides, WOU is the only approximation whose accuracy improves as time increases (above a certain local maximum in the KL divergence), whereas the E and SO pseudo-tpds either deteriorate or stabilize as time increases. The exception is the scenario with low diffusivity where SO is almost equal to WOU (and both are close to the true tpd). E is systematically behind SO in performance. Further empirical results in [11] corroborate this pattern.

We compare now the efficiency of WOU, SO, and E in estimating the unknown parameters of the WN process in $p = 2$ from a sample of $N = 250$ points. We consider $\Delta = 0.05, 0.20, 0.50, 1.00$ and four representative parameter choices for the WN process. The trajectories are simulated using the E method with time step 0.001 and then subsampled for given Δ's. $\mathbf{\Sigma}$ is assumed to be known to avoid the inherent unidentifiabilities of $\mathbf{A}$ and $\mathbf{\Sigma}$ when

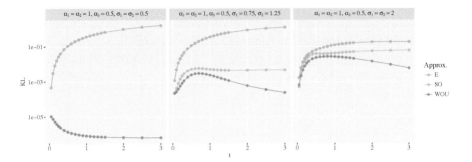

FIGURE 4.7

D^A_{t,σ^2_0} curves of the WN process with $p = 2$. Note the vertical log-scale. From left to right, panels represent small, moderate, and high diffusivities. The PDE was solved with $M_x = M_y = 240$, $M_t = \lceil 1500t \rceil$, and $\sigma_0 = 0.1$.

Δ is large and the tpd converges to the sdi. To summarize the overall performance of the three estimators (E, SO, and WOU) of the 5-variate parameter $\boldsymbol{\lambda} = (\alpha_1, \alpha_2, \alpha_3, \mu_1, \mu_2)$, a global measure of relative performance is considered. This measure is the componentwise average of Relative Efficiency (RE), where the RE is measured with respect to the best estimator at a given component in terms of Mean Squared Error (MSE). Hence, if $\hat{\boldsymbol{\lambda}}_j$ ($j \in \{E, SO, WOU\}$) is the best estimator for all the components of $\boldsymbol{\lambda}$, then $\text{RE}(\hat{\boldsymbol{\lambda}}_j) = 1$. $\text{RE}(\hat{\boldsymbol{\lambda}}_j)$ is estimated with 1000 Monte Carlo replicates and $\hat{\boldsymbol{\lambda}}_j$ is obtained by maximizing the approximate likelihood with a common optimization procedure that employs stationary estimates as starting values.

Table 4.1 gives the REs for E, SO, and WOU in $p = 2$. When averaging across scenarios and discretization times, the global ranking of performance is:

TABLE 4.1

Relative efficiencies for the WN process with $p = 2$, $\boldsymbol{\mu} = \left(\frac{\pi}{2}, -\frac{\pi}{2}\right)$, $\alpha_1 = \alpha_2 = \alpha$, $\alpha_3 = \frac{\alpha}{2}$, and $\boldsymbol{\Sigma} = \sigma^2 \mathbf{I}$. Bold font denotes the most efficient estimator for each row and scenario.

	$\alpha = 1, \quad \sigma = 1$			$\alpha = 2, \quad \sigma = 1$		
Δ	E	SO	WOU	E	SO	WOU
0.05	**0.9765**	0.9244	0.8999	**0.9920**	0.8452	0.8460
0.20	**0.9985**	0.8214	0.8229	0.7234	0.9978	**0.9993**
0.50	0.5679	0.9868	**0.9972**	0.4370	**1.0000**	0.9980
1.00	0.4296	0.9872	**0.9998**	0.3467	**1.0000**	0.9970
	$\alpha = 1, \quad \sigma = 2$			$\alpha = 2, \quad \sigma = 2$		
Δ	E	SO	WOU	E	SO	WOU
0.05	0.9297	**1.0000**	0.9422	**0.9635**	0.8752	0.8793
0.20	0.8249	0.9573	**0.9916**	0.6017	0.7333	**1.0000**
0.50	0.6050	0.6607	**1.0000**	0.3797	0.6406	**1.0000**
1.00	0.5254	0.5432	**1.0000**	0.2690	0.4214	**1.0000**

WOU (0.9608), SO (0.8372), and E (0.6607). E is the best performing method for $\Delta = 0.05$ but its relative efficiency quickly decays as Δ increases. SO and WOU perform similarly for low diffusive scenarios ($\sigma = 1$), but for $\sigma = 2$ WOU significantly outperforms SO for $\Delta = 0.20, 050, 1.00$, a fact explained by the proneness of the tpd to be multimodal in those situations. The competitive performance of WOU under all scenarios and Δ's, in addition to its affordable computational cost, places it as the preferred estimation method for the WN process with $p = 2$. Similar empirical results hold for $p = 1$, see [11] for details.

As a conclusion, the WN process is seen to be a suitable toroidal diffusion for the needs of ETDBN: OU-like toroidal diffusion, with known sdi, time-reversible, and with tractable inference.

4.3 ETDBN: An Evolutionary Model for Protein Pairs

Shortly stated, ETDBN [13] is a generative evolutionary model of protein sequence and structure evolution that accounts for evolutionary dependencies due to shared ancestry, dependencies between sequence and structure, and local dependencies between aligned sites. Throughout this section we detail structure, training, and benchmarking of ETDBN.

4.3.1 Hidden Markov Model Structure

ETDBN is a dynamic Bayesian network model for a pair of aligned homologous proteins which can be viewed as a Hidden Markov Model (HMM; see Figure 4.8). Each hidden node corresponds to an aligned site in a sequence alignment and adopts an evolutionary hidden state specifying a distribution over three different observation pairs: a pair of amino acid characters, a pair of dihedral angles, and a pair of secondary structure classifications. A transition probability matrix specifies neighboring dependencies between adjacent evolutionary states along the alignment. For example, a transition from a hidden state encoding predominantly α-helix evolution to a hidden state also encoding α-helix evolution is expected to occur more frequently than a transition to a hidden state encoding β-sheet evolution.

The sequence of hidden nodes in the HMM is denoted as $\mathbf{H} := (H^1, H^2, \ldots, H^m)$, where m is the length of the sequence alignment $\mathbf{M}_{ab} := (M_{ab}^1, \ldots, M_{ab}^m)$. Each hidden node H^i in the HMM corresponds to a *site observation* pair $\left(\mathbf{P}_a^{x(i)}, \mathbf{P}_b^{y(i)}\right)$ at an aligned site i in $\mathbf{M}_{ab}$ of the two homologous proteins $\mathbf{P}_a = \left(\mathbf{P}_a^1, \ldots, \mathbf{P}_a^{|\mathbf{P}_a|}\right)$ and $\mathbf{P}_b = \left(\mathbf{P}_b^1, \ldots, \mathbf{P}_b^{|\mathbf{P}_b|}\right)$. For $i = 1, \ldots, m$, $M_{ab}^i \in \left\{\binom{x(i)}{y(i)}, \binom{x(i)}{-}, \binom{-}{y(i)}\right\}$ specifies the homology relationship at position i of the alignment: homologous (no insertions or deletions), deletion with respect to $\mathbf{P}_a$, and insertion with respect to $\mathbf{P}_a$, respectively. $x(i) \in \{1, \ldots, |\mathbf{P}_a|\}$ and

$y(i) \in \{1, \ldots, |\mathbf{P}_b|\}$ specify the indices of the positions in $\mathbf{P}_a$ and $\mathbf{P}_b$, respectively, and $|\mathbf{P}|$ denotes the number of amino acids of protein $\mathbf{P}$. To simplify the discussion that follows, we treat the sequence alignment $\mathbf{M}_{ab}$ as given *a priori*, but in practice we extend the HMM to marginalize out an unobserved alignment (see [13] for more details). Therefore, we exclude $\mathbf{M}_{ab}$ in the equations that follow to make our descriptions more concise.

ETDBN is parametrized by h hidden states. Thus, every hidden node H^i corresponding to an aligned site i can take h possible hidden states and the HMM transition matrix is $h \times h$. Each hidden state specifies a distribution over a *site-class* pair (r_a^i, r_b^i) that is controlled by the evolutionary time t_{ab}. A site-class pair consists of two site-classes, r_a^i and r_b^i, each of the two site-classes taking two integer values, i.e., $(r_a^i, r_b^i) \in R := \{(1,1), (1,2), (2,1), (2,2)\}$. Briefly stated, these site-classes serve for encoding two types of evolution. This is discussed further in Section 4.3.2.

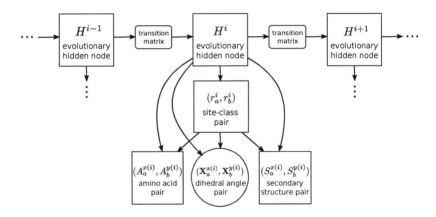

FIGURE 4.8
HMM architecture of ETDBN. Every hidden node corresponds to a sequence aligned pair of amino acid positions of two homologous proteins $\mathbf{P}_a$ and $\mathbf{P}_b$. The horizontal edges between evolutionary hidden nodes encode neighboring dependencies between aligned amino acid positions. The arrows between the evolutionary hidden nodes and site-class pair nodes encode the conditional independence between the observation pair variables $\left(A_a^{x(i)}, A_b^{y(i)} \right)$ (amino acid site pair), $\left(\mathbf{X}_a^{x(i)}, \mathbf{X}_b^{y(i)} \right) = \left(\langle \phi_a^{x(i)}, \psi_a^{x(i)} \rangle, \langle \phi_b^{y(i)}, \psi_b^{y(i)} \rangle \right)$ (dihedral angle site pair), and $\left(S_a^{x(i)}, S_b^{y(i)} \right)$ (secondary structure class site pair). The circles represent continuous variables and the rectangles represent discrete variables.

The state of H^i, together with the site-class pair, and the evolutionary time separating proteins $\mathbf{P}_a$ and $\mathbf{P}_b$, t_{ab}, specify a distribution over three conditionally independent stochastic processes describing each of the three types of site observation pairs: $\left(A_a^{x(i)}, A_b^{y(i)} \right)$, $\left(\mathbf{X}_a^{x(i)}, \mathbf{X}_b^{y(i)} \right)$, and $\left(S_a^{x(i)}, S_b^{y(i)} \right)$. This conditional independence structure allows the likelihood of a site observation pair at an aligned site i to be written as follows:

$$
p\big(\mathbf{P}_a^{x(i)}, \mathbf{P}_b^{y(i)} \mid H^i, r_a^i, r_b^i, t_{ab}\big) := \overbrace{p\big(A_a^{x(i)}, A_b^{y(i)} \mid H^i, r_a^i, r_b^i, t_{ab}\big)}^{\text{amino acid evolution}}
$$

$$
\times \overbrace{p\big(\mathbf{X}_a^{x(i)}, \mathbf{X}_b^{y(i)} \mid H^i, r_a^i, r_b^i, t_{ab}\big)}^{\text{dihedral angle evolution}}
$$

$$
\times \overbrace{p\big(S_a^{x(i)}, S_b^{y(i)} \mid H^i, r_a^i, r_b^i, t_{ab}\big)}^{\text{secondary structure evolution}}, \qquad (4.13)
$$

where $p(X)$ denotes either the pdf or the probability mass function of the random variable X. The assumption of conditional independence provides computational tractability, allowing us to avoid costly marginalization when certain combinations of data are missing (e.g., amino acid sequences present, but secondary structures and dihedral angles missing).

The amino acid and secondary structure evolution terms in Equation (4.13) are modelled using time-reversible Continuous-Time Markov Chains (CTMC). CTMCs are standard tools for modelling the evolution of discrete states, such as amino acid characters or secondary structure classes. The parameters of the amino acid and secondary structure CTMCs are specified by the hidden state, H^i, and evolutionary site-classes (r_a^i, r_b^i). Each CTMC shares a symmetric rate matrix across all hidden states and stationary frequencies specific to each value of H^i and (r_a^i, r_b^i) (see details in [13]). The amino acid CTMC contains 20 states and the secondary structure classes has 3: Helix (H), Sheet (S), and Coil (C). The joint time-dependent pdfs in (4.13) are obtained using the pulley principle (Figure 4.3) and the transition probabilities of the CTMCs. The dihedral angle evolution is modelled using the bivariate WN process introduced in (4.12) and, using the approximate tpd given in (4.12), $p\big(\mathbf{X}_a^{x(i)}, \mathbf{X}_b^{y(i)} \mid H^i, r_a^i, r_b^i, t_{ab}\big) := p_{t_{ab}}^{\text{WOU}}\big(\mathbf{X}_b^{y(i)} \mid \mathbf{X}_a^{x(i)}; \mathbf{A}^\ell, \boldsymbol{\mu}^\ell, \boldsymbol{\Sigma}^\ell\big) f_{\text{WN}}\big(\mathbf{X}_a^{x(i)}; \mathbf{A}^\ell, \boldsymbol{\mu}^\ell, \boldsymbol{\Sigma}^\ell\big)$, where ℓ depends on the value of (H^i, r_a^i, r_b^i). The time-reversibility of p_t^{WOU} (Corollary 1) implies that the roles of a and b are exchangeable in the above expression.

Finally, note that the HMM structure and the sequence alignment entail that the joint pdf of a pair of related proteins $\mathbf{P}_a = (\mathbf{A}_a, \mathbf{X}_a, \mathbf{S}_a)$ and $\mathbf{P}_b = (\mathbf{A}_b, \mathbf{X}_b, \mathbf{S}_b)$ is given by

$$
p(\mathbf{P}_a, \mathbf{P}_b \mid t_{ab}) = \sum_{\mathbf{H}} p(\mathbf{P}_a, \mathbf{P}_b \mid \mathbf{H}, t_{ab}) p(\mathbf{H})
$$

$$
= \sum_{\mathbf{H}} p\big(\mathbf{P}_a^{x(1)}, \mathbf{P}_b^{y(1)} \mid H^1, t_{ab}\big) p(H^1)
$$

$$
\times \prod_{i=2}^{m} p\big(\mathbf{P}_a^{x(i)}, \mathbf{P}_b^{y(i)} \mid H^i, t_{ab}\big) p(H^i \mid H^{i-1}), \qquad (4.14)
$$

where the sum is carried over the h^m possible sequences $\mathbf{H}$. The factor $p(\mathbf{P}_a^{x(i)}, \mathbf{P}_b^{y(i)} \mid H^i, t_{ab})$ is given in terms of (4.13) in (4.17). The summation in (4.14) can be efficiently computed using the HMM forward algorithm, and hidden node sequences can be efficiently sampled from (4.14) using the Forward Filtering Backward Sampling (FFBS) algorithm.

4.3.2   Site-Classes: Constant Evolution and Jump Events

We now turn to the meaning of the site-class pairs. Two modes of evolution are modelled: *constant evolution* and *jump events*. Constant evolution occurs when the site-class starting in protein $\mathbf{P}_a$ at aligned site i, namely r_a^i, is the same as the site-class ending in protein $\mathbf{P}_b$ at aligned site i, r_b^i, i.e., $r_a^i = r_b^i$.

As already stated, a site-class specifies the parameters of the three conditionally independent stochastic processes describing evolution. A limitation of constant evolution is that the coupling between the three stochastic processes is somewhat weak. This in part stems from the time-reversibility of the stochastic processes – swapping the order of one of the three observation pairs at a homologous site, e.g., (glycine, proline) instead of (proline, glycine), does not alter the likelihood in Equation (4.13). Alternatively restated: a "directional coupling" of an amino acid interchange does not inform the *direction* of change in dihedral angle or secondary structure. For example, replacing a glycine in an α-helix in one protein with a proline at the homologous position in a second protein is expected to break the α-helix and to strongly inform the plausible dihedral angle conformations in the second protein.

Ideally, we would consider a model in which the underlying site-classes were not fixed over the evolutionary trajectory separating the two proteins, as in the case of constant evolution as described above, but instead were able to "evolve" in time. This would allow occasional switches in the underlying site-class at a particular homologous site, which would create a stronger dependency between amino acid, dihedral angle, and secondary structure evolution that captures the directional coupling we desire. To approximate this "ideal" model in a computationally efficient manner we introduce the notion of a jump event. A jump event occurs when $r_a^i \neq r_b^i$. Whereas constant evolution is intended to capture angular drift (changes in dihedral angles localized to a region of the *Ramachandran plot* – see Figure 4.9), a jump event is intended to create a directional coupling between amino acid and structure evolution, and is also expected to capture angular shift (large changes in dihedral angles, possibly between distant regions of the Ramachandran plot).

The hidden state at node H^i, together with the evolutionary time t_{ab} separating proteins $\mathbf{P}_a$ and $\mathbf{P}_b$, provides a distribution over a site-class pair:

$$p(r_a^i, r_b^i \mid H^i, t_{ab}) = p(r_a^i \mid H^i, r_b^i, t_{ab})p(r_b^i \mid H^i), \qquad (4.15)$$

where we consider

$$p(r_a^i \mid H^i, r_b^i, t_{ab}) := \begin{cases} e^{-\gamma_{H^i} t_{ab}} + \pi_{H^i, r_b^i}(1 - e^{-\gamma_{H^i} t_{ab}}), & \text{if } r_a^i = r_b^i, \\ \pi_{H^i, r_b^i}(1 - e^{-\gamma_{H^i} t_{ab}}), & \text{if } r_a^i \neq r_b^i, \end{cases}$$

with $p(r_a^i \mid H^i) := \pi_{H^i, r_a^i}$ and $p(r_b^i \mid H^i) := \pi_{H^i, r_b^i}$. Both π_{H^i, r_a^i} and π_{H^i, r_b^i} are model parameters specifying the stationary probability of starting in site-class r_a^i and r_b^i, respectively, and therefore are not conditional on the evolutionary time t_{ab}. $\gamma_{H^i} > 0$ is a parameter specific to hidden state H^i specifying the jump rate. The site-class jump probabilities have been chosen so that time-reversibility holds; in other words it happens that: $p(r_a^i \mid H^i, r_b^i, t_{ab})p(r_b^i \mid H^i) = p(r_b^i \mid H^i, r_a^i, t_{ab})p(r_a^i \mid H^i)$.

The hidden state at node H^i, together with a site-class pair (r_a^i, r_b^i) and the evolutionary time t_{ab}, specifies the likelihood over site observation pairs:

$$p\big(\mathbf{P}_a^{x(i)}, \mathbf{P}_b^{y(i)} \mid H^i, r_a^i, r_b^i, t_{ab}\big)$$
$$:= \begin{cases} p\big(\mathbf{P}_a^{x(i)}, \mathbf{P}_b^{y(i)} \mid H^i, r_c^i, t_{ab}\big), & \text{if } r_a^i = r_b^i = r_c^i, \\ p\big(\mathbf{P}_a^{x(i)} \mid H^i, r_a^i\big)p\big(\mathbf{P}_b^{y(i)} \mid H^i, r_b^i\big), & \text{if } r_a^i \neq r_b^i. \end{cases} \tag{4.16}$$

Constant evolution is considered *constant* because each observation type at an aligned site i is drawn from the same stochastic process given by H^i and r_c^i. Note that the strength of the evolutionary dependency within an observation pair depends on t_{ab}.

In the case of a jump event, the evolutionary processes are, after the evolutionary jump, restarted independently in the sdi of the new site-class. Thus the site observations $\mathbf{P}_a^{x(i)}$ and $\mathbf{P}_b^{y(i)}$ are assumed to be drawn from the sdis of two separate stochastic processes corresponding to site-classes r_a^i and r_b^i, respectively. This implies that, conditional on a jump, the likelihood of the observations is no longer dependent on t_{ab}. A jump event can therefore best be thought of as an abstraction that captures the end-points of the evolutionary process, but ignores the potential evolutionary trajectory linking the two site observations. The advantage of abstracting the evolutionary trajectory is that there is no need to perform a computationally expensive integration of all possible trajectories. The likelihood of an observation pair is now simply

$$p\big(\mathbf{P}_a^{x(i)}, \mathbf{P}_b^{y(i)} \mid H^i, t_{ab}\big)$$
$$= \sum_{(r_a^i, r_b^i) \in R} p\big(\mathbf{P}_a^{x(i)}, \mathbf{P}_a^{y(i)} \mid H^i, r_a^i, r_b^i, t_{ab}\big)p(r_a^i, r_b^i \mid H^i, t_{ab}), \tag{4.17}$$

where (4.16) and (4.15) provide the terms in (4.17).

4.3.3 Model Training

Training and Test Datasets

In order to train and evaluate ETDBN, a training dataset of 1200 protein pairs (2400 proteins; 417870 site observation pairs) and a test dataset of 38 protein pairs (76 proteins; 14125 site observation pairs) were assembled from 1032 protein families in the HOMSTRAD database [28] – a database of homologous protein structures. Dihedral angles were computed from the PDB coordinates of each protein structure using the `BioPython.PDB` package [17]. The secondary structure at each amino acid position in every protein was annotated using the `DSSP` software [33].

Model Training

Stochastic Expectation-Maximization (StEM, [12]) was used to train the model. StEM is a stochastic version of the well-known Expectation-Maximization iterative algorithm [12], which is commonly used for fitting the parameters of HMMs. Its distinguishing feature is that the E-step consists of filling in the values of the latent variables using sampling. StEM is attractive due to its computational efficiency and its tendency to avoid getting stuck in local minima [12].

We describe in what follows the E- and M-steps. To that end, let us denote by $\Psi^{(k)}$ the model parameters at the k-th iteration. FFBS was used in the E-step to sample hidden node sequences $\mathbf{H}$ and site-classes $(\mathbf{r}_a, \mathbf{r}_b)$. The Metropolis–Hastings algorithm was used to sample t_{ab} for each protein pair in the training dataset conditional on the observations and parameters $\Psi^{(k)}$ (see (4.18) and the description that follows for details). In other words, at iteration k for each pair of aligned observation sequences $\mathbf{P}_a$ and $\mathbf{P}_b$ samples were drawn from the following joint distribution:

$$\left(\mathbf{H}, \mathbf{r}_a, \mathbf{r}_b, t_{ab}\right)^{(k)} \sim p\left(\mathbf{H}, \mathbf{r}_a, \mathbf{r}_b, t_{ab} \mid \mathbf{P}_a, \mathbf{P}_b, \Psi^{(k)}\right).$$

In the M-step the samples from the previous E-step were used to update the hidden node parameters ($\Psi^{(k+1)}$) using Efficient Sufficient Statistics (ESSs). For example, the MLE of the transition probability matrix at a particular step can be obtained by calculating the proportion of hidden state transitions within the sampled hidden sequences. Where ESSs were not used, the COBYLA optimization algorithm [29] in the `NLOpt` library [20] was used to update the parameters.

Model Selection

Models with hidden states ranging from 8 to 112 were trained until convergence for varying numbers of repetitions (2 to 4) using different initial random number seeds. The highest log-likelihood model of each repetition was selected for downstream analysis. Following that, marginal likelihoods $p(\text{data} \mid \text{fitted model})$ and corresponding Bayesian Information Criterion (BIC) scores were computed under each model by fixing the alignments to the respective HOMSTRAD alignments. The alignments were fixed *a priori* to make computation of the marginal likelihoods computationally tractable. The 64 hidden state model was selected as the best model, as it presented the lowest BIC. Additionally, predictive accuracies under a homology modelling scenario, $p(\mathbf{X}_b \,|\, \mathbf{A}_a, \mathbf{A}_b, \mathbf{X}_a,$ fitted model), were calculated for each of the 38 protein pairs in the test dataset for each of the models. The chosen 64 hidden state model had predictive accuracies comparable to the model with the highest predictive accuracies (a model with 32 hidden states). Further details about the model selection can be found in [13].

4.3.4 Benchmarks

In this section we perform a series of benchmarks that examine the performance of ETDBN. First, we test how well the model reproduces the empirical distributions of dihedral angles. Next we compare how adding increasingly informative conditioning observations affects uncertainty in the estimates of evolutionary times and improves levels of accuracy in the prediction of dihedral angles. These analyses are facilitated by the conditional independence structure in (4.13), which enables computationally efficient posterior inference under different combinations of observed or missing data.

Dihedral Angles Distribution

The generative nature of ETDBN allows the model to be easily interrogated. In this benchmark, we compare the histograms of dihedral angle pairs present in real data with the dihedrals sampled from ETDBN. Figure 4.9 presents this comparison for the 417870 dihedral angle pairs present in the training dataset and 19324 dihedral angle pairs associated to proline. These distributions are shown in the Ramachandran plots and are a useful tool for visualizing the conformational possibilities associated with different amino acids. There is a close correspondence between dihedral angles sampled under the model (Figure 4.9, left) and the empirical distributions (Figure 4.9, right). This serves only as a partial validation of ETDBN, since it was expected that the empirical dihedral angle distributions be well-modelled, given that ETDBN is effectively a mixture model with a large number of mixture components.

Posterior Inference: Evolutionary Times

The posterior distribution of t_{ab}, conditional on the observed dihedral angles $\mathbf{X}_a$ and $\mathbf{X}_b$, is given by Bayes' theorem as

$$
\begin{aligned}
p(t_{ab} \,|\, \mathbf{X}_a, \mathbf{X}_b) &\propto p(\mathbf{X}_a, \mathbf{X}_b \,|\, t_{ab}) p(t_{ab}) \\
&= p(t_{ab}) \sum_{\mathbf{H}} p\big(\mathbf{X}_a^{x(1)}, \mathbf{X}_b^{y(1)} \,|\, H^1, t_{ab}\big) p(H^1) \\
&\quad \times \prod_{i=2}^{m} p\big(\mathbf{X}_a^{x(i)}, \mathbf{X}_b^{y(i)} \,|\, H^i, t_{ab}\big) p(H^i \,|\, H^{i-1}).
\end{aligned}
\tag{4.18}
$$

The prior distribution over $p(t_{ab})$ is given by an exponential distribution with $\lambda = \frac{1}{10}$, chosen for its biological plausibility. The Metropolis–Hastings algorithm can be used to sample the posterior distribution,

$p(t_{ab} \mid \mathbf{X}_a, \mathbf{X}_b)$, using the associated prior and likelihood in (4.18) (obtained by the HMM structure as in (4.14)), which can be efficiently computed using the HMM forward algorithm. Analogous expressions hold for $p(t_{ab} \mid \mathbf{A}_a, \mathbf{A}_b)$ and $p(t_{ab} \mid \mathbf{A}_a, \mathbf{A}_b, \mathbf{X}_a, \mathbf{X}_b)$.

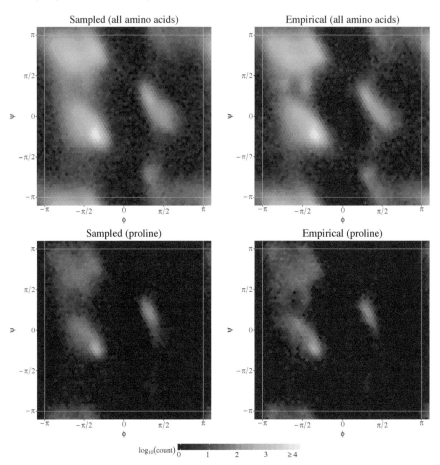

FIGURE 4.9

Ramachandran plots depicting histograms of the model-sampled dihedral angles (left column) and the dihedral angles in the training dataset (right column). The top row shows the distributions for all amino acids (417870) and the bottom for proline only (19324). The number of angles in the left and right columns is the same.

The left half of Figure 4.10 depicts the boxplots of posterior standard deviations of the evolutionary times inferred for the proteins in the testing dataset under three different conditions (sequence only, angles only, and both sequence and angles). A one-sided Wilcoxon signed-rank test was used to test whether the standard deviations were significantly smaller when using both types of observations, compared to using only sequence or dihedral angle information.

As might be expected, the standard deviations of the posterior evolutionary times were significantly smaller when both the sequence and dihedral angle observations were used for posterior inference, compared to using only sequences (p-value $= 0.00014$) or only dihedral angles (p-value $= 0.00002$). Hence, the model's accuracy on the posterior of t_{ab} improves as more conditioning information is considered. Finally, note also that the sequence alone provides sharper posteriors than the dihedrals alone, a fact likely explained by the direct relation between t_{ab} and $(\mathbf{A}_a, \mathbf{A}_b)$.

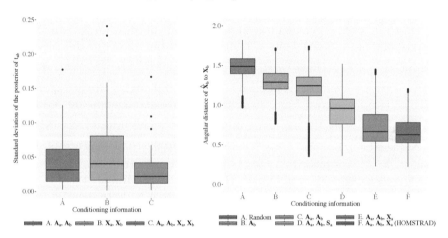

FIGURE 4.10

Left: box plots depicting the distribution of posterior standard deviations for the protein pairs in the test dataset under three different scenarios. The mean standard deviations were 0.43, 0.58, and 0.33 for A, B, and C, respectively. Right: comparison of mean predictive accuracy (measured using angular distance, lower is better) for the 38 protein pairs in the test dataset, giving a representative view of predictive accuracy under six different combinations of observations.

Posterior Inference: Structure

ETDBN can be used to sample (i.e., predict) the dihedral angles of a protein b, $\mathbf{X}_b$, from $\mathbf{A}_b$, $\mathbf{A}_a$, $\mathbf{X}_a$, $\mathbf{S}_b$, $\mathbf{S}_a$, or any combination of them. We describe in the next three paragraphs the procedure for achieving so.

Conditional on having observed $(\mathbf{A}_a, \mathbf{A}_b, \mathbf{X}_a)$, the sequence of missing dihedral angles $\mathbf{X}_b$ can be sampled using the following procedure. First, similar to (4.18), the likelihood $p(t_{ab} \mid \mathbf{A}_a, \mathbf{A}_b, \mathbf{X}_a)$ and the prior $p(t_{ab})$ are used to sample a single evolutionary time, $t_{ab}{}^{(k)}$, using Metropolis–Hastings:

$$t_{ab}{}^{(k)} \sim p(t_{ab} \mid \mathbf{A}_a, \mathbf{A}_b, \mathbf{X}_a)p(t_{ab}). \tag{4.19}$$

When only a single observation is present for a particular observation type (e.g., only $\mathbf{X}_a$ in $p(t_{ab} \mid \mathbf{A}_a, \mathbf{A}_b, \mathbf{X}_a)$), it is assumed that the observation is drawn from the sdi of the corresponding stochastic process. Conditional on the observations and newly sampled evolutionary time, $t_{ab}^{(k)}$, the forward probabilities already calculated in (4.19) can be used to perform FFBS algorithm to draw a sequence of hidden states, $\mathbf{H}^{(k)} \sim p(\mathbf{H} \mid \mathbf{A}_a, \mathbf{A}_b, \mathbf{X}_a, t_{ab}^{(k)})$.

For each hidden state $H^{i\,(k)}$ in the newly sampled hidden sequence $\mathbf{H}^{(k)}$, a site-class pair $(r_a^i, r_b^i)^{(k)}$ taking on one of four possible values is sampled with probability proportional (recall an application of Bayes' theorem in the first factor) to $p(A_a^{x(i)}, A_b^{y(i)}, \mathbf{X}_a^{x(i)} \mid H^{i\,(k)}, r_a^i, r_b^i, t_{ab}^{(k)}) p(r_a^i, r_b^i \mid H^{i\,(k)}, t_{ab}^{(k)})$.

If $r_a^{i\,(k)} \neq r_b^{i\,(k)}$, a jump is implied and a dihedral angle pair $\mathbf{X}_b^{y(i)\,(k)}$, corresponding to aligned site i, is drawn from the sdi of the diffusion whose parameters are specified by $r_b^{i\,(k)}$: $\mathbf{X}_b^{y(i)\,(k)} \sim p(\mathbf{X}_b^{y(i)} \mid H^{i\,(k)}, r_b^{i\,(k)})$. If $r_a^{i\,(k)} = r_b^{i\,(k)}$, constant evolution is implied and a dihedral angle pair corresponding to aligned site i is drawn from the diffusion whose parameters are specified by $r_a^{i\,(k)} = r_b^{i\,(k)}$ using the sampling procedure obtained from (4.12): $\mathbf{X}_b^{y(i)\,(k)} \sim p(\mathbf{X}_b^{y(i)} \mid H^{i\,(k)}, r_b^{i\,(k)}, \mathbf{X}_a^{x(i)}, t_{ab}^{(k)})$. Doing so produces a sequence of dihedral angles, $\mathbf{X}_b^{(k)}$, sampled from the correct posterior distribution.

For each of the 38 protein pairs $(\mathbf{P}_a, \mathbf{P}_b)$, the dihedral angles of $\mathbf{P}_b$ in each pair were treated as missing, and 5000 sequences of $\mathbf{X}_b$ were sampled under six different combinations of observations for each. Following that, predictive accuracy was measured using the average angular distance between the sampled $(\hat{\mathbf{X}}_b)$ and known $(\mathbf{X}_b)$ dihedral angles. Figure 4.10 shows the results. In Figure 4.10A, no data was used for prediction, hence the predicted angles correspond to random draws from the model. In Figure 4.10B–4.10E, the following observations are used, respectively: $\mathbf{A}_b$, $(\mathbf{A}_b, \mathbf{A}_a)$, $(\mathbf{A}_a, \mathbf{A}_b, \mathbf{S}_a)$, and $(\mathbf{A}_a, \mathbf{A}_b, \mathbf{X}_a)$. Finally, in Figure 4.10F the same combination of observations was used as in Figure 4.10E, but the sequence alignment was treated as known *a priori* instead of marginalized. These results show that increasingly informative observations lead to better predictive accuracy, which is consistent with what we expect.

In the right half of Figure 4.11 we provide a detailed graphical example of a single pair of homologous annexin proteins (PDB 1ala and PDB 1ann) for which sampling was performed under four different conditions. PDB 1ala and PDB 1ann are moderately diverged, having an amino acid sequence identity of 56%. While the full protein (PDB 1ann) was sampled, only a 35 amino acid fragment is depicted. The reason is that ETDBN is considered a local model and is not designed to capture the global properties of protein structure. For example, proteins have a strong tendency to be globular (compact) in nature – a global, coarse-grained property not enforced by our model. Therefore ETDBN does not constitute a complete homology modelling method in itself. Rather, it can be used as a building block (much like fragment libraries model local structure [30]) in protein structure prediction and homology modelling methods. Fine-grained distributions, such as ETDBN, can be combined with

with coarse-grained distributions (that capture the globular nature of proteins, for example) in a statistically principled manner using a method known as the *reference ratio method* [10, 16], an extension out of the scope of this work.

Despite these caveats, it is clear once again in Figure 4.11 that introducing increasingly informative sequence and structural observations leads to better predictions. As more informative data is added the distribution of samples tends to concentrate around the native structure. Increasingly informative conditioning observations also improve the prediction of the secondary structure elements, as can be seen by comparing the patterns of cartoon helices and coils in the centroid and reference structures.

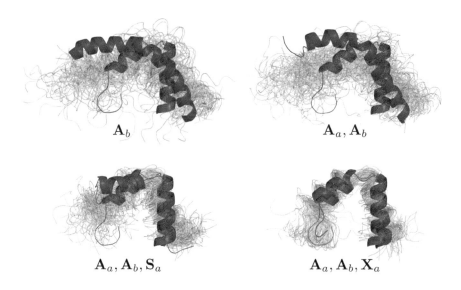

FIGURE 4.11
Samples from ETDBN of a 35 amino acid fragment of an annexin protein under increasingly informative conditioning observations (from left to right and up to down). In red is the native protein structure (protein b, corresponding to positions 100–135 of PDB 1ala) and in gray is 100 samples from the model corresponding to the same region. In blue is the centroid structure (sample with the lowest average RMSD to all other samples) based on 2000 samples from the model. The homologous protein used for the purposes of prediction (protein a) was PDB 1ann.

4.4 Case Study: Detection of a Novel Evolutionary Motif

A benefit of ETDBN is that the 64 evolutionary hidden states learned during the training phase are interpretable. We give an example of a hidden state encoding a jump event detected in a number of protein pairs in our test and training datasets, suggesting that this hidden state encodes an *evolutionary motif* – a common pattern of sequence-structure evolution.

Evolutionary Hidden State 13 (Figure 4.12), referred to in the sequel as EHS13, was selected from the 64 hidden states due to its encoding of a jump event with a significant angular shift (a large change in dihedral angle). A notable feature of this EHS13 is that the dihedral angles corresponding to site-classes r_1 and r_2, respectively, are associated with moderately different amino acid distributions. This is expected to be informative of a specific directional transition in dihedral angles corresponding to a particular directional exchange between amino acids.

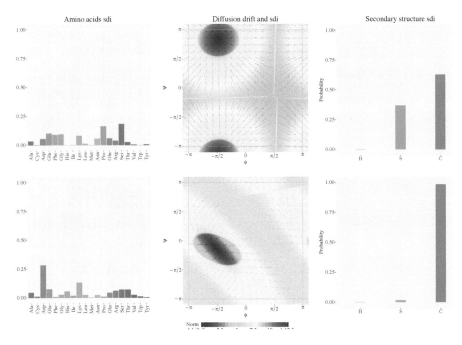

FIGURE 4.12

Depiction of EHS13 and its associated evolutionary structural motif. This hidden state was sampled at 2.59% of sites (the average was 1.56%). The equilibrium frequencies of r_1 and r_2 were $\pi_1 = 0.181$ and $\pi_2 = 0.819$, respectively, and the jump rate was $\gamma = 1.32$. The first and second rows depict the parameters encoded by the two site-classes r_1 and r_2, respectively. The first and third columns show the estimated sdis for the amino acid and secondary structure stochastic processes, respectively. The second column depicts the drift vector fields of the WN processes. The color gradient represents the intensity of the drift, measured as the norm of the arrows, and with shading proportional to the stationary density.

We performed a search of protein pairs containing the jump represented by ESH13. Posterior inference was performed conditioned on the amino acid sequence and dihedral angles, $(\mathbf{A}_a, \mathbf{A}_b, \mathbf{X}_a, \mathbf{X}_b)$, of 238 protein pairs. This was done to identify aligned sites encoding jump events corresponding to potential evolutionary motifs. Aligned sites pertaining to a single hidden state and with evidence of a jump event $(r_a^i \neq r_b^i)$ at posterior probability > 0.99 were identified, that is, the i's such that $p(H^i, r_a^i \neq r_b^i \mid \mathbf{A}_a, \mathbf{A}_b, \mathbf{X}_a, \mathbf{X}_b) > 0.99$.

Twelve aligned sites in eleven different protein pairs corresponding to $H^i = 13$ (EHS13) were identified. A homologous pair of annexin proteins, PDB 1ann (from *Bos taurus*) and PDB 1ala (from *Gallus gallus*), was selected out of the 11 pairs with a jump corresponding to EHS13 (Figure 4.13) at aligned site E161/P163. Most aligned sites in the annexin pair had low posterior jump probabilities (≈ 0.0), with the exception of three successive aligned sites starting with the aligned site of interest, E161/P163, which had high posterior jump probabilities (≈ 1.0). Site-classes $\mathbf{r}_1$ and $\mathbf{r}_2$ indicated that an exchange between a proline and a glutamate at positions 161 and 163, respectively, is informative of the dihedral angle change specified by EHS13.

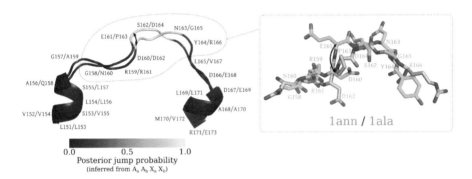

FIGURE 4.13
Depiction of two homologous annexin proteins, PDB 1ann and 1ala, superimposed. Left: cartoon representation of the two proteins corresponding to regions L151–R171 and L153–E173, respectively, with posterior jump probabilities at each position overlaid. Right: ball-and-stick representation giving atomic detail for a seven amino acid region (G158–Y164 and N160–R166, respectively). The exchange between a glutamate (E161 in 1ann) and a proline (P163 in 1ala) is associated with a large change in dihedral angle as indicated by the curved arrows.

It is unknown whether this apparent evolutionary motif has functional consequences, however, the associated local structural and sequence changes between 1ann and 1ala appear significant. The three successive aligned sites with high posterior jump probabilities have dihedral transitions with large angular distances of 1.97, 1.20, and 1.96, respectively, in comparison to the mean angular distance between the dihedral angles of 1ann and 1ala (mean 0.288 and 90%-confidence interval for the mean $(0.034, 0.806)$). All three positions involve an exchange of amino acids. Glutamate to proline (E161/P163) in the first, serine to aspartic acid (S162/D164) in the second, and asparagine to glycine (N163/G165) in the third. The involvement of proline at the first of the three positions in a large conformational shift is unsurprising as proline adopts distinctive dihedral conformations compared to the other amino acids. Likewise, the involvement of a glycine at the third position is also unsurprising given that it is the smallest and most flexible amino acid in terms of the dihedral angle conformations it can adopt. The transition to two such flexible amino acids in PDB 1ala may be indicative of positive selection acting to confer a beneficial structural conformation, although more evidence is required to substantiate this conclusion.

The presence of the identified jump in 11 protein pairs with high posterior probabilities suggests that this may represent a common evolutionary motif. Conjecturally, the identification of evolutionary motifs together with improved modelling may in the future prove useful for several reasons: (*i*) improvement in homology modelling predictions due to the more accurate prediction of large conformational changes; (*ii*) improved estimates of evolutionary parameters, such as evolutionary times, which may be inflated when large conformation shifts are not modelled; and (*iii*) help in identifying classes of functionally relevant positions that are potential drug targets, given that large changes in dihedral angles might be associated with consequential and predictable functional changes.

4.5 Conclusions

We have shown that the WN process, an ergodic, time-reversible, and tractable Ornstein–Uhlenbeck toroidal analogue, is a cornerstone in the development of ETDBN, a tractable, generative, and interpretable probabilistic model of protein sequence and structure evolution on a local scale. The probabilistic nature of ETDBN allows rigorous statements about uncertainty to be made. The ability to infer various quantities of interest (such as evolutionary times or missing structures) and to interpret the parameters of the model demonstrates ETDBN's usefulness in gaining biological insights. Many existing computational models of biological structures lack statistical rigor, relying on heuristic techniques that do not provide any quantification of uncertainty. We envisage that the use of problem-adapted statistical methods, like the toroidal diffusions considered in this chapter, will grow in importance as practitioners increasingly demand a rigorous understanding of the underlying assumptions and inferences made in the construction and application of their models.

Acknowledgments

This work is part of the Dynamical Systems Interdisciplinary Network, University of Copenhagen, and was funded by the University of Copenhagen 2016 Excellence Programme for Interdisciplinary Research (UCPH2016-DSIN), and by project MTM2016-76969-P from the Spanish Ministry of Economy, Industry and Competitiveness, and European Regional Development Fund (ERDF).

Bibliography

[1] K. Arnold, L. Bordoli, J. Kopp, and T. Schwede. The SWISS-MODEL workspace: a web-based environment for protein structure homology modelling. *Bioinformatics*, 22(2):195–201, 2006.

[2] D. S. Bernstein and W. So. Some explicit formulas for the matrix exponential. *IEEE Trans. Automat. Control*, 38(8):1228–1232, 1993.

[3] W. Boomsma, K. V. Mardia, C. C. Taylor, J. Ferkinghoff-Borg, A. Krogh, and T. Hamelryck. A generative, probabilistic model of local protein structure. *Proc. Natl. Acad. Sci. U.S.A.*, 105(26):8932–8937, 2008.

[4] W. Boomsma, P. Tian, J. Frellsen, J. Ferkinghoff-Borg, T. Hamelryck, K. Lindorff-Larsen, and M. Vendruscolo. Equilibrium simulations of proteins using molecular fragment replacement and NMR chemical shifts. *Proc. Natl. Acad. Sci. U.S.A.*, 111(38):13852–13857, 2014.

[5] C. J. Challis and S. C. Schmidler. A stochastic evolutionary model for protein structure alignment and phylogeny. *Mol. Biol. Evol.*, 29(11):3575–3587, 2012.

[6] K. A. Dill. Polymer principles and protein folding. *Protein Sci.*, 8(6):1166–1180, 1999.

[7] K. A. Dill and J. L. MacCallum. The protein-folding problem, 50 years on. *Science*, 338(6110):1042–1046, 2012.

[8] J. Felsenstein. Evolutionary trees from DNA sequences: a maximum likelihood approach. *J. Mol. Evol.*, 17(6):368–376, 1981.

[9] J. Felsenstein. Phylogenies and the comparative method. *Am. Nat.*, 125(1):1–15, 1985.

[10] J. Frellsen, K. V. Mardia, M. Borg, J. Ferkinghoff-Borg, and T. Hamel-ryck. Towards a general probabilistic model of protein structure: the reference ratio method. In T. Hamelryck, K. V. Mardia, and J. Ferkinghoff-Borg, editors, *Bayesian Methods in Structural Bioinformatics*, Statistics for Biology and Health. Springer, Berlin, 2012.

[11] E. García-Portugués, M. Sørensen, K. V. Mardia, and T. Hamelryck. Langevin diffusions on the torus: estimation and applications. *Stat. Comput.*, to appear.

[12] W. R. Gilks, S. Richardson, and D. J. Spiegelhalter, editors. *Markov Chain Monte Carlo in Practice*. Interdisciplinary Statistics. Chapman & Hall, London, 1996.

[13] M. Golden, E. García-Portugués, M. Sørensen, K. V. Mardia, T. Hamel-ryck, and J. Hein. A generative angular model of protein structure evolution. *Mol. Biol. Evol.*, 34(8):2085–2100, 2017.

[14] N. V Grishin. Estimation of evolutionary distances from protein spatial structures. *J. Mol. Evol.*, 45(4):359–369, 1997.

[15] A. M. Gutin and A. Y. Badretdinov. Evolution of protein 3D structures as diffusion in multidimensional conformational space. *J. Mol. Evol.*, 39:206–209, 1994.

[16] T. Hamelryck, M. Borg, M. Paluszewski, J. Paulsen, J. Frellsen, C. Andreetta, W. Boomsma, S. Bottaro, and J. Ferkinghoff-Borg. Potentials of mean force for protein structure prediction vindicated, formalized and generalized. *PLoS One*, 5(11):e13714, 2010.

[17] T. Hamelryck and B. Manderick. PDB file parser and structure class implemented in python. *Bioinformatics*, 19(17):2308–2310, 2003.

[18] J. L. Herman, C. J. Challis, Á. Novák, J. Hein, and S. C. Schmidler. Simultaneous Bayesian estimation of alignment and phylogeny under a joint model of protein sequence and structure. *Mol. Biol. Evol.*, 31(9):2251–2266, 2014.

[19] S. R. Jammalamadaka and A. SenGupta. *Topics in Circular Statistics*, volume 5 of *Series on Multivariate Analysis*. World Scientific Publishing, River Edge, 2001.

[20] S. G. Johnson. The NLopt nonlinear-optimization package, 2014.

[21] R. P. Joosten, T. A. H. te Beek, E. Krieger, M. L. Hekkelman, R. W. W. Hooft, R. Schneider, C. Sander, and G. Vriend. A series of PDB related databases for everyday needs. *Nucleic Acids Res.*, 39(suppl 1):D411–D419, 2011.

[22] J. Kent. Discussion of paper by K. V. Mardia. *J. Roy. Statist. Soc. Ser. B*, 37(3):377–378, 1975.

[23] J. Kent. Time-reversible diffusions. *Adv. in Appl. Probab.*, 10(4):819–835, 1978.

[24] K. V. Mardia. *Statistics of Directional Data*, volume 13 of *Probability and Mathematical Statistics*. Academic Press, London, 1972.

[25] K. V. Mardia and J. Frellsen. Statistics of bivariate von Mises distributions. In T. Hamelryck, K. V. Mardia, and J. Ferkinghoff-Borg, editors, *Bayesian Methods in Structural Bioinformatics*, Statistics for Biology and Health. Springer, Berlin, 2012.

[26] K. V. Mardia and J. Frellsen. Directional statistics in bioinformatics. In C. Ley and T. Verdebout, editors, *Applied Directional Statistics: Modern Methods and Case Studies*. Chapman & Hall/CRC, 2018.

[27] D. S. Marks, L. J. Colwell, R. Sheridan, T. A. Hopf, A. Pagnani, R. Zecchina, and C. Sander. Protein 3D structure computed from evolutionary sequence variation. *PLoS One*, 6:e28766, 2011.

[28] K. Mizuguchi, C. M. Deane, T. L. Blundell, and J. P. Overington. HOMSTRAD: a database of protein structure alignments for homologous families. *Protein Sci.*, 7(11):2469–2471, 1998.

[29] M. J. D. Powell. A direct search optimization method that models the objective and constraint functions by linear interpolation. In S. Gomez and J.-P. Hennart, editors, *Advances in Optimization and Numerical Analysis*. Springer Netherlands, 1994.

[30] C. Rohl, C. Strauss, K. Misura, and D. Baker. Protein structure prediction using Rosetta. *Methods Enzymol.*, 383:66–93, 2004.

[31] I. Shoji and T. Ozaki. A statistical method of estimation and simulation for systems of stochastic differential equations. *Biometrika*, 85(1):240–243, 1998.

[32] J. L. Thorne, H. Kishino, and J. Felsenstein. Inching toward reality: an improved likelihood model of sequence evolution. *J. Mol. Evol.*, 34(1):3–16, 1992.

[33] W. G. Touw, C. Baakman, J. Black, T. A. H. te Beek, E. Krieger, R. P. Joosten, and G. Vriend. A series of PDB-related databanks for everyday needs. *Nucleic Acids Res.*, 43(D1):D364–D368, 2015.

[34] J. B. Valentin, C. Andreetta, W. Boomsma, S. Bottaro, J. Ferkinghoff-Borg, J. Frellsen, K. V. Mardia, P. Tian, and T. Hamelryck. Formulation of probabilistic models of protein structure in atomic detail using the reference ratio method. *Proteins: Struct., Funct., Bioinf.*, 82(2):288–299, 2014.

5

Noisy Directional Data

Thanh Mai Pham Ngoc

Université Paris-Sud
Université Paris-Saclay

CONTENTS

5.1 Introduction 95
5.2 Some Preliminaries about Harmonic Analysis on SO(3) and $\mathbb{S}^2$ 96
5.3 Model and Assumptions 98
    5.3.1 Null and Alternative Hypotheses 98
    5.3.2 Noise Assumptions 99
5.4 Test Constructions 100
5.5 Numerical Illustrations 102
    5.5.1 The Testing Procedures 102
    5.5.2 Alternatives 103
    5.5.3 Simulations 104
    5.5.4 Real Data: Paleomagnetism 107
    5.5.5 Real Data: UHECR 108
    Bibliography 109

5.1 Introduction

We have at our disposal noisy directional observations Z_i's which have the following form:

$$Z_i = \varepsilon_i X_i, \quad i = 1, \ldots, N \tag{5.1}$$

where the ε_i's are i.i.d. random variables of SO(3) the rotation group in $\mathbb{R}^3$ and the X_i's are i.i.d. random variables on $\mathbb{S}^2$, the unit sphere in $\mathbb{R}^3$. We suppose that X_i and ε_i are independent. We also assume that the distributions of Z_i and X_i are absolutely continuous with respect to the uniform measure on $\mathbb{S}^2$ and we set f_Z and f the densities of Z_i and X_i, respectively. The distribution of ε_i is absolutely continuous with respect to the Haar probability measure on SO(3) and we will denote the density of the ε_i's by f_ε. We assume that f_ε is known. We would like to construct nonparametric goodness-of-fit tests from the $Z_i's$ to test $f = f_0$ in an adaptive way. This means that the procedure does

95

not depend on the particular f under the alternative and will automatically adapt itself to regularity of f. Model (5.1) is the spherical convolution model. Indeed we have

$$f_Z = f_\varepsilon * f,$$

where $*$ denotes the convolution product which is defined below in (5.8). Roughly speaking, the spherical convolution model provides a setup where each genuine observation X_i is contaminated by a small random rotation. And the smoother the density of the noise ε, the harder the problem becomes as the convolution effect is increased.

Considering goodness-of-fit testing in the spherical convolution model not only finds its interest in some important applications, but it also fills a gap both in the noisy setup testing literature and the spherical convolution one. Indeed, convolution models have been extensively studied in the Euclidean setting and more recently in other geometric frameworks, like the hyperbolic plane (see [6]) or the sphere. However, so far, only estimation has been treated in the spherical convolution model. For the nonparametric estimation problem with noisy directional data in model (5.1), one is interested in recovering the underlying density f from the Z_i' s. The pioneer works of [5], [8], [9] introduced a minimax estimation procedure based on the Fourier basis of $\mathbb{L}_2(\mathbb{S}^2)$. Recently, [7] proposed an optimal and adaptive hard thresholding estimation procedure based on needlets.

This chapter is organized as follows. Section 1.2 gives a brief overview about harmonic analysis on $SO(3)$ and $\mathbb{S}^2$ which will be necessary to construct the testing procedure. In Section 1.3 we introduce the test hypotheses and the regularity assumptions on f and the noise ε. Section 1.4 introduces the testing procedure and finally Section 1.5 presents numerical results with simulations and real data in astrophysics and paleomagnetism.

5.2 Some Preliminaries about Harmonic Analysis on $SO(3)$ and $\mathbb{S}^2$

Our test procedure is constructed via a spectral approach. So before introducing our testing procedure, we shall give first a brief overview of Fourier analysis on $SO(3)$ to deal with the noise and on $\mathbb{S}^2$ for the unknown density. Most of the material can be found in expanded form in [5], [8], [12], [13], [16].

Let $\mathbb{L}_2(SO(3))$ denote the space of square integrable functions on $SO(3)$, that is, the set of measurable functions f on $SO(3)$ for which

$$\|f\|_2 = \left(\int_{SO(3)} |f(g)|^2 dg \right)^{\frac{1}{2}} < \infty,$$

where dg is the Haar measure on $SO(3)$.

Let D^l_{mn} for $-l \le m$, $n \le l$, $l = 0, 1, \ldots$ be the eigenfunctions of the Laplace Beltrami operator on SO(3), hence, $\sqrt{2l+1}D^l_{mn}$, $-l \le m$, $n \le l$, $l = 0, 1, \ldots$ is a complete orthonormal basis for $\mathbb{L}_2(\text{SO}(3))$ with respect to the probability Haar measure. The D^l_{mn}'s are called *rotational harmonics*. Explicit formulae of the rotational harmonics D^l_{mn} in terms of Euler angles exist but we do not need them here. Next, for $f \in \mathbb{L}_2(\text{SO}(3))$, we define the rotational Fourier transform on SO(3) by the $(2l+1) \times (2l+1)$ matrices f^{*l} with entries

$$f^{*l}_{mn} = \int_{\text{SO}(3)} f(g) D^l_{mn}(g) dg. \tag{5.2}$$

The rotational inversion can be obtained by

$$f(g) = \sum_{l \ge 0} \sum_{-l \le m, \, n \le l} f^{*l}_{mn}(2l+1)\overline{D^l_{mn}(g)}. \tag{5.3}$$

(5.3) is to be understood in $\mathbb{L}_2$-sense although with additional smoothness conditions, it can hold pointwise.

A parallel spherical Fourier analysis is available on $\mathbb{S}^2$. Any point on $\mathbb{S}^2$ can be represented by

$$\omega = (\cos\phi\sin\theta, \sin\phi\sin\theta, \cos\theta)^t,$$

with $\phi \in [0, 2\pi)$, $\theta \in [0, \pi)$. We also define the functions:

$$Y^l_m(\omega) = Y^l_m(\theta, \phi) := \sqrt{\frac{(2l+1)}{4\pi}\frac{(l-m)!}{(l+m)!}} P^l_m(\cos\theta) e^{im\phi}, \tag{5.4}$$

for $-l \le m \le l$, $l = 0, 1, \ldots$, $\phi \in [0, 2\pi)$, $\theta \in [0, \pi)$ where the P^l_m's are the associated Legendre functions. The functions Y^l_m obey

$$Y^l_{-m}(\theta, \phi) = (-1)^m \overline{Y^l_m(\theta, \phi)}. \tag{5.5}$$

Let $\mathbb{L}_2(\mathbb{S}^2)$ denote the space of square integrable functions on $\mathbb{S}^2$, that is, the set of measurable functions f on $\mathbb{S}^2$ for which

$$\|f\|_2 = \left(\int_{\mathbb{S}^2} |f(x)|^2 dx\right)^{\frac{1}{2}} < \infty,$$

where dx is the Lebesgue measure on the sphere $\mathbb{S}^2$. It is well-known that $\mathbb{L}_2(\mathbb{S}^2)$ is a Hilbert space with the inner product

$$\langle f, g \rangle_{\mathbb{L}_2} = \int_{\mathbb{S}^2} f(x)\overline{g(x)} dx, \quad f, g \in \mathbb{L}_2(\mathbb{S}^2).$$

The set $\{Y^l_m, -l \le m \le l, l = 0, 1, \ldots\}$ is forming an orthonormal basis of $\mathbb{L}_2(\mathbb{S}^2)$, generally referred to as the spherical harmonic basis. Again, as above, for $f \in \mathbb{L}_2(\mathbb{S}^2)$, we define the spherical Fourier transform on $\mathbb{S}^2$ by

$$f_m^{\star l} = \int_{\mathbb{S}^2} f(x)\overline{Y_m^l(x)}dx. \tag{5.6}$$

We think of (5.6) as the vector entries of the $(2l + 1)$ vector

$$f^{\star l} = [f_m^{\star l}]_{-l \le m \le l}, \; l = 0, \; 1, \ldots$$

The spherical inversion can be obtained by

$$f(\omega) = \sum_{l \ge 0} \sum_{-l \le m \le l} f_m^{\star l} Y_m^l(\omega). \tag{5.7}$$

The bases detailed above are important because they realize a singular value decomposition of the convolution operator created by our model. We define for $f_\varepsilon \in \mathbb{L}_2(\mathrm{SO}(3))$, $f \in \mathbb{L}_2(\mathbb{S}^2)$ the convolution by the following formula:

$$f_\varepsilon * f(\omega) = \int_{\mathrm{SO}(3)} f_\varepsilon(u) f(u^{-1}\omega) du, \tag{5.8}$$

and we have for all $-l \le m \le l$, $l = 0, \; 1, \ldots$ that

$$(f_\varepsilon * f)_m^{\star l} = \sum_{n=-l}^{l} (f_\varepsilon^{\star l})_{mn} f_n^{\star l} = (f_\varepsilon^{\star l} f^{\star l})_m. \tag{5.9}$$

We finally define the operator norm

$$\|f_\varepsilon^{\star l}\|_{op} = \sup_{h \ne 0, h \in \mathbb{H}_l} \frac{\|f_\varepsilon^{\star l} h\|_2}{\|h\|_2}, \tag{5.10}$$

where $\mathbb{H}_l$ is the vector spanned by $\{Y_m^l, -l \le m \le l\}$, for each $l = 0, 1, \ldots$. The operator norm (5.10) is used in the next section to assess the regularity of the noise ε and hence to quantify the difficulty of the convolution problem.

5.3   Model and Assumptions

5.3.1   Null and Alternative Hypotheses

Our nonparametric alternatives are expressed in terms of Sobolev classes of order s on $\mathbb{S}^2$ (see, e.g., [5] for a definition on the sphere) $W_s(\mathbb{S}^2, R)$. It means that the derivative of order s of the density f is square-integrable and bounded by R.

For the uniform density of probability on the sphere namely $f_0 = (4\pi)^{-1} \mathbb{1}_{\mathbb{S}^2}$, we want to test the hypothesis

$$H_0 : \quad f = f_0,$$

from observations $Z_1, \ldots, Z_N$ given by model (5.1). We consider the alternative

$$H_1 : f \in W_s(\mathbb{S}^2, R) \text{ and } \|f - f_0\|_2^2 \geq \mathcal{C}\psi_N, \tag{5.11}$$

where $\mathcal{C}$ is a constant and ψ_N is the testing rate.

5.3.2 Noise Assumptions

Depending on the regularity of the noise ε, the testing problem is more or less difficult. More precisely, the testing problem is strongly related to the rate of decrease of the Fourier transform of the noise density f_ε. The smoother f_ε the quicker is the rate of decay of $\|f_\varepsilon^{\star l}\|_{op}$ and the slower the testing rate. We shall now see how to assess this regularity in terms of rotational Fourier coefficients decay of f_ε. We will say that the distribution of ε is ordinary smooth of order ν if the rotational Fourier transform of f_ε satisfies the following assumption:

Assumption 1 *For all $l \geq 0$, the matrix $f_\varepsilon^{\star l}$ is invertible and there exist positive constants d_0, d_1, ν such that*

$$\|f_{\varepsilon^{-1}}^{\star l}\|_{op} \leq d_0^{-1} l^\nu \quad \text{and} \quad \|f_\varepsilon^{\star l}\|_{op} \leq d_1 l^{-\nu},$$

where we have denoted the matrix $(f_\varepsilon^{\star l})^{-1}$ by $f_{\varepsilon^{-1}}^{\star l}$.

Recall that we assume that f_ε is known, consequently d_0 and ν are also considered known. Roughly speaking, Fourier transform of ordinary smooth noise has a polynomial decay.

An example of ordinary smooth noise is given by the Rotational Laplace distribution. The latter is the rotational analogue of the well-known Euclidean Laplace distribution (known also as double exponential distribution). It has been discussed in depth in [5]. The expanded form of the Rotational Laplace distribution in terms of rotational harmonics is the following:

$$f_\varepsilon = \sum_{l \geq 0} \sum_{m=-l}^{l} (1 + \sigma^2 l(l+1))^{-1} (2l+1) \overline{D_{mm}^l}, \tag{5.12}$$

for some $\sigma^2 > 0$ which is a variance parameter. Hence we have

$$(f_\varepsilon^{\star l})_{mn} = (1 + \sigma^2 l(l+1))^{-1} \delta_{mn},$$

for $l = 0, 1, \ldots$ and where $\delta_{mn} = 1$ if $m = n$ and is 0 otherwise. The Rotational Laplace distribution is ordinary smooth with a smoothness index $\nu = 2$.

We may have a more regular noise, called supersmooth noise. This kind of noise is of interest since it includes the Gaussian distribution. We will say that the distribution of ε is supersmooth of order ν if the rotational Fourier transform of f_ε satisfies

Assumption 2 *For all $l \geq 0$, the matrix f_ε^{*l} is invertible and there exist real numbers $\nu_1 \leq \nu_0$, and positive constants d_0, d_1, δ, β such that*

$$\|f_{\varepsilon-1}^{*l}\|_{op} \leq d_0^{-1} l^{-\nu_0} \exp(l^\beta/\delta) \quad and \quad \|f_\varepsilon^{*l}\|_{op} \leq d_1 l^{\nu_1} \exp(-l^\beta/\delta).$$

Again as we assume that f_ε is known, consequently d_0 and ν_0, δ and β are also considered known. Roughly speaking, Fourier transform of a supersmooth noise has an exponential decay.

The Gaussian distribution on SO(3) can be written as follows (see [8]):

$$f_\varepsilon = \sum_{l \geq 0} \sum_{m=-l}^{l} \exp(-\sigma^2 l(l+1)/2)(2l+1)\overline{D_{mm}^l}, \quad (5.13)$$

for $\sigma > 0$. This is an example of a supersmooth distribution with $\delta = 2/\sigma^2$ and $\beta = 2$.

5.4   Test Constructions

As the alternative H_1 (5.11) is expressed in terms of the $\mathbb{L}_2$-distance $\| \cdot \|_2$, to build a test statistic, we first have to construct an unbiased estimator of the quadratic functional $\int_{\mathbb{S}^2} (f - f_0)^2 = \|f - f_0\|_2^2$. To do so, we remark that thanks to the Parseval equality

$$\int_{\mathbb{S}^2} (f - f_0)^2 = \sum_{l \geq 0} \sum_{m=-l}^{l} |f_m^{*l} - f_{0m}^{*l}|^2 = \sum_{l \geq 1} \sum_{m=-l}^{l} |f_m^{*l}|^2,$$

where the last equality comes from the fact that $(f_0)_m^{*l} \neq 0$ only for $(l, m) = (0, 0)$. Consequently, it goes back to find an estimator for the Fourier coefficient f_m^{*l}. First, we need to isolate f_m^{*l}. By inverting the convolution product we get

$$f^{*l} = f_{\varepsilon-1}^{*l} f_Z^{*l}, \text{ for } l = 0, 1, \dots$$

Second, as the noise $f_{\varepsilon-1}^{*l}$ is supposed to be known, it remains to find an estimate for f_Z^{*l}. Due to the law of large numbers, a natural and consistent estimator of the Fourier coefficient $(f_Z^{*l})_n$ is given by the empirical mean

$$\frac{1}{N} \sum_{i=1}^{N} \overline{Y_n^l(Z_i)}.$$

Hence a natural estimator of f_m^{*l} is given by

$$\hat{f}_m^{*l} = \frac{1}{N} \sum_{i=1}^{N} \sum_{n=-l}^{l} (f_{\varepsilon-1}^{*l})_{mn} \overline{Y_n^l(Z_i)}.$$

If we denote by $\Phi_{lm}(x) = \sum_{n=-l}^{l} (f_{\varepsilon-1}^{*l})_{mn} \overline{Y_n^l}(x)$ then an estimator of $\|f - f_0\|_2^2$ is given by

$$T_L = \sum_{l=1}^{L} \sum_{m=-l}^{l} \frac{2}{N(N-1)} \sum_{i_1 < i_2} \Phi_{lm}(Z_{i_1}) \overline{\Phi_{lm}(Z_{i_2})}, \qquad (5.14)$$

where the choice of maximal level L is crucial and will be solved next when defining the testing procedures.

We are now in position to define our test procedure D_N from the statistic T_L. As we face noisy observations, we have to distinguish two cases according to the regularity of the noise ε. If the distribution of ε is ordinary smooth of order ν then

$$D_N = \mathbb{1}_{\left\{ \max_{L \in \mathcal{L}} (|T_L|/t_L^2) > \sqrt{2/K_0} \right\}}, \qquad (5.15)$$

with $\mathcal{L} = \{2^{j_0}, \ldots, 2^{j_m}\}$ a family of levels and $\sqrt{2/K_0}$ the quantile of the statistics $\max_{L \in \mathcal{L}} (|T_L|/t_L^2)$. The minimum level is set to $j_0 = \lceil \log_2(\log \log N) \rceil$ and the maximum level to $j_m = \lceil \log_2(N^{1/3} (\log \log N)^{-3/2}) \rceil$. The threshold t_L^2 is set to

$$t_L^2 = L^{2\nu+1} \sqrt{\log \log N}/N.$$

The computation of K_0 is done through Monte Carlo runs and will be specified in the numerical section.

If the distribution of ε is supersmooth of order ν then the test statistic is similar but with a different threshold t_L. Furthermore, it is sufficient to consider only one level L^* instead of a maximum

$$D_N = \mathbb{1}_{\left\{ |T_{L^*}|/t_{L^*}^2 > K_0 \right\}},$$

with $L^* = \lfloor (\delta \log(N)/8)^{\frac{1}{\beta}} \rfloor$ and

$$t_{L^*}^2 = L^{*-2\nu_0+1} \exp(2L^{*\beta}/\delta)/N.$$

The quantile K_0 is again computed with Monte Carlo runs and will be specified in the numerical section.

In both cases, thresholds t_L^2 and t_L^{*2} are of order of the variance of T_L and T_{L^*}, respectively, under H_0. Furthermore thresholds t_L^2 and t_L^{*2} do not depend on any unknown parameters.

From a theoretical point of view, the tests D_N are adaptive for the Sobolev classes $W_s(\mathbb{S}^2, R)$. They also reach optimal testing rates ψ_N in the minimax sense over $W_s(\mathbb{S}^2, R)$, for more details see [10]. In summary, our tests do not need extra knowledge of the regularity s to achieve the minimax rates.

5.5   Numerical Illustrations

In this section, we highlight the numerical performances of our test procedure D_N and compare them with other well-known directional test procedures. We both deal with simulations and real data in astrophysics and paleomagnetism.

5.5.1   The Testing Procedures

We shall now explain the various directional testing procedures we consider in this numerical section.

Let us start with our adaptive testing procedure that will be denoted throughout this section by SHT (as Spherical Harmonics Test). To compute the quantile K_0 appearing in the definition of D_N, we generate 1000 times N observations uniformly under H_0. Then, we compute by 1000 Monte Carlo runs the 5% quantile of the statistics $\max_{L \in \mathcal{L}}(|T_L|/t_L^2)$ or $|T_{L^*}|/t_{L^*}^2$. We point out that our numerical procedure is notably fast all the more so as we are in dimension 2. Furthermore, we do not have any tuning parameter to calibrate.

To compare our results, we have implemented two other procedures. The first one is called *the Nearest Neighbor test* and was proposed by [11]. It will be denoted by NN in the sequel. For each observation Z_i, one must compute the distance Y_i to its nearest neighbor. The Wilcoxon test statistic is

$$W = \sqrt{12N}\left(\frac{1}{2} - \frac{1}{N}\sum_{i=1}^{N}\phi(Y_i)\right),$$

where $\phi(z) = 1 - [(1 + \cos z)/2]^{N-1}$. The distribution of W is asymptotically standard Gaussian. Notice that this test was designed for non-noisy data.

The second procedure was introduced by [1] and [4]. The test statistic is

$$F_N = \frac{3N}{2} - \frac{4}{N\pi}\sum_{i=1}^{N-1}\sum_{j=i+1}^{N}(d(Z_i, Z_j) + \sin(d(Z_i, Z_j))),$$

where $d(Z_i, Z_j) = \arccos\langle Z_i, Z_j\rangle$ is the spherical distance between Z_i and Z_j, and the quantiles are computed via simulations under H_0. Again, this test was designed for non-noisy data.

So as not to leave anything out, we mention two other testing procedures based on needlets which can be seen as the wavelets on the sphere. In [2], the authors implement two procedures called Multiple and PlugIn for the noise free case. The Multiple test is based on a family of linear estimators of the density f_Z whereas the PlugIn test considers a hard thresholding procedure on needlets.

5.5.2 Alternatives

We have investigated the performances of the different testing procedures described above for two kinds of alternatives. These alternatives aim at describing different relevant scenarios in practice.

The first family of alternatives is nonisotropic, unimodal with a Gaussian shape. More precisely, it is a mixture of a Gaussian-like density with the uniform density f_0. We will denote this alternative by H_1^a. The H_1^a density has the form

$$f(x) = (1 - \delta)f_0(x) + \delta h_\gamma(x),$$

where $h_\gamma(x) := C_\gamma \exp(-d(x, x_0)^2/(2\gamma^2))$, d is the spherical distance, C_γ is a normalization constant such that $\int_{\mathbb{S}^2} f(x)dx = 1$ and x_0 is $(\pi/2, 0)$ in spherical coordinates. In the sequel, we chose $\delta = 0.08$ and $\gamma = 5\pi/180$, i.e., $\gamma = 5°$. Remark that with this choice of parameters, the dose of uniformness injected in H_1^a is high and complicates the detection of the alternative from the null hypothesis.

This density is particularly meaningful in the field of astrophysics since very often one seeks for some departure from isotropy and some principal direction. As [2] also considered H_1^a, this will permit us to compare the performances of our test to the Multiple and PlugIn procedures of [2]. Figure 5.1 allows us to visualize this alternative. The density is represented in spherical coordinates as a surface $z = f(\theta, \phi)$. To visualize points on the sphere, we use Hammer projection, because of its equal-area property.

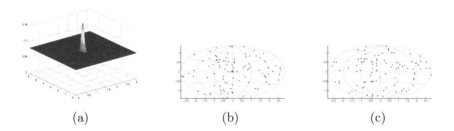

| (a) | (b) | (c) |

FIGURE 5.1
(a) Representation of the H_1^a density in spherical coordinates. (b) 100 random draws X_i from the H_1^a distribution, (c) 100 random draws Z_i from H_1^a convolved with a Rotational Laplace noise with variance 0.1.

The second alternative we consider and which is denoted by H_1^b is the Watson distribution [17]. Its density is

$$f(\theta, \phi) = C \exp(-2\cos^2(\theta)),$$

with C such that $\int_0^{2\pi} \int_0^\pi f(\theta, \phi) \sin(\theta) d\theta d\phi = 1$. This distribution has a girdle form, distributed around the equator. This choice is motivated by two reasons: first, this gives an alternative very different from H_1^a; second, it plays a role in applications. For example, in the case of gamma-ray bursts (see [15]), many theories assumed that the sources of those flashes were located around the galactic plane (then a girdle distribution), whereas other proposed that gamma-ray bursts come from beyond the Milky Way (rather a uniform distribution). Figure 5.2 presents this alternative. Notice that the presence of noise (Figure 5.2 c) prevents us from seeing the equatorial nature of the distribution.

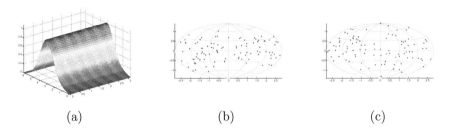

(a)   (b)   (c)

FIGURE 5.2
(a) Representation of the Watson density in spherical coordinates. (b) 100 random draws X_i from the Watson distribution, (c) 100 random draws Z_i from the Watson distribution convolved with a Gaussian noise with variance 0.2.

5.5.3   Simulations

For the two alternatives, we computed the test power for a prescribed level of 5%, for our test procedure (denoted by SHT), for the nearest neighbor test (denoted by NN) and the Beran–Giné test (denoted by BG). Tables 5.1–5.4 give the results in percent for different kinds of noisy data: no noise, Laplace noise with variance 0.05, 0.1, 0.2, and Gaussian noise with variance 0.05, 0.1, 0.2.

As expected, the increase of the size sample improves the power, whereas presence of noise reduces it.

For the H_1^a alternative, in the absence of noise, our procedure performs better than the two others. When adding some noise, our procedure has better results than the NN procedure but slightly worse than the BG procedure. Note that it is only possible to compare our results with those of [2] (see their Figure 7) for the noise-free case and for $N = 100$ observations. Their Multiple and PlugIn procedures for a resolution level equal to 3 perform better than the three other procedures presented here. Nonetheless, it is important

TABLE 5.1
Test powers for $N = 100$ and H_1^a

Noise type variance	No noise	Laplace			Gaussian		
		0.05	0.1	0.2	0.05	0.1	0.2
SHT	53	18	13	10	13	8	8
NN	19	10	10	8	9	7	7
BG	30	19	18	10	18	12	11
Multiple (J=3)	62						
PlugIn (J=3)	63						

TABLE 5.2
Test powers for $N = 250$ and H_1^a

Noise type variance	No noise	Laplace			Gaussian		
		0.05	0.1	0.2	0.05	0.1	0.2
SHT	95	45	35	19	20	19	12
NN	26	11	11	8	11	8	9
BG	61	43	38	24	41	30	20

TABLE 5.3
Test powers for $N = 100$ and H_1^b

Noise type variance	No noise	Laplace			Gaussian		
		0.05	0.1	0.2	0.05	0.1	0.2
SHT	100	98	83	49	45	31	21
NN	64	34	19	10	30	17	10
BG	93	48	27	15	32	19	10

TABLE 5.4
Test powers for $N = 250$ and H_1^b

Noise type variance	No noise	Laplace			Gaussian		
		0.05	0.1	0.2	0.05	0.1	0.2
SHT	100	100	100	95	73	86	51
NN	91	52	34	16	46	26	10
BG	100	99	89	41	99	71	15

to notice that our procedure SHT is entirely data-driven and has no tuning parameter, contrary to the one of [2]. In addition, [2] has not dealt with the noise scenario and their procedures are rather complicated and lengthy for practical purposes.

When the alternative is the Watson distribution, our procedure performs pretty well and is clearly better than the others. Indeed the test powers for our procedure are often twice higher.

We also computed the ROC curves for the three methods for different noise and numbers of observations settings. Let us recall that the *Receiver Operating Characteristic* curves allow us to illustrate the performance of a test by plotting the true positive rate versus the false positive rate, at various threshold settings. Roughly speaking, the greater the area under the ROC curve, the better the test.

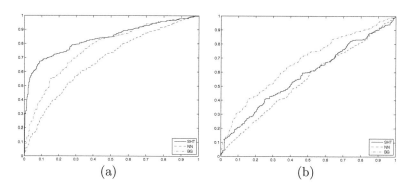

FIGURE 5.3
ROC curves for the three methods and for the alternative H_1^a: (a) No noise and $N = 100$. (b) Rotational Laplace noise with variance 0.1 and $N = 100$.

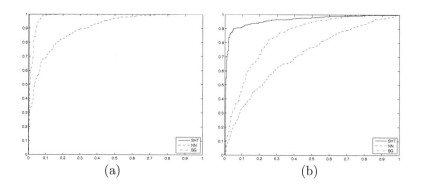

FIGURE 5.4
ROC curves for the three methods and for the alternative H_1^b: (a) No noise and $N = 100$. (b) Rotational Laplace noise with variance 0.1 and $N = 100$.

We point out that on Figure 5.4a the solid line corresponding to the performance of our procedure SHT is mixed up with the axes passing through the points $(0,0), (0,1)$, and $(1,1)$.

5.5.4 Real Data: Paleomagnetism

In paleomagnetism, some minerals in rocks have the particular property of conserving the direction of magnetic field when they cool down. This allows geologists to retrieve the direction of the Earth's magnetic field in the past ages, and this also provides information about the past location of tectonic plates. Since these records of magnetic field are directions, it is about spherical data. According to the accuracy of the measuring devices, the measurements can be more or less noisy. Hence the measurements of those noisy directions can be modeled by (5.1).

To illustrate our method, we use the data given by [3]: these are 52 measurements of magnetic remanence from specimens of red beds from the Bowen Basin (Queensland, Australia), after thermal demagnetization to 670°C. Demagnetization is a process which is used in order to eliminate unwanted magnetic fields. For our illustration, the advantage of this data set is that the possible anisotropy is not visible to the naked eye, contrary to a lot of data sets in paleomagnetism, see Figure 5.5.

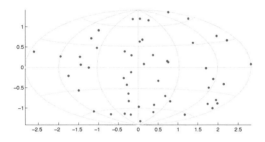

FIGURE 5.5
Representation of the 52 measurements of magnetic remanence (Hammer projection).

The obtained p-values for this sample are given in Table 5.5, assuming different kinds of possible noise. Whatever the possible noise we consider, there is no statistical evidence of anisotropy, which indicates that the demagnetization process succeeded.

TABLE 5.5
p-values for the magnetism data.

Noise type	No noise	Laplace			Gaussian		
variance		0.05	0.1	0.2	0.05	0.1	0.2
p-value	0.90	0.81	0.79	0 .77	0.73	0.63	0.74

5.5.5   Real Data: UHECR

In astrophysics, a burning issue consists in understanding the behavior of the so-called Ultra High Energy Cosmic Rays (UHECR). The latter are cosmic rays with an extreme kinetic energy (of the order of 10^{19} eV) and the rarest particles in the universe. The source of those most energetic particles remains a mystery and the stake lies in finding out their origins and which process produces them. Astrophysicists have at their disposal directional data which are measurements of the incoming directions of the UHECR on Earth. Needless to say that finding out more about the law of probability of those incoming directions is crucial to gain an insight into the mechanisms generating the UHECR. Faÿ et al. [2] recently developed isotropy goodness-of-fit tests based on the so-called needlets for the nonperturbated case. Their study is focused on the practical aspect with nice simulations connected to realistic cosmic rays scenarios. But the difficulty lies in the fact that the observed UHECR do not come necessarily from the genuine direction as specified by [2]. Their trajectories are deflected by galactic and intergalactic fields. As this deflection is inevitable in the measurements, it is quite challenging and essential to take into account this uncertainty in the statistical modelling. A first way to model the deflection in the incoming directions can be done thanks to the model (5.1) with random rotations. Several hypotheses are made about the underlying probability of the incoming directions. A uniform density would suggest that the UHECR are generated by cosmological effects, such as the decay of relic particles from the Big Bang. On the contrary, if these UHECR are generated by astrophysical phenomena (such as acceleration into Active Galactic Nuclei [AGN]), then we should observe a density function which is highly non-uniform and tightly correlated with the local distribution of extragalactic supermassive black holes at the center of nearby galaxies (AGN). First results seemed to favor a non-uniform density but as underlined by [2], a more recent analysis based on 69 observations of UHECR softens this conclusion of anisotropy. To this prospect, these relevant considerations lead naturally to goodness-of-fit testing on the uniform density in the noisy model (5.1).

To apply our procedure to UHECR data of the observatory Pierre Auger [14], we need to take into account the observatory exposure. Indeed, only cosmic rays with zenith angle of arrival less than $60°$ can be observed. Then, a coverage function over the years of observation can be computed from geometrical considerations and it is displayed in Figure 5.6.

In addition to the noise due to extragalactic magnetic fields, a selection is done depending on whether the ray is in the observation area. Denoting the coverage density by g_0, the observations are now $V_1, \ldots, V_N$ where the density of V is proportional to g_0 times f_Z: $f_V = cg_0 f_Z$, with c such that f_V is a density. The relevant test is then $f = f_0 \Leftrightarrow f_V = g_0$. Hence we implement an extended method. Our initial test procedure is based on the estimation of $(f_Z)_n^{\star l}$ by $N^{-1} \sum_i \overline{Y_n^l(Z_i)}$. Then it is sufficient to apply the same procedure but with the estimator $N^{-1} \sum_{i=1}^N (\overline{Y_n^l}/(cg_0))(V_i)$. Indeed this quantity approx-

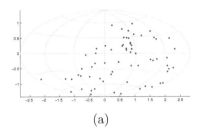
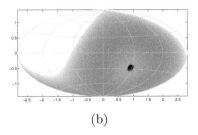

(a)   (b)

FIGURE 5.6
(a) Representation of the 69 arrival directions of highest energy cosmic rays
(Pierre Auger data). (b) Coverage function g_0 for the Pierre Auger observatory
(the darker the more observed, white area non-observed).

imates $\int f_V \overline{Y_n^l}/(cg_0) = \int f_Z \overline{Y_n^l} = (f_Z)_n^{\star l}$. Using this method, we obtain the
p-values given in Table 5.6, assuming different kinds of possible noise.

TABLE 5.6
p-values for the Pierre Auger data

Noise type	No noise	Laplace			Gaussian		
variance		0.05	0.1	0.2	0.05	0.1	0.2
p	0.003	0.014	0.034	0.092	0.016	0.001	0.076

Then our method confirms what was already noticed by astrophysicists:
there seems to be some kind of anisotropy in the UHECR phenomenon.

Bibliography

[1] R. J. Beran. Testing for uniformity on a compact homogeneous space. *J. Appl. Probability*, 5:177–195, 1968.

[2] G. Faÿ, J. Delabrouille, G. Kerkyacharian, and D. Picard. Testing the isotropy of high energy cosmic rays using spherical needlets. *Ann. Appl. Stat.*, (2):1040–1073, 2013.

[3] N. I. Fisher, T. Lewis, and B. J. J. Embleton. *Statistical Analysis of Spherical Data.* Cambridge University Press, Cambridge, 1987.

[4] E. Giné. Invariant tests for uniformity on compact Riemannian manifolds based on Sobolev norms. *Ann. Statist.*, 3:1243–1266, 1975.

[5] D. M. Healy, Jr., H. Hendriks, and P. T. Kim. Spherical deconvolution. *J. Multivariate Anal.*, 67:1–22, 1998.

[6] S. F. Huckemann, P. T. Kim, J.-Y. Koo, and A. Munk. Möbius deconvolution on the hyperbolic plane with application to impedance density estimation. *Ann. Statist.*, 38:2465–2498, 2010.

[7] G. Kerkyacharian, T. M. Pham Ngoc, and D. Picard. Localized spherical deconvolution. *Ann. Statist.*, 39:1042–1068, 2011.

[8] P. T. Kim and J.-Y. Koo. Optimal spherical deconvolution. *J. Multivariate Anal.*, 80:21–42, 2002.

[9] P. T. Kim, J.-Y. Koo, and H. J. Park. Sharp minimaxity and spherical deconvolution for super-smooth error distributions. *J. Multivariate Anal.*, 90:384–392, 2004.

[10] C. Lacour and T. M. Pham Ngoc. Goodness-of-fit test for noisy directional data. *Bernoulli*, 20:2131–2168, 2014.

[11] J. M. Quashnock and D. Q. Lamb. Evidence for the galactic origin of gamma-ray bursts. *M.N.R.A.S.*, 265:L45–L50, 1993.

[12] J. D. Talman. *Special Functions: A Group Theoretic Approach*. Based on lectures by Eugene P. Wigner. With an introduction by Eugene P. Wigner. W. A. Benjamin, Inc., New York-Amsterdam, 1968.

[13] A. Terras. *Harmonic Analysis on Symmetric Spaces and Applications. I.* Springer-Verlag, New York, 1985.

[14] The Pierre AUGER Collaboration. Update on the correlation of the highest energy cosmic rays with nearby extragalactic matter. *Astroparticle Physics*, 34:314–326, 2010.

[15] G. Vedrenne and J.-L. Atteia. *Gamma-Ray Bursts: The Brightest Explosions in the Universe*. Springer/Praxis Books, 2009.

[16] N. Ja. Vilenkin. *Special Functions and the Theory of Group Representations*. Translated from the Russian by V. N. Singh. Translations of Mathematical Monographs, Vol. 22. American Mathematical Society, Providence, R. I., 1968.

[17] G. S. Watson. Equatorial distributions on a sphere. *Biometrika*, 52:193–201, 1965.

6

On Modeling of $SE(3)$ Objects

Louis-Paul Rivest

Department of Mathematics and Statistics, Université Laval, Québec, Canada

Karim Oualkacha

Department of Mathematics, Université du Québec À Montréal, Montréal, Canada

CONTENTS

6.1 Introduction 111
6.2 The One Axis Model in $SE(3)$ 113
    6.2.1 Rotation Matrices and Cardan Angles in $SO(3)$ 113
    6.2.2 A Geometric Construction of the One Axis Model 113
6.3 Modeling Data from $SE(3)$ 115
6.4 Estimation of the Parameters 116
    6.4.1 The Rotation Only Estimator of the Rotation Axis A_3 and B_3 116
    6.4.2 The Translation Only Estimator of the Parameters 117
    6.4.3 The Rotation-Translation Estimator of the Parameters 118
6.5 Numerical Examples 119
    6.5.1 Simulations 119
    6.5.2 Data Analysis 121
6.6 Discussion 123
    Bibliography 124

6.1 Introduction

This chapter considers the modeling of rotation translation pairs, (R, t), where R is an 3×3 rotation matrix and $t \in \Re^3$ is a 3-dimensional translation vector. In rigid body kinematics, such a pair is used to characterize the position of a rigid object; t and R, respectively, give its location and its orientation. The set of all these pairs is denoted $SE(3)$, see [19]. There are special cases of interest, the pure rotation $(R, 0)$ and the pure translation (I_3, t), where I_3 is the 3×3 identity matrix. The set of 3×3 rotation matrices is denoted $SO(3)$.

First, it is convenient to review some mathematical properties of $SE(3)$. A pair (R, t) represents the position of a rigid body in three dimensions as it allows to express the position $\mathbf{X} = (X_1, X_2, X_3)^T$ of a point in the rigid body in the global reference frame in terms of its coordinates $\mathbf{x} = (x_1, x_2, x_3)^T$ in a local reference frame attached to the body. This transformation is given by

$$\mathbf{X} = R\mathbf{x} + t. \tag{6.1}$$

The $SE(3)$ object (R, t) can be expressed as the matrix

$$M = (R, t) = \begin{pmatrix} R & t \\ 0_3^T & 1 \end{pmatrix}, \quad R \in SO(3), \quad t \in \Re^3.$$

The transformation (6.1) can be written in terms of the matrix M as $(\mathbf{X}^T, 1)^T - M(\mathbf{x}^T, 1)^T$. So $SE(3)$ can be seen as the set of 4×4 matrices M. Each object has an inverse $M^{-1} = (R^T, -R^T t) \in SE(3)$ giving the position of the global reference frame with respect to the local one. $SE(3)$ is closed under product operation as $(R_1, t_1)(R_2, t_2) := M_1 M_2 = (R_1 R_2, R_1 t_2 + t_1)$. If M_1 gives the position of object $\mathcal{B}$ with respect to object $\mathcal{A}$ and M_2 gives that of $\mathcal{C}$ with respect to $\mathcal{B}$, then $M_1 M_2$ gives the position of $\mathcal{C}$ with respect to $\mathcal{A}$. The identity element is $I_4 = (I_3, 0_3)$. The set $SE(3)$ is a Lie group, see [19] for details.

A SE(3) data point gives the position (R, t) of a rigid body $\mathcal{A}$ at a given time. It can be measured using k markers attached to $\mathcal{A}$. First body $\mathcal{A}$ is put in a reference position and the 3-D coordinates $\mathbf{x}_{01}^{\mathcal{A}}, \ldots, \mathbf{x}_{0k}^{\mathcal{A}}$ of the markers are recorded; this defines the local reference frame attached to the body. To measure a sample position i, the 3-D coordinates of the markers $\mathbf{x}_{i1}^{\mathcal{A}}, \ldots, \mathbf{x}_{ik}^{\mathcal{A}}$ are recorded in that position. Considering (6.1), $(R_i^{\mathcal{A}}, t_i^{\mathcal{A}})$ is estimated by minimizing $\sum_j \|\mathbf{x}_{ij}^{\mathcal{A}} - t_i^{\mathcal{A}} - R_i^{\mathcal{A}} \mathbf{x}_{0j}^{\mathcal{A}}\|^2$. Methods for minimizing this expression are discussed in [34]; one needs $k \geq 3$ for the estimate of $(R_i^{\mathcal{A}}, t_i^{\mathcal{A}})$ to be uniquely defined. The capture of the knee motion data considered in Section 6.5.2 used 8 markers, 4 attached to the thigh segment and 4 to the lower leg. As the subject was walking, the 3-D positions of the 8 markers were recorded by a camera system at a frequency of 50 Hz, that is 50 times per second. At each time point i, pairs $(R_i^{\mathcal{A}}, t_i^{\mathcal{A}})$ and $(R_i^{\mathcal{B}}, t_i^{\mathcal{B}})$ were calculated to characterize the position of the thigh segment and of the lower leg segment, respectively. The relative position of the lower leg with respect to the thigh, giving the knee posture at time i, is $(R_i, t_i) = (R_i^{\mathcal{A}}, t_i^{\mathcal{A}})^{-1}(R_i^{\mathcal{B}}, t_i^{\mathcal{B}})$. The data set analyzed in Section 6.5.2 is $\{(R_i, t_i)\}$. The estimation of the $SE(3)$ position of a rigid body is important in biomechanics, see for instance [13] for a recent discussion of this problem.

Probabilistic models for $SE(3)$ objects have received limited attention in the literature. Proposals can however be found in [4, sec. 14.5], [22], and [15]. For rotation matrices, the classical reference is [6]. Relevant advances have thereafter been made by [11], [12], with additional work done by [1], [3], [17], [18], [21], [24], and [25]. In this work we use a very simple error model where the translation and the rotation components are independent. The translation errors are assumed to follow a $N_3(0, \sigma^2 I_3)$ distribution, where $\sigma^2 > 0$ is the error variance, while the rotation errors are assumed to follow a symmetric [6] model indexed by a positive concentration parameter κ. The rest of the chapter is organized as follows: the parametrization of rotation matrices is reviewed, and the predicted values under a one axis model are calculated. Methods for estimating the parameters are then presented. The application section then presents the results of a Monte Carlo study and the analysis of gait analysis data on the motion of the knee joint.

6.2 The One Axis Model in $SE(3)$

6.2.1 Rotation Matrices and Cardan Angles in $SO(3)$

For the application considered in this work it is convenient to parameterize rotation matrices using Cardan angles. Three Cardan angles are needed to construct an $SO(3)$ object. Here we use the so-called XYZ convention; it represents a 3×3 rotation matrix R using the angles $\rho \in [-\pi, \pi), \xi \in [-\pi/2, \pi/2)$ and $\theta \in [-\pi, \pi)$ as

$$
\begin{aligned}
R &= R(\rho, X)R(\xi, Y)R(\theta, Z) \qquad\qquad (6.2)\\
&= \begin{pmatrix} 1 & 0 & 0 \\ 0 & c(\rho) & -s(\rho) \\ 0 & s(\rho) & c(\rho) \end{pmatrix} \begin{pmatrix} c(\xi) & 0 & s(\xi) \\ 0 & 1 & 0 \\ -s(\xi) & 0 & c(\xi) \end{pmatrix} \begin{pmatrix} c(\theta) & -s(\theta) & 0 \\ s(\theta) & c(\theta) & 0 \\ 0 & 0 & 1 \end{pmatrix} \\
&= \begin{pmatrix} c(\theta)c(\xi) & -s(\theta)c(\xi) & s(\xi) \\ c(\theta)s(\xi)s(\rho) + s(\theta)c(\rho) & -s(\theta)s(\xi)s(\rho) + c(\theta)c(\rho) & -s(\rho)c(\xi) \\ -c(\theta)s(\xi)c(\rho) + s(\rho)s(\theta) & s(\theta)s(\xi)c(\rho) + s(\rho)c(\theta) & c(\rho)c(\xi) \end{pmatrix},
\end{aligned}
$$

where $c(.)$ and $s(.)$ denote the cosine and the sine functions, respectively. Thus R is expressed as a rotation with respect to the X axis followed by rotations with respect to the Y and the Z axes. Conversely, given a rotation $R \in SO(3)$, one can calculate the three Cardan angles according to the (XYZ) convention as $\xi = \arcsin(R_{13}) \in [-\pi/2, \pi/2]$, $\rho = \arctan(-R_{23}, R_{33}) \in [-\pi, \pi]$ and $\theta = \arctan(-R_{12}, R_{11}) \in [-\pi, \pi]$, where R_{ij} denotes the entry in the i th row and the j th column of R and $\arctan(a, b)$ is equal to the angle x such that $\sin(x) = a/(a^2 + b^2)^{1/2}$ and $\cos(x) = b/(a^2 + b^2)^{1/2}$. In a similar way, one can obtain the Cardan angles for other conventions, see [16] for details.

6.2.2 A Geometric Construction of the One Axis Model

The goal of this section is to provide an algebraic representation of the relative motion of one rigid body with respect to another when both rotate about the same axis, a line in $\Re^3$ that is called $\mathcal{L}_R$. First a laboratory coordinate system of axes is set up with the following specifications: (i) the direction of $\mathcal{L}_R$ is the

Z-axis of the coordinate system and (*ii*) the origin of the coordinate system belongs to $\mathcal{L}_R$. In the laboratory coordinate system, $\mathcal{L}_R = \{z(0,0,1)^\top \mid z \in \Re\}$. Let $(A^\top, -t_a)$ be an $SE(3)$ object giving the reference position of $\mathcal{A}$. Define (A_1, A_2, A_3) as the three columns of rotation matrix A and (t_{a1}, t_{a2}, t_{a3}) as the three entries of $t_a^\top$. Thus if $x \in \Re^3$ is the local coordinate of point P in $\mathcal{A}$, then the coordinate of P in the laboratory system of axes is $-t_a + A^\top x$, when rigid body $\mathcal{A}$ is in its reference position. In $\mathcal{A}$'s reference frame, $\mathcal{L}_R$ consists of the points x such that $-t_a + A^\top x = z(0,0,1)^\top$ for some $z \in \Re$. This is equivalent to $x = zA_3 + At_a$. Thus, A_3 is the direction of $\mathcal{L}_R$ in $\mathcal{A}$'s system of axes and $At_a = A_1 t_{a1} + A_2 t_{a2} + A_3 t_{a3}$, gives the coordinates of the lab origin, a point of $\mathcal{L}_R$, in $\mathcal{A}$'s system of axes. Thus in $\mathcal{A}$, $\mathcal{L}_R$ is determined by its direction A_3 and one of its points, $A_1 t_{a1} + A_2 t_{a2}$. A rotation of θ_a around $\mathcal{L}_R$ is described by the $SE(3)$ object $(R(\theta_a, Z), 0)$ in the global reference frame. In $\mathcal{A}$'s coordinate system this becomes $\{R(\theta_a, Z), 0\}(A^\top, -t_a) = \{R(\theta_a, Z)A^\top, -R(\theta_a, Z)t_a\}$. As θ_a varies, this sequence of $SE(3)$ describes the movement of $\mathcal{A}$ when it rotates around $\mathcal{L}_R$.

Consider a second rigid body, say $\mathcal{B}$, with local coordinate system defined by $(B^\top, -t_b) \in SE(3)$. The $SE(3)$ object giving $\mathcal{B}$'s position once it has undergone a rotation of θ_b around $\mathcal{L}_R$ is $\{R(\theta_b, Z)B^\top, -R(\theta_b, Z)t_b\}$. The relative position of $\mathcal{B}$ with respect to $\mathcal{A}$ when both rotate about $\mathcal{L}_R$ is $\{R(\theta_a, Z)A^\top, -R(\theta_a, Z)t_a\}^{-1}\{R(\theta_b, Z)B^\top, R(\theta_b, Z)t_b\}$. This is equal to $\{AR(\theta_b - \theta_a, Z)B^\top, At_a - AR(\theta_b - \theta_a, Z)t_b\}$. Notice that t_{a3} and t_{b3} depend on the position of the laboratory origin on $\mathcal{L}_R$. These two parameters are not identifiable, only their difference, $t_z = t_{a3} - t_{b3}$, is. The five translation parameters are put in two vectors, $t_a = (t_{a1}, t_{a2}, t_z)^T$ and $t_b = (t_{b1}, t_{b2}, 0)^T$. If $\theta_i = \theta_b - \theta_a$, then the relative position of $\mathcal{B}$ with respect to $\mathcal{A}$ is

$$D_i = \begin{pmatrix} \Psi_i & d_i \\ 0^T & 1 \end{pmatrix} = \begin{pmatrix} AR(\theta_i, Z)B^T & At_a - AR(\theta_i, Z)t_b\} \\ 0^T & 1 \end{pmatrix}. \quad (6.3)$$

To parameterize the rotations matrices A and B we use the (XYZ) Cardan angles parametrization. Since the third angle θ in (6.2) for the rotation about Z is confounded with θ_i in (6.3), A and B are parameterized using only two angles each, say (a_x, a_y, b_x, b_y). Considering (6.2), in this parametrization the rotation axis is expressed as

$$A_3 = \begin{pmatrix} \sin(a_y) \\ -\sin(a_x)\cos(a_y) \\ \cos(a_x)\cos(a_y) \end{pmatrix}, \quad B_3 = \begin{pmatrix} \sin(b_y) \\ -\sin(b_x)\cos(b_y) \\ \cos(b_x)\cos(b_y) \end{pmatrix},$$

in the local coordinate system for $\mathcal{A}$ and $\mathcal{B}$, respectively. Without loss of generality we assume that $a_x, a_y, b_x, b_y \in (-\pi/2, \pi/2)$. Thus we use a parametrization where the third entries of the two rotation axes are positive.

6.3 Modeling Data from $SE(3)$

The goal of the analysis is to estimate a vector of nine parameters, $\beta = (t_{a1}, t_{a2}, t_{b1}, t_{b2}, t_z, a_x, a_y, b_x, b_y)$, using the data set $\{M_i = (R_i, t_i) : i = 1, \ldots, n\}$, where M_i is equal to D_i perturbed by experimental errors with D_i defined in (6.3). The rotation angles θ_i are regarded as nuisance parameters that need to be estimated as well. One can model the data point M_i as

$$M_i = D_i F_i, \qquad F_i = (E_i, e_i) \in SE(3), \tag{6.4}$$

where D_i is given by (6.3) and F_i, the experimental error, is distributed according to the SE(3) density, with respect to $SE(3)$ Haar measure, proposed by [22],

$$f(F; \kappa, \sigma^2) = \frac{\exp\{\operatorname{tr}(\kappa E) - \|e\|^2/(2\sigma^2)\}}{c_\kappa (2\pi\sigma^2)^{3/2}}, \qquad F = (E, e) \in SE(3), \tag{6.5}$$

where κ describes the scatter of E around the 3×3 identity matrix and σ^2 is the variance of the translation experimental errors. In general κ is large and we use the following approximation, $c_\kappa \approx e^{3\kappa}(\pi/\kappa)^{3/2}$.

Since the Jacobian of the transformation $M = DF$ is 1, [4, sec. 6.4], the proposed density for M_i is given by

$$f(M; D_i, \beta, \kappa, \sigma^2) = \frac{\exp\{\operatorname{tr}(\kappa R^T \Psi_i) - \|t - d_i\|^2/(2\sigma^2)\}}{c_\kappa (2\pi\sigma^2)^{3/2}}, \quad M = (R, t) \in SE(3). \tag{6.6}$$

This density describes the scatter of a random SE(3) object M_i around its modal value D_i. It is the product of the Fisher–von Mises–Langevin density for R_i with modal value Ψ_i, and of the normal distribution density $N_3(d_i, \sigma^2 I_3)$ for t_i. Thus R_i and t_i are independent.

The log-likelihood for β, κ, σ^2 and $\theta_i, i = 1, \ldots, n$ is given by

$$l_{rt}(\beta, \kappa, \sigma^2) = l_r(\beta, \kappa) + l_t(\beta, \sigma^2), \tag{6.7}$$

where

$$l_r(\beta, \kappa) = -n \log(c_\kappa) + \kappa \sum_{i=1}^{n} \operatorname{tr}\left\{ B^T R_i^T A R\left(\theta_i, Z\right) \right\}, \tag{6.8}$$

and

$$l_t(\beta, \sigma^2) = -\frac{3n}{2} \log(2\pi\sigma^2) \tag{6.9}$$

$$-\frac{1}{2\sigma^2} \sum_{i=1}^{n} \|t_i - A\left\{t_a - N_{t_b} R\left(\theta_i + \alpha_b, Z\right)(1, 0, 0)^T\right\}\|^2,$$

are the log-likelihoods of the rotation and the translation marginal models. Note that the translation log-likelihood uses a reparameterization of β where (t_{b1}, t_{b2}) is expressed in polar coordinates, $(t_{b1}, t_{b2}) = N_{t_b}\{\cos(\alpha_b), \sin(\alpha_b)\}$, for $\alpha_b \in (0, 2\pi)$ and $N_{t_b} > 0$.

An alternative expression for (6.6) is

$$f(M; D_i, \beta, \kappa, \sigma^2) = \exp\{-\|M_i - D_i\|_W^2/2\}/\{\exp(-3\kappa)c_\kappa(2\pi\sigma^2)^{3/2}\}$$

where $\|M_i - D_i\|_W^2 = \mathrm{tr}\{(M_i - D_i)W(M_i - D_i)^\top\}$ and $W = \mathrm{diag}(\kappa, \kappa, \kappa, 1/\sigma^2)$, is a least square distance between M_i and D_i. Thus, when the variance components (κ, σ^2) are known, maximizing (6.7) amounts to estimating the parameters using a least square criterion; this is discussed in [22].

6.4   Estimation of the Parameters

This section discusses the estimation of the parameters using the marginal rotation model, the marginal translation model and the combined model given in (6.6). The estimation strategy is first to get rid of the nuisance angles θ_i by maximizing over θ_i when all the other parameters are kept fixed. The number of θ parameters is equal to the number of data points n. This is a situation analogous to that described in [20]: consistent estimators of the other parameters can be constructed, in either a large sample size or a large concentration setting. Some care however is needed with the maximum likelihood method since it then gives biased estimators of the parameters κ and σ^2 for the error variances; this is called the Neyman–Scott paradox. These estimators need to be corrected for the estimation of the angles θ_i; such corrections are provided for the three models considered in this section.

6.4.1   The Rotation Only Estimator of the Rotation Axis A_3 and B_3

When the only data available is the rotation data $\{R_i\}$, one can only estimate the directions, A_3 and B_3, of the fixed rotation axis in the reference frame attached to $\mathcal{A}$ and $\mathcal{B}$, respectively. This problem is considered in [28]. The rotation log-likelihood (6.8) depends on the rotation parameters $\beta_r = (a_x, a_y, b_x, b_y)^\top$, on the angles $\theta_i, i = 1 \ldots, n$, and on κ. Thus, $(n + 5)$ parameters need to be estimated. The problem of maximizing $l_r(\beta_r, \kappa)$ with respect to $\theta_i, \ i = 1, \ldots, n$, when A, B and κ are fixed is first considered. The log-likelihood is a linear function of $\cos(\theta_i)$ and $\sin(\theta_i)$,

$$l_r(\beta_r, \kappa) = -n\log(c_\kappa) + \kappa\sum_{i=1}^{n}\{\cos(\theta_i)\,(Q_{i11} + Q_{i22}) + \sin(\theta_i)\,(Q_{i21} - Q_{i12}) + Q_{i33}\},$$

where $Q_i = B^T R_i^T A$ is a rotation matrix. As such, it can be shown to satisfy $(Q_{i11} + Q_{i22})^2 + (Q_{i21} - Q_{i12})^2 = (1 + Q_{i33})^2$ and an alternative expression for the likelihood is

$$l_r(\beta_r, \kappa) = -n \log(c_\kappa) + \kappa \sum_{i=1}^{n} \{(1 + Q_{i33}) \cos(\theta_i - \theta_{iR}) + Q_{i33}\}, \quad (6.10)$$

where, for $i = 1, \ldots, n$,

$$\theta_{iR} = \arctan(Q_{i21} - Q_{i12}, Q_{i11} + Q_{i22}). \quad (6.11)$$

Likelihood (6.10) is maximum for $\theta_i = \theta_{iR}$ and the profile likelihood for A and B is given by

$$
\begin{aligned}
l_r^p(\beta_r, \kappa) &= -n \log(c_\kappa) + \kappa \sum_{i=1}^{n} (1 + 2Q_{i33}) \\
&= -n \log(c_\kappa) + \kappa \sum_{i=1}^{n} (1 + 2A_3^T R_i B_3).
\end{aligned}
$$

The estimators of A_3 and B_3 are, respectively, the left and the right singular vectors of the largest singular value, λ_1, of $\bar{R} = \sum_{i=1} R_i/n$.

The large concentration maximum likelihood estimator (MLE) of κ is derived easily using $c_\kappa \approx e^{3\kappa}(\pi/\kappa)^{3/2}$. Equating the derivative of $l_r^p(\beta_r, \kappa)$, with respect to κ, to zero leads to $\hat{\kappa}_{MLE} = 3n/\{4n(1 - \lambda_1)\}$. Notice that the MLE estimator, $\hat{\kappa}_{MLE}$, is positively biased as it does not take into account the $(n + 4)$ degrees of freedom for the model parameters $\beta_r, \theta_i, i = 1, \ldots, n$. A biased corrected estimator is $\hat{\kappa}_r = (3n - n - 4)/\{4n(1 - \lambda_1)\}$.

6.4.2 The Translation Only Estimator of the Parameters

Translation data is available when a single marker is attached to one of the rigid bodies, say $\mathcal{B}$. This does not allow the construction of local reference frame in $\mathcal{B}$ so that the parameters b_x, b_y, α_b are not estimable using this data. The translation log-likelihood $l_t(\beta, \sigma^2)$, see (6.9), involves a vector of six structural parameters $\beta_t = (t_{a1}, t_{a2}, N_{tb}, t_z, a_x, a_y)^T$, the angles $\theta_i, i = 1 \ldots, n$, and σ^2. Thus, $(n + 7)$ parameters need to be estimated.

Some simple manipulations lead to

$$
l_t(\beta_t, \sigma^2) = -\frac{3n}{2} \log(2\pi\sigma^2)
$$
$$
-\frac{1}{2\sigma^2} \sum_{i=1}^{n} \left\{ ||A_{(-3)}^T t_i - t_{a(-3)}^T + N_{tb}\{\cos(\theta_i), \sin(\theta_i)\}^T||^2 + (A_3^T t_i - t_z)^2 \right\},
$$

where $A_{(-3)}$ denotes the 3×2 matrix obtained by deleting the third column of the 3×3 matrix A and $x_{(-3)} = (x_1, x_2)^T$ for all 3×1 vector $x = (x_1, x_2, x_3)^T$.

The maximization of $l_t(\beta_t, \sigma^2)$ with respect to θ_i, $i = 1, \ldots, n$, when A, t_{a1}, t_{a2}, N_{t_b}, t_z and σ^2 are fixed, is first considered. Since $l_t(\beta_t, \sigma^2)$ is a linear function of $\cos(\theta_i)$ and $\sin(\theta_i)$, the maximum is obtained at the value

$$\theta_{it} \;=\; \arctan\left(t_{a2} - A_2^T t_i, t_{a1} - A_1^T t_i\right), \quad \text{for} \quad i = 1, \ldots, n. \quad (6.12)$$

Replacing θ_i by the above expression in (6.9) yields the profile log-likelihood

$$l_t^p(\beta_t, \sigma^2) = -\frac{3n}{2}\log(2\pi\sigma^2) - \frac{1}{2\sigma^2}SSE_t(\beta), \qquad (6.13)$$

where $SSE_t(\beta_t)$ is given by

$$SSE_t(\beta_t) = \sum_{i=1}^{n}\left\{\left[N_{t_b} - ||t_{a_{(-3)}} - A_{(-3)}^T t_i||\right]^2 + (A_3^T t_i - t_z)^2\right\}.$$

Maximizing the log-likelihood with respect to N_{t_b} and t_z gives

$$\hat{N}_{t_b} = \frac{1}{n}\sum_{i=1}^{n}||t_{a_{(-3)}} - A_{(-3)}^T t_i|| \quad \text{and} \quad \hat{t}_z = A_3^T \bar{t},$$

and so, one can write the profile log-likelihood of A_3 and t_a by replacing $SSE_t(\beta_t)$ in (6.13) by

$$\sum_{i=1}^{n}\left[||t_{a_{(-3)}} - A_{(-3)}^T t_i|| - \frac{1}{n}\sum_{i=1}^{n}||t_{a_{(-3)}} - A_{(-3)}^T t_i||\right]^2 + \sum_{i=1}^{n}A_3^T(t_i - \bar{t})(t_i - \bar{t})^T A_3.$$

$$(6.14)$$

The maximization of $l_t^p(\beta_t, \sigma^2)$ with respect to A_3 and t_a can be done using a general optimization program. An estimator of σ^2, corrected for the estimation of the angles θ_i, is $\hat{\sigma}_t^2 = SSE_t(\hat{\beta}_t)/(3n - n - 6)$.

From a geometrical point of view, as $\mathcal{B}$ rotates around $\mathcal{A}$, $\mathcal{B}$'s marker trajectory is a circle in a plane orthogonal to the rotation axis $\mathcal{L}_R$. Minimizing the second part of (6.14), $\sum A_3^T(t_i - \bar{t})(t_i - \bar{t})^T A_3$, yields estimator $\hat{A}_{3s}$, the eigenvector corresponding to the smallest eigenvalue of $\sum(t_i - \bar{t})(t_i - \bar{t})^T$. This is a well-known estimator discussed in [26], [27], [29], [30] and by [8] and [9] in a biomechanical setting. It does not use the fact that the marker's trajectory is a circle; this information is brought in by the first component of (6.14).

6.4.3   The Rotation-Translation Estimator of the Parameters

This section uses the complete data $\{(R_i, t_i)\}$ to estimate the model parameters. We assume that σ^2 goes to 0 and that κ goes to ∞ in such a way that $\sigma^2\kappa = O(1)$. The constant C, defined as the limit of $1/(\sigma^2\kappa)$, is assumed to be known. It represents the weight of the translation log-likelihood in the joint log-likelihood $l_{rt}(\cdot)$. Now $l_{rt}(\cdot)$ depends on the vector parameters $\beta = (t_{a1}, t_{a2}, N_{t_b}, \alpha_b, t_z, a_x, a_y, b_x, b_y)_{9\times1}^T$, the angles θ_i, $i = 1\ldots, n$, and κ. Thus, $(n + 10)$ parameters need to be estimated.

Simple manipulations of the rotation-translation log-likelihood given by (6.7) lead to

$$l_{rt}(\beta, \kappa, \theta_i) = \tag{6.15}$$

$$\kappa \sum_{i=1}^{n} tr\left\{B^T R_i^T A R\left(\theta_i, Z\right)\right\} + CN_{t_b}\{\cos(\theta_i + \alpha_b), \sin(\theta_i + \alpha_b)\}^T \left\{t_{a(-3)} - A_{(-3)}^T t_i\right\}$$

$$- n\log(c_\kappa) - \frac{3n}{2}\log(\frac{2\pi}{C\kappa}) - \frac{C\kappa}{2}\sum_{i=1}^{n}\left\{\left(A_3^T t_i - t_z\right)^2 + N_{t_b}^2 + ||t_{a(-3)} - A_{(-3)}^T t_i||^2\right\}.$$

The problem of maximizing the log-likelihood with respect to θ_i, $i = 1, \ldots, n$, when A, B, t_a, t_b, t_z and κ are fixed, is first considered. As before, (6.15) is a linear function for $\cos(\theta_i)$ and $\sin(\theta_i)$. Thus, for $i = 1, \ldots, n$, the maximum value of (6.15) is obtained at

$$\theta_i = \arctan(G_{2i}, G_{1i}), \tag{6.16}$$

where

$$G_{i1} = (1 + Q_{i33})\cos(\theta_{iR}) + CN_{t_b}||t_{a(-3)} - A_{(-3)}^T t_i|| \cos(\theta_{it} - \alpha_b),$$

$$G_{i2} = (1 + Q_{i33})\sin(\theta_{iR}) + CN_{t_b}||t_{a(-3)} - A_{(-3)}^T t_i|| \sin(\theta_{it} - \alpha_b),$$

$Q_i = B^T R_i^T A$, θ_{iR} is given by (6.11) and θ_{it} is the translation angle defined by (6.12). Using the approximation $c_\kappa \approx e^{3\kappa}(\pi/\kappa)^{3/2}$ and replacing θ_i in (6.15) by (6.16) gives the following profile log-likelihood

$$l_{rt}^p(\beta, \kappa) = -\frac{3n}{2}\log\left(\frac{\pi}{\kappa}\right) - \frac{3n}{2}\log\left(\frac{2\pi}{C\kappa}\right) - \frac{\kappa}{2}SSE_{rt}(\beta),$$

where

$$SSE_{rt}(\beta) = \tag{6.17}$$

$$\sum_{i=1}^{n}\left[6 - 2\sqrt{G_{1i}^2 + G_{2i}^2} - 2Q_{i33} + C\left\{N_{t_b}^2 + ||t_{a(-3)} - A_{(-3)}^T t_i||^2 + \left(A_3^T t_i - t_z\right)^2\right\}\right].$$

The maximization of $l_{rt}^p(\beta, \kappa)$ (i.e., minimization of $SSE_{rt}(\beta)$), with respect to β, does not depend on κ and can be carried out in R. A bias-corrected estimator for κ is $\hat{\kappa}_{rt} = (6n - n - 9)/SSE_{rt}(\hat{\beta})$.

6.5 Numerical Examples

6.5.1 Simulations

Monte Carlo simulations were carried out to evaluate the sampling properties of the estimators obtained with the models considered in this work. Errors

F_i were generated from (6.5). The true value of A_3 and B_3 is $(0, 0, 1)^T$ with $a_x = a_y = b_x = b_y = 0$. The true translation components are $t_a = (60, 90, 0)$, $t_b = (100, 120, 0)$ and $t_z = 100$. We considered two sample sizes $n = 30$ and 100, two sets of error parameters $(\kappa = 700, \sigma = 5)$ and $(\kappa = 700, \sigma = 10)$ and two standard deviations for the angle θ_i, $sd_\theta = 0.1$ and 0.5. For each combination of n, σ and sd_θ, 1000 simulations were performed. The translation weight constant, C, was estimated for each replication b ($b = 1, \ldots, 1000$) as $\hat{C}_{(b)} = 1/\{\hat{\kappa}_{r(b)}\hat{\sigma}^2_{t(b)}\}$, where $\hat{\kappa}_{r(b)}$ and $\hat{\sigma}^2_{t(b)}$ are the unbiased estimators of κ and σ^2 from the rotation only and the translation only models, respectively. Root mean squared errors (r.m.s.e.) and biases are reported in Tables 6.1 and 6.2.

Table 6.1 provides r.m.s.e. for the three estimators of $\sin(a_x)$ and $\sin(a_y)$ using the rotation only (R), the translation only (T) and the combined rotation-translation (R-T) models. The parameters a_x and a_y are the two Cardan angles of A. Table 6.1 shows an important gain in precision of the combined estimator when compared to the rotation only and the translation only estimators. The biases of the estimators of $\sin(a_x) \sin(a_y)$ are not reported here since they are small ($\leq 10^{-3}$). The rotation only estimators of a_x and a_y are better than the translation only estimators, especially when sd_θ is small and σ is large. Note also that the root mean squared errors are inversely proportional to sd_θ.

Table 6.2 gives the estimations of t_a and t_z obtained with the rotation-translation and the translation only models. When the standard deviation of θ_i is small, the translation only estimators of t_a and t_z have large biases, they are bigger than 15%. The biases of the rotation-translation estimators of the translation parameters are always small; they are less than 7%. Also the r.m.s.e. decreases when either the sample size or the standard deviation of the rotation angles increases.

TABLE 6.1

Root mean squared errors for the rotation only (R), the translation only (T), and the rotation-translation (R-T) estimators of $\sin(a_x)$ and $\sin(a_y)$, where a_x and a_y are the two Cardan angles parametrization of A.

			$\sin(\hat{a}_x)$			$\sin(\hat{a}_y)$		
σ	n	sd_θ	R	T	R-T	R	T	R-T
	30	.1	0.054	0.580	0.049	0.053	0.403	0.046
5	30	.5	0.011	0.036	0.009	0.011	0.031	0.009
	100	.1	0.029	0.431	0.025	0.029	0.311	0.024
	100	.5	0.006	0.018	0.005	0.006	0.016	0.005
	30	.1	0.054	0.653	0.053	0.052	0.454	0.051
10	30	.5	0.011	0.075	0.011	0.011	0.062	0.011
	100	.1	0.028	0.627	0.027	0.028	0.437	0.027
	100	.5	0.006	0.035	0.005	0.006	0.030	0.005

TABLE 6.2

Biases and root mean squared errors in parenthesis for the translation only (T) and the rotation-translation (R-T) estimators of t_{a1}, t_{a2}, and t_z.

			$\hat{t}_{a1}$		$\hat{t}_{a2}$		$\hat{t}_z$	
σ	n	sd_θ	T	R-T	T	R-T	T	R-T
	30	.1	-51.8(89)	-1.17(11.8)	-53.6(150.1)	-0.39(12.8)	-28.2(50.2)	-0.25(2.8)
5	30	.5	-0.09(5.9)	-0.07(2.5)	0.14(6.92)	0.04(2.5)	-0.11(1.8)	0.003(0.9)
	100	.1	-11(64.7)	-0.4(6.3)	-0.13(109.3)	-0.04(6.6)	-14.3(34.5)	-0.03(1.4)
	100	.5	-0.19(2.9)	-0.06(1.3)	-0.11(3.2)	0.01(1.3)	-0.03(0.9)	0.003(0.5)
	30	.1	-81.0(93.3)	-4.93(22.9)	-101(138.6)	-6.45(24.0)	-35.8(56.9)	-4.16(3.3)
10	30	.5	-0.18(12.1)	0.07(4.2)	-0.09(14.1)	-0.13(4.3)	-0.57(3.9)	-0.10(1.9)
	100	.1	-72.8(94.8)	-0.1(11.1)	-90.2(149.1)	-0.34(11.5)	-34.1(56.3)	-0.12(1.7)
	100	.5	0.35(5.6)	-0.02(2.2)	0.48(6.5)	-0.03(2.3)	-0.01(1.7)	0.02(1.1)

6.5.2 Data Analysis

We fit the proposed model to the data collected in an experiment carried out by Michael Pierrynowski at the School of Rehabilitation of McMaster University, Canada. The sample consists of $n = 1000$ $SE(3)$ data points, M_i, that give the position of the right tibia with respect to the right femur in gait analysis. The subject is walking on a treadmill at 2.5 mph and the data was collected at 50 Hz for 20 seconds. To avoid the effect of skin movement artifact, the data was recorded with the markers screwed in the right femur and the right tibia of the experimental subject. When standing, the leg and the thigh are in a vertical position, the local x, y, and z axes point forward, inward and up respectively.

The estimates of κ and σ from the rotation only and the translation only models are given by $\hat{\kappa}_r = 836.63$ and $\hat{\sigma}_t = 4.68$ mm. In the rotation translation calculations we used $C = 1/(\hat{\kappa}_r \hat{\sigma}_t^2) = 5.46 \times 10^{-5}$. The estimates of the rotation axis in the local reference frame attached to the femur from respectively the rotation only, the translation only and the rotation-translation models are

$$\hat{A}_3^r = \begin{pmatrix} 0.137 \\ -0.990 \\ 0.001 \end{pmatrix}, \quad \hat{A}_3^t = \begin{pmatrix} 0.326 \\ -0.940 \\ 0.098 \end{pmatrix}, \quad \hat{A}_3^{rt} = \begin{pmatrix} 0.155 \\ -0.988 \\ 0.013 \end{pmatrix}.$$

The estimates of the translation parameters from the rotation-translation model are given by

$$\hat{t}_{a(-3)} = (-27.95, -430.04)^T, \quad \hat{t}_{b(-3)} = (-90.62, 35.66)^T, \quad \hat{t}_z = -34.36.$$

Thus the knee rotates about a y axis that points outward. To interpret the translation parameters note that $\hat{A}_1^{rt} = (0.988, 0.155, -0.002)^T$ and $\hat{A}_2^{rt} = (0, 0.03, 0.999)^T$; thus t_{a1} is a coordinate in the x-axis that goes forward and

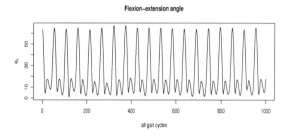

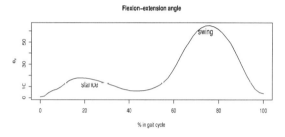

FIGURE 6.1
The flexion-extension angle $\hat{\theta}_i$ over all gait cycles and for a single cycle.

t_{a2} refers to the z-axis that goes up. The z Cardan angle of $\hat{A}^T R_i \hat{B}$, $\hat{\theta}_i$, is the flexion-extension angle around the rotation axis $\hat{A}_3^{rt}$. These angles are presented in Figure 6.1 which shows that the data set contains 18 complete gait cycles. A gait cycle is comprised of a stance and a swing phase, see [23]. A stance is the period during which the foot is on the ground (from 0% to 60% of the gait cycle). A swing phase corresponds to the time when the foot is in the air (from 60% to 100% of the gait cycle), see also [32].

As observed by [2], [14] and [31], a knee's motion cannot be completely described by the one axis model. Even if it gives a good global description of the knee motion, model (6.6) is not completely accurate. Additional analysis will be carried out with this data set. First we investigate whether the orientation of the knee rotation axis changes between the stance and the swing phase. Then a change in the location of the rotation axis will be studied.

To investigate whether the rotation axis changes between the two phases, separate estimates of the stance and the swing rotation axes were calculated for the 18 gait cycles of the data set. To test the null hypothesis of equality $\mathcal{H}_0$, we used the rotation only estimates of A_3, because the translation model in (6.6) is not completely true for the knee. We used the two-sample Watson–William test given by [18], whose approximate null distribution is a $F_{2,68}$, the Fisher distribution with 2 and 68 degrees of freedom. One has $F_{obs} = 35.75$, ($p\_value = 0$), the hypothesis $\mathcal{H}_0$ is rejected. The estimates of A_3 for the stance and the swing phases are given by $\hat{\bar{A}}_{3st}^r = (-0.047, -0.994, 0.096)^T$ and $\hat{\bar{A}}_{3sw}^r = (0.131, -0.991, 0.021)^T$ respectively. Thus a rotation of $(\bar{\hat{A}}_{3st}^r, \widehat{\bar{\hat{A}}_{3sw}^r}) = 11.08 \, degree$ maps $\bar{\hat{A}}_{3st}^r$ to $\bar{\hat{A}}_{3sw}^r$ in the plane orthogonal to the vector $(0.386, 0.07, 0.92)^T$. To our knowledge such a change has not been noted in the biomechanical literature. See for instance [31] for a detailed discussion.

To examine how the location of the rotation axis changes along one gait cycle, one can calculate local estimates of the translations t_a along the gait cycle using the local likelihood functions as defined in [5] and [33]. Minus the local likelihood at $100 \times p\%$ of the gait cycle is $\sum_i w_i(p) SSE_i(\beta_{rt})$, where $p \in [0,1]$, $SSE_i(\beta_{rt})$ is the i th term in Equation (6.17),

$$w_i(p) = \max \left\{ 0, 1 - \left[\frac{1 - \cos\left(2\pi(p - p_i)\right)}{h} \right]^2 \right\},$$

is the Epanechnikov kernel, [7], and $p_i \in [0,1]$ is the fraction in the gait cycle for observation i, calculated from $\hat{\theta}_i$, and h is a smoothing parameter which is equal to 2.3 in the calculation. One can calculate the local MLE of the model parameters at different points along the gait cycle. The estimates of the points $(\hat{t}_{a1j}, \hat{t}_{a2j})$, $j = 1, \ldots, 100$, and the pathway of the position of the rotation axis in the plane spanned by $(\hat{A}_1^{rt}, \hat{A}_2^{rt})$ along the gait cycle are given in Figure 6.2. Since the orientation of the rotation axis does not change much over a gait cycle (i.e., $\hat{A}_{kj}^{rt} \approx \hat{A}_k^{rt}$, for $k = 1, 2$), we interpret the pair $(\hat{t}_{a1j}, \hat{t}_{a2j})$ as giving, respectively, the forward and the upward position of the rotation axis at $100 \times p_j\%$ of the gait cycle in a plane orthogonal to that axis.

The rotation axis goes forward by $4 \, mm$ from 0% to 40% of the cycle, thus most of the motion occurs in the stance. At the end of the swing phase, the leg is no longer bearing weight so the thigh extends slightly and the rotation axis goes down 2 mm, back to its original position. This displacement of the rotation axis, as measured in the thigh reference frame, can possibly be explained by the displacement of the femur on the tibial plateau. Figure 6.2 shows that the rotation axis moves forward at the beginning of the stance; this is similar to the findings presented in Figure 2 of [10]. In [14]'s investigation of the gait stance phase, Figure 4 shows an anterior-posterior motion (that is along $\hat{A}_1^{rt}$) and a distal-proximal motion (that is along $\hat{A}_2^{rt}$) of the femur with respect to the tibia of a same order of magnitude as the displacement of the rotation axis reported in Figure 6.2. Consequently, the data analysis presented in this section allows a precise characterization of the knee's motion.

6.6 Discussion

This chapter has shown how $SE(3)$ data is collected. It has proposed a one axis model to analyze this data, motivated by an investigation of the motion of the knee. The gains of efficiency associated with using the combined rotation-

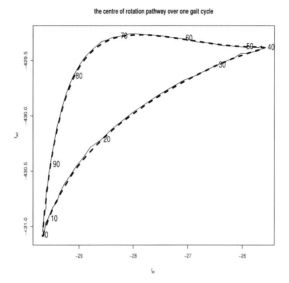

FIGURE 6.2
The instantaneous center of rotation pathway of the knee joint, the numbers are sequential positions along the gait cycle.

translation data rather than modeling the marginal rotation or translation data have been measured in a simulation study.

This work has used a combined estimation criterion that minimizes $\sum \|M_i - D_i\|_W^2$. An alternative criterion, minimizing $\sum \|M_i^{-1} - D_i^{-1}\|_W^2$, leads to different estimators for β. Which one should be used? The determination of a proper estimation criterion is still an open problem in $SE(3)$ which is discussed in [22]. One could also argue that the proposed error model, see (6.5), is too simplistic. In (6.4), one could have both right-hand side and left-hand side errors. Kume et al. [15] also noted some error heterogeneity in rotation data sets; thus the proposed homoscedastic error model for the rotation component in (6.5) might fail and extensions are needed to properly describe experimental data. In summary, the selection of an estimation criterion and of a probability model for an $SE(3)$ data set raises interesting foundational issues that will be the focus of future work.

Bibliography

[1] M. A. Bingham, D. J. Nordman, and S. B. Vardeman. Modeling and inference for measured crystal orientations and a tractable class of symmetric distributions for rotations in three dimensions. *Journal of the American Statistical Association*, 104(488):1385–1397, 2009.

[2] A. M. J. Bull and A. A. Amis. Knee joint motion: Description and measurement. *Proceedings of the Institution of Mechanical Engineers, Part H: Journal of Engineering in Medicine*, 212(5):357–372, 1998. doi: 10.1243/0954411981534132.

[3] Y. Chikuse. *Statistics on Special Manifolds*. Springer, 2003.

[4] G. S. Chirikjian and A. Kyatkin. *Engineering Applications of Noncommutative Harmonic Analysis: With Emphasis on Rotation and Motion Groups*. CRC Press: Boca Raton, FL, Abingdon, 2001.

[5] J. B. Copas. Local likelihood based on kernel censoring. *Journal of the Royal Statistical Society. Series B (Methodological)*, 57(1):221–235, 1995.

[6] T. D. Downs. Orientation statistics. *Biometrika*, 59(3):665–676, 1972.

[7] V. A. Epanechnikov. Non parametric estimation of a multivariate probability density. *Theory of Probability and Its Applications*, 14(1):153–158, 1969.

[8] S. S. Gamage and J. Lasenby. New least squares solutions for estimating the average centre of rotation and the axis of rotation. *Journal of Biomechanics*, 35(1):87–93, 2002.

[9] M. Halvorsen, K. Lesser and A. Lundberg. A new method for estimating the axis of rotation and the center of rotation. *Journal of Biomechanics*, 32:1221–1227, 1999.

[10] J. H. Hollman, R. H. Deusinger, L. R. Van Dillen, and M. J. Matava. Knee joint movements in subjects without knee pathology and subjects with injured anterior cruciate ligaments. *Physical Therapy*, 82(10):960–972, 2002.

[11] P. E. Jupp and K. V. Mardia. Maximum Likelihood Estimators for the Matrix Von Mises-Fisher and Bingham Distributions. *The Annals of Statistics*, 7(3):599–606, 1979.

[12] C. G. Khatri and K. V. Mardia. The Von Mises-Fisher Matrix Distribution in Orientation Statistics. *Journal of the Royal Statistical Society. Series B (Methodological)*, 39(1):95–106, 1977.

[13] M. Klous and S. Klous. Marker-based reconstruction of the kinematics of a chain of segments: A new method that incorporates joint kinematic constraints. *Journal of Biomechanical Engineering*, 132(7):074501 (7 pages), 2010.

[14] M. Kozanek, A. Hosseini, F. Liu, S. K. Van de Velde, T. J. Gill, H. E. Rubash, and G. Li. Tibiofemoral kinematics and condylar motion during the stance phase of gait. *Journal of Biomechanics*, 42(12):1877 – 1884, 2009.

[15] A. Kume, S. P. Preston, and A. T. A. Wood. Saddlepoint approximations for the normalizing constant of fisher bingham distributions on products of spheres and stiefel manifolds. *Biometrika*, 100(4):971–984, 2013.

[16] S. M. LaValle. *Planning Algorithms*. Cambridge University Press, 2006.

[17] C. A. León, J.-C. Massé, and L.-P. Rivest. A statistical model for random rotations. *Journal of Multivariate Analysis*, 97(2):412–430, 2006.

[18] K. V. Mardia and P. E. Jupp. *Directional Statistics*. Wiley, 2000.

[19] J. M. McCarthy. *Introduction to Theoretical Kinematics*. The MIT Press, Cambridge, MA, 1990.

[20] J. Neyman and E. Scott. Consistent estimates based on partially consistent observations. *Econometrica*, 16:1–32, 1948.

[21] K. Oualkacha and L.-P. Rivest. A new statistical model for random unit vectors. *Journal of Multivariate Analysis*, 100(1):70–80, 2009. doi: 10.1016/j.jmva.2008.03.004.

[22] K. Oualkacha and L.-P. Rivest. On the estimation of an average rigid body motion. *Biometrika*, 99(3):585–598, 2012. doi: 10.1093/biomet/ass020.

[23] J. Perry. *Gait Analysis: Normal and Pathological Function*. Delmar Learning, 1992.

[24] M. J. Prentice. Orientation statistics without parametric assumptions. *Journal of the Royal Statistical Society: Series B (Methodological)*, 48(2): 214–222, 1986.

[25] D. Rancourt, L.-P. Rivest, and J. Asselin. Using orientation statistics to investigate variations in human kinematics. *Journal of the Royal Statistical Society: Series C (Applied Statistics)*, 49(1):81–94, 2000.

[26] L.-P. Rivest. Some local linear models for the assessment of geometric integrity. *Statistica Sinica*, 5:204–210, 1995.

[27] L.-P. Rivest. Some linear model techniques for analyzing small circle spherical data. *Revue Canadienne de Statistique*, 27(3):623–638, 1999.

[28] L.-P. Rivest. A directional model for the statistical analysis of movement in three dimensions. *Biometrika*, 88(3):779–791, 2001.

[29] C. Sheringer. A model for fitting a plane to a set of points by least squares. *Acta Crystallographica*, 27:1470–1472, 1971.

[30] V. Shomaker, J. Waser, R. E. Marsh, and G. Bergman. To fit a plane or a line to a set of points by least squares. *Acta Crystallographica*, 12: 600–604, 1959.

[31] K. M. Smith, P. N. Refshauge and J. M. Scarvell. Development of the concepts of the knee kinematics. *Archives of Physical Medicine and Rehabilitation*, 84:1895–1902, 2003.

[32] D. H. Sutherland, K. R. Kaufman, and J. R. Moitoza. Kinematics of normal human walking. J. Rose and J.G. Gamble, (eds.), *Human Walking*, Baltimore, Williams and Wilkins, pages 23–44, 1994.

[33] R. Tibshirani and T. Hastie. Local likelihood estimation. *Journal of the American Statistical Association*, 82:559–567, 1987.

[34] F. E. Veldpaus, H. J. Woltring, and L. J. M. G. Dortmans. A least-squares algorithm for the equiform transformation from spatial marker co-ordinates. *Journal of Biomechanics*, 21:45–54, 1988.

7

Spatial and Spatio-Temporal Circular Processes with Application to Wave Directions

Giovanna Jona-Lasinio

Sapienza University of Rome

Alan E. Gelfand

Duke University

Gianluca Mastrantonio

Polytechnic of Turin

CONTENTS

7.1	Introduction		130
7.2	The Wrapped Spatial and Spatio-Temporal Process		131
	7.2.1	Wrapped Spatial Gaussian Process	132
	7.2.2	Wrapped Spatio-Temporal Process	133
	7.2.3	Kriging and Forecasting	133
	7.2.4	Wave Data for the Examples	134
	7.2.5	Wrapped Skewed Gaussian Process	136
	7.2.5.1	Space-Time Analysis Using the Wrapped Skewed Gaussian Process	140
7.3	The Projected Gaussian Process		141
	7.3.1	Univariate Projected Normal Distribution	141
	7.3.2	Projected Gaussian Spatial Processes	142
	7.3.3	Model Fitting and Inference	145
	7.3.4	Kriging with the Projected Gaussian Processes	146
	7.3.4.1	An Example Using the Projected Gaussian Process	147
	7.3.5	The Space-Time Projected Gaussian Process	149
	7.3.6	A Separable Space-Time Wave Direction Data Example	149
7.4	Space-Time Comparison of the WN and PN Models		151
7.5	Joint Modeling of Wave Height and Direction		153
7.6	Concluding Remarks		157
	Bibliography		158

7.1   Introduction

An under-investigated area in directional data is the development of space
and space time processes where directional observations are taken at spatial
locations, usually over time as well, and proximity in space and time affects
dependence between directions. Such data take us to the world of geostatistical
modeling [6], introducing structured dependence in space and time.

Wave direction data provide an illustration of this setting. In the sea or
ocean, there is a conceptual wave direction at every location and time. Thus,
the challenge is to introduce structured dependence into angular data. In
particular, circular data have no magnitudes. One value is neither larger nor
smaller than another; angular values are only assigned given an orientation.
Our approach begins with models for linear variables over space and time
using Gaussian processes. Then, we use either wrapping or projection to obtain
Gaussian processes for circular data. Hence, circular dependence structure is
inherited from that of the linear process.

So, in this chapter, we propose statistical tools to model observed wave
data. We are interested in modeling wave directions alone as well as significant wave heights and directions jointly. Values are available at multiple locations and time points. Specifically, we work with the output of a deterministic
wave forecast model implemented by Istituto Superiore per la Protezione e la
Ricerca Ambientale (ISPRA).

Many univariate distributions have been proposed to model circular data,
see for example [31], while few multivariate extensions, generally bivariate, are
available. Some interesting examples are the bivariate circular distributions
proposed by [28], [32], and [42], the bivariate wrapped normal of [19], the
multivariate von Mises of [30], and the multivariate wrapped normal of [5]
and from them new complex models have been built. Moreover, the analysis
of circular data has become less descriptive and more inferential, addressing
problems well beyond i.i.d. samples from a particular distribution. Examples
include linear models [15, 23], spatial models [35], temporal models [8, 17],
and spatio-temporal models [25] (cross-reference to F. Lagona's Chapter 3).

Here, we review work developed by the authors in a series of papers published over the last few years focusing on spatial and spatio-temporal modeling
for wave directions in the spirit noted above. All are cast as hierarchical models, with fitting and inference within a Bayesian inference framework. The
hierarchical specification is vital; latent variables are introduced to facilitate
passage from linear variables to circular variables. The Bayesian framework is
arguably most attractive for this setting. We obtain full posterior inference including uncertainties, we avoid potentially inappropriate asymptotics, and we
enable routine prediction (kriging) over space and time. Furthermore, model
fitting is straightforwardly implemented using Markov chain Monte Carlo [6].
Specifically, Gibbs sampling provides loops, updating parameters given latent

variables and data and, then, updating latent variables given parameters and data [20].

We start with a purely spatial setting, as initially introduced in [20]. The authors proposed the wrapped spatial Gaussian process, induced from a linear spatial Gaussian process. They explored dependence structure and showed how to implement spatial prediction of mean directions and concentrations in this setting. In follow-on work space time modeling was considered [34] with further extension to skew normal processes in [33]. In parallel, space and space-time models based on a projected Gaussian process were explored in [38, 46]. Also, wave height was considered in the context of a joint response model over space and time in [48], modeling wave height as a linear variable and then wave direction given wave height, introducing the latter as a linear variable in a circular regression. Altogether, the current state of the art for modeling circular space-time data includes the wrapped Gaussian process and the projected Gaussian process.

Again, our running example involves the main outputs of marine forecasts: outgoing wave directions, measured directionally in degrees relative to a fixed orientation, and wave height. Numerical models for weather and marine forecasts need statistical post-processing. Wave heights, like wind speed, being linear variables can be treated in several ways [21, 22, 49], while wave directions, being circular variables, cannot be treated according to standard post-processing techniques (see [7, 10] and references therein). For example, in [7] bias correction and ensemble calibration forecasts of surface wind direction are proposed. The authors use circular-circular regression as in [23] for bias correction and Bayesian model averaging with the von Mises distribution for ensemble calibration. However, their approach does not explicitly account for spatial and temporal structure.

The chapter is organized as follows. Section 7.2 presents the wrapped spatial and spatio-temporal Gaussian process. Section 7.3 provides the projected Gaussian analogues. Section 7.4 offers a spatio-temporal comparison between the wrapped Gaussian process and the projected Gaussian process. Section 7.5 takes up the joint modeling of directions and heights with Section 7.6 offering concluding remarks.

7.2 The Wrapped Spatial and Spatio-Temporal Process

We briefly review the wrapped modeling approach to circular data. We provide the spatial and then the space-time setting; they are similar in terms of modeling and implementation. Let $Y \in \mathbb{R}$ be a random variable on the real line and let $g(y)$ and $G(y)$ be respectively its probability density function and cumulative distribution function. The random variable

$$\theta = Y \bmod 2\pi, \; \theta \in [0, 2\pi), \tag{7.1}$$

is the wrapped version of Y having period 2π. The probability density function of θ, $f(\theta)$, is obtained by wrapping the probability density function of Y, $g(y)$, around a circle of unit radius via the transformation $Y = \theta + 2\pi K$, yielding the form

$$f(\theta) = \sum_{k=-\infty}^{\infty} g(\theta + 2\pi k). \tag{7.2}$$

That is, $K \in \mathbb{Z} \equiv \{0, \pm 1, \pm 2, \ldots\}$, resulting in the doubly infinite sum in (7.2).

Equation (7.2) shows that $g(\theta + 2\pi k)$ is the joint distribution of (θ, K). Hence, the marginal distribution of K is $P(K = k) = \int_0^{2\pi} g(\theta + 2\pi k) d\theta$ while the conditional distributions are $P(K = k|\theta) = g(\theta + 2\pi k)/\sum_{j=-\infty}^{\infty} g(\theta + 2\pi j)$, and $f(\theta|K = k) = g(\theta + 2\pi k)/\int_0^{2\pi} g(\theta + 2\pi k) d\theta$. The introduction of K as a latent variable facilitates model fitting as shown in [20] and initially introduced in [9]. Extension of the wrapping approach to multivariate distributions is direct. Let $\mathbf{Y} = (Y_1, Y_2, \ldots, Y_p) \sim g(\cdot)$, with $g(\cdot)$ a p-variate distribution on $\mathbb{R}^p$ indexed by say $\boldsymbol{\eta}$ and let $\mathbf{K} = (K_1, K_2, \ldots, K_p)$ be such that $\mathbf{Y} = \boldsymbol{\Theta} + 2\pi \mathbf{K}$. Then the distribution of $\boldsymbol{\Theta}$ is

$$f(\boldsymbol{\theta}) = \sum_{k_1=-\infty}^{+\infty} \sum_{k_2=-\infty}^{+\infty} \cdots \sum_{k_p=-\infty}^{+\infty} g(\boldsymbol{\theta} + 2\pi \mathbf{k}). \tag{7.3}$$

From (7.3) we see, as in the univariate case, that the joint density of $(\boldsymbol{\Theta}, \mathbf{K})$ is $g(\boldsymbol{\theta} + 2\pi \mathbf{k})$. If $g(\cdot; \boldsymbol{\eta})$ is a p-variate normal density, with $\boldsymbol{\eta} = (\boldsymbol{\mu}, \boldsymbol{\Sigma})$, then $\boldsymbol{\Theta}$ has a p-variate wrapped normal distribution with parameters $(\boldsymbol{\mu}, \boldsymbol{\Sigma})$. Here, we introduce the latent random vector of winding numbers $\mathbf{K}$ to facilitate model fitting. Mardia and Jupp [31] point out that only a few values of K are needed to obtain a reasonable approximation of the wrapped distribution and Lasinio et al. [20] show, when $g(\cdot; \boldsymbol{\eta})$ is Gaussian, how to choose the set of values of K based on the variance of the associated conditional distribution.

7.2.1   Wrapped Spatial Gaussian Process

Let $Y(\mathbf{s})$ be a Gaussian process (GP) with $\mathbf{s} \in \mathbb{R}^2$, mean function $\mu(\mathbf{s})$ and covariance function say $\sigma^2 \rho(||\mathbf{s}_i - \mathbf{s}_j||; \boldsymbol{\psi})$. Here, $\rho(\cdot)$ is a valid correlation function providing the correlation between variables at locations $\mathbf{s}_i$ and $\mathbf{s}_j$, indexed by parameters $\boldsymbol{\psi}$ with σ^2 being the homogeneous spatial variance. For locations $\mathbf{s}_1, \mathbf{s}_2, \ldots, \mathbf{s}_n$, $\mathbf{Y} = (Y(\mathbf{s}_1), Y(\mathbf{s}_2), \ldots, Y(\mathbf{s}_n)) \sim N(\boldsymbol{\mu}, \sigma^2 R(\boldsymbol{\psi}))$, where $\boldsymbol{\mu} = (\mu(\mathbf{s}_1), \ldots, \mu(\mathbf{s}_n))$ and $R(\boldsymbol{\psi})_{ij} = \rho(\mathbf{s}_i - \mathbf{s}_j; \boldsymbol{\psi})$. As a consequence, the wrapped random process is such that $\boldsymbol{\Theta} \sim WrapN(\boldsymbol{\mu}, \sigma^2 R(\boldsymbol{\psi}))$ [20], where $WrapN(\cdot, \cdot)$ indicates the wrapped normal distribution. Differently from [9], where multivariate wrapped modeling employs replications to learn about a general $\boldsymbol{\Sigma}$ for the multivariate model, we do not require replications due to the structured spatial dependence introduced through the GP.

7.2.2 Wrapped Spatio-Temporal Process

Turning to space and time, suppose we seek to model $\{\theta(\mathbf{s}, t) \in [0, 2\pi), \mathbf{s} \in \mathcal{S} \subseteq \mathbb{R}^2, t \in \mathcal{T} \subseteq \mathbb{R}^+\}$, a spatio-temporal process of angular variables. We can model $\theta(\mathbf{s}, t)$ as a spatio-temporal wrapped Gaussian process through its linear counterpart $Y(\mathbf{s}, t)$, a spatio-temporal Gaussian process. We write the linear GP $Y(\mathbf{s}, t)$ as $Y(\mathbf{s}, t) = \mu_Y + \omega_Y(\mathbf{s}, t) + \tilde{\varepsilon}_Y(\mathbf{s}, t)$ where μ_Y is a constant mean function, $\omega_Y(\mathbf{s}, t)$ is a zero mean space-time GP with covariance function $\sigma^2 \rho(\mathbf{h}, u)$ with ρ now a valid space-time correlation function, $\mathbf{h}, u$ denoting the spatial and temporal separations, respectively, and $\tilde{\varepsilon}(\mathbf{s}, t) \overset{iid}{\sim} N(0, \phi_Y^2)$ is pure error. It is convenient to work with the marginalized model where we integrate over all of the $\omega_Y(\mathbf{s}, t)$, see [6], that is,

$$Y(\mathbf{s}, t) = \mu_Y + \varepsilon_Y(\mathbf{s}, t). \tag{7.4}$$

Then, $\varepsilon_Y(\mathbf{s}, t)$ is a zero mean Gaussian process with covariance function

$$\mathrm{Cov}(\varepsilon_Y(\mathbf{s}_i, t_j), \varepsilon_Y(\mathbf{s}_{i'}, t_{j'})) = \sigma_Y^2 \mathrm{Cor}(\mathbf{h}_{i,i'}, u_{j,j'}) + \phi_Y^2 1_{(i=i')} 1_{(j=j')}.$$

7.2.3 Kriging and Forecasting

A customary objective with spatial and spatio-temporal processes is prediction, known as *kriging*. For example, with wave data, we seek to predict wave direction at a new location and time, say $(\mathbf{s}_0, t_0)$, given what we have observed. Following [34], the Bayesian framework enables a full predictive distribution. Let $\mathcal{D} \subset \mathbb{R}^2 \times \mathbb{R}^+$ be the set of n observed points. Let $\boldsymbol{\Theta} = \{\theta(\mathbf{s}, t), (\mathbf{s}, t) \in \mathcal{D}\}$ be the vector of observed circular variables. Let $\mathbf{Y} = \{Y(\mathbf{s}, t), (\mathbf{s}, t) \in \mathcal{D}\}$ be the associated linear ones and let $\mathbf{K} = \{K(\mathbf{s}, t), (\mathbf{s}, t) \in \mathcal{D}\}$ be the associated vector of winding numbers. We seek the predictive distribution of $\theta(\mathbf{s}_0, t_0)|\boldsymbol{\Theta}$, that is not known in closed form. We use composition sampling within Markov chain Monte Carlo (MCMC) to obtain samples from it. Here, again we move from the circular process to the linear one, i.e., a sample from the distribution of $Y(\mathbf{s}_0, t_0)|\boldsymbol{\Theta}$ can be considered as a sample from $\theta(\mathbf{s}_0, t_0), K(\mathbf{s}_0, t_0)|\boldsymbol{\Theta}$. If we let $\boldsymbol{\Psi}_Y$ be the vector of all parameters, we can write

$$g(\theta(\mathbf{s}_0, t_0), K(\mathbf{s}_0, t_0)|\boldsymbol{\Theta})$$

$$= \sum_{K(s,t),(s,t) \in \mathcal{D}} \int_{\boldsymbol{\Psi}_Y} g(\theta(\mathbf{s}_0, t_0), K(\mathbf{s}_0, t_0)|\boldsymbol{\Psi}_Y, \mathbf{K}, \boldsymbol{\Theta}) g(\boldsymbol{\Psi}_Y, \mathbf{K}|\boldsymbol{\Theta}) d\boldsymbol{\Psi}_Y. \tag{7.5}$$

So, suppose, for each posterior sample in $\{\mathbf{K}_l^*, \boldsymbol{\Psi}_{Y,l}^*, l = 1, 2, \ldots, L\}$ we generate a realization from the distribution of $\theta(\mathbf{s}_0, t_0), K(\mathbf{s}_0, t_0)| \boldsymbol{\Psi}_{Y,l}^*, \mathbf{K}, \boldsymbol{\Theta}$. Then, we will obtain the set of posterior samples $\{\theta_l^*(\mathbf{s}_0, t_0), K_l^*(\mathbf{s}_0, t_0), l = 1, 2, \ldots, L\}$ from $\theta(\mathbf{s}_0, t_0), K(\mathbf{s}_0, t_0)|\boldsymbol{\Theta}$. If we retain the set $\{\theta_l^*(\mathbf{s}_0, t_0), l = 1, 2, \ldots, L\}$, we will have samples from the desired predictive distribution. The requisite distribution theory is presented in [34]. We omit the details here.

7.2.4 Wave Data for the Examples

We model wave directions using data obtained as outputs from a deterministic computer model implemented by Istituto Superiore per la Protezione e la Ricerca Ambientale (ISPRA). The computer model starts from a wind forecast model predicting the surface wind over the entire Mediterranean. The hourly evolution of sea wave spectra is obtained by solving energy transport equations using the wind forecast as input. Wave spectra are locally modified using a source function describing the wind energy, the energy redistribution due to nonlinear wave interactions, and energy dissipation due to wave fracture. The model produces estimates every hour on a grid with 10×10 km cells [44, 45]. The ISPRA dataset has forecasts for a total of 4941 grid points over the Italian Mediterranean. We restrict ourselves to the Adriatic Sea area where there are 1494 points (see Figure 7.1). We select several time points from the large wave-database such that three sea states, *calm, transition*, and *storm*, are present. The sea state is usually defined through the wave height (which is also supplied by the computer model output); when this height is below 1 meter, we have *calm*, when it is between 1 and 2 meters we have *transition* (between calm and storm), and when it is greater than 2 meters we have a *storm*. Wave directions vary more in calm than in storm.

To evaluate model performance, following [20, 34], we hold out a validation dataset from the model fitting. Then, we use *average prediction error* (APE), defined as the average circular distance across pairs of held out observations and associated predictive values where, for a given pair, we adopt as circular distance $d(\alpha, \beta) = 1 - \cos(\alpha - \beta)$ [18, p. 15] (see [34] for details).

However, a stronger predictive goal regards the performance of entire predictive distribution compared with the held out observation at a given location. For this we use the *continuous ranked probability score* (CRPS) for circular variables as defined in [14]:

$$CRPS(F, \delta) = \int_0^{2\pi} (F(y) - \mathbb{1}(y - \delta))^2 \, dy, \tag{7.6}$$

where $\mathbb{1}(\cdot)$ is the Heaviside step function, F is a predictive distribution and δ is a holdout observation. Equation (7.6) measures the integrated square distance between the predictive distribution and the degenerate cumulative distribution associated with the holdout value. An elegant computational result shows that, with circular distance as above, (7.6) can be written as

$$CRPS(F, \delta) = E(d(\Delta, \delta)) - \frac{1}{2}E(d(\Delta, \Delta^*)), \tag{7.7}$$

where Δ and Δ^* are independent copies of a circular variable with distribution F. In our setting, samples from the posterior predictive distribution associated with any δ can be generated. Hence, Monte Carlo integration for (7.7) is readily available. CRPS would be calculated as an average over the hold out observations; small values are preferred.

FIGURE 7.1
Adriatic map, dark points highlight the study area.

In the study area, storms are more often produced by North-East winds (see Figure 7.2 (a)) and induce a very homogeneous wave field over the entire area, while the calm state can be highly variable (see Figure 7.2 (c-d)). In what follows we show results of fitting models to data from the ISPRA database during the months of April and May 2010. In Figure 7.3 the rose diagrams of the two months of data are reported.

Wave Data: Spatial Estimation

We estimate the wrapped spatial Gaussian process for a portion of the Adriatic Sea (see Figure 7.4). The selected area includes 378 grid points of which 100 are kept aside for validation (the black dots in Figure 7.4). The chosen region is often highly variable during calm and transition so it becomes a challenging area for estimation. We run an increasing number of MCMC iterations (from 22,000 to 35,000) going from the storm (less variable) to the calm state (most variable) keeping 2000 iterations for inference. The prior structure is as follows: for the mean we use a wrapped Gaussian distribution with mean π and variance 1, for the decay parameter a $\Gamma(1.3, 100)$, for σ^2 an Inv$\Gamma(100, 10)$; for the decay parameter and σ^2 we use an adaptive Metropolis update starting from initial guesses. In Table 7.1 APE and CRPS are computed under the three sea states; the increasing variability in wave direction induces a larger uncertainty in the point estimates. A space-time nonseparable example is presented in Section 7.4 also with these data.

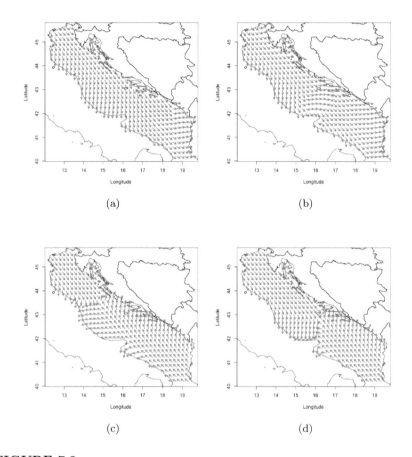

(a)   (b)

(c)   (d)

FIGURE 7.2

Time windows showing different sea states. The four panels represent the observed wave direction over the entire area at: (a) 12:00 on 5/5/2010 (storm); (b) 00:00 on 6/5/2010 (transition between storm and calm); (c) 00:00 on 7/5/2010 (calm); (d) 12:00 on 7/5/2010 (calm).

7.2.5   Wrapped Skewed Gaussian Process

To allow the model to capture asymmetric circular phenomena we propose the wrapped version of the skewed Gaussian. By now, there is a fairly rich literature on skew multivariate normal models [2, 3, 4] but all are *inline*. The first attempt to wrap the skew normal distribution for circular data can be found in [38] where its basic properties are derived. Follow-on work appears in [39, 41]. The first extension to multivariate wrapped skew distributions, in particular, to a spatial and spatio-temporal setting has been proposed in [33]

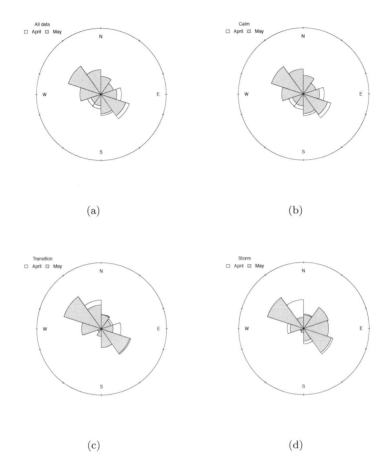

FIGURE 7.3
Rose diagrams of wave directions for April and May 2010 overall and by state,
(a) overall, (b) calm, (c) transition, (d) storm.

and is the focus of this subsection.

The Univariate Case

We begin with the univariate wrapped skew normal distribution. Let D and
W be two independent standard normal variables, let $\mu \in \mathbb{R}^1$, $\sigma^2 \in \mathbb{R}^+$, and
$\lambda \in \mathbb{R}$. Then, the random variable

$$Y = \mu + \frac{\sigma\lambda}{\sqrt{1+\lambda^2}}|D| + \frac{\sigma}{\sqrt{1+\lambda^2}}W - \frac{\sigma\lambda\sqrt{2}}{\sqrt{\pi(1+\lambda^2)}} \tag{7.8}$$

FIGURE 7.4
Map of the portion of the Adriatic Sea selected for the examples using the wrapped spatial Gaussian process, in black the selected validation sites.

TABLE 7.1
Average prediction error (APE) and continuous ranked probability scores for the wrapped spatial Gaussian process estimated on wave directions by sea state in the north Adriatic Sea

	Storm	Transition	Calm
APE	2×10^{-4}	0.1275	0.2805
CRPS	6.1×10^{-5}	0.0307	0.1533

is said to be distributed as a skew normal variable [1] with parameters μ, σ^2 and λ; i.e., $Y|\Psi \sim SN(\mu, \sigma^2, \lambda)$, where Ψ denotes the vector of parameters. Let $\phi(\cdot)$ and $\Phi(\cdot)$ be the probability density function (pdf) and the cumulative density function (cdf), respectively, of a standard normal. Then, the pdf of $Z|\Psi$ is

$$\frac{2}{\sigma}\phi\left(\frac{y - \mu + \frac{\sigma\lambda\sqrt{2}}{\sqrt{\pi(1+\lambda^2)}}}{\sigma}\right)\Phi\left(\lambda\left(\frac{y - \mu + \frac{\sigma\lambda\sqrt{2}}{\sqrt{\pi(1+\lambda^2)}}}{\sigma}\right)\right) \quad (7.9)$$

The mean of Y is μ (the definition in (7.8) was made in order to center Y at μ) and the variance is $\sigma^2\lambda^2/(1 + \lambda^2)(1 - 2/\pi) + \sigma^2/(1 + \lambda^2)$.

With the transformation $\Theta = Y \bmod 2\pi$, implying $\Theta \in [0, 2\pi)$, we obtain a random variable with support on the unit circle. We can express the inline

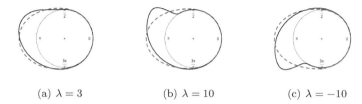

(a) $\lambda = 3$ (b) $\lambda = 10$ (c) $\lambda = -10$

FIGURE 7.5
Densities of the wrapped skew normal (solid line) with $\mu = \pi$, $\sigma^2 = 1$ and different values of λ along with the associated densities of the wrapped normal (dashed line) having the same circular mean and variance.

variable as $Y = \Theta + 2\pi K$. This transformation defines what is called a *wrapped skew normal* (WSN) distribution, as introduced in [38].

The pdf of $\Theta | \Psi$ is

$$
\sum_{k \in \mathbb{Z}} \frac{2}{\sigma} \phi \left(\frac{\theta + 2\pi k - \mu + \frac{\sigma \lambda \sqrt{2}}{\sqrt{\pi(1+\lambda^2)}}}{\sigma} \right) \Phi \left(\lambda \left(\frac{\theta + 2\pi k - \mu + \frac{\sigma \lambda \sqrt{2}}{\sqrt{\pi(1+\lambda^2)}}}{\sigma} \right) \right).
$$

$$(7.10)$$

The infinite sum in (7.10) cannot be evaluated but, to display the density, as with the wrapped normal case, we can obtain an accurate approximation by appropriately truncating the sum. Figure 7.5 illustrates the effect of introduction of skewness into the wrapped normal density.

To obtain a sample from a wrapped skew normal we first obtain a sample from the skew normal and then transform it as above to a circular variable. If we let K be a random variable, the density inside the sum in (7.10) is the joint density of $(\Theta, K | \Psi)$ where Ψ denotes all of the parameters in (7.10). Then, we marginalize over K to obtain the density of the circular variable. Pewsey [38] gives the fundamental properties of the WSN along with closed forms for the cosine and sine moments.

Indeed, from (7.8) we can see that

$$
Y | D, \Psi \sim N \left(\mu + \frac{\sigma \lambda}{\sqrt{1+\lambda^2}} |D| - \frac{\sigma \lambda \sqrt{2}}{\sqrt{\pi(1+\lambda^2)}}, \frac{\sigma^2}{1+\lambda^2} \right) \qquad (7.11)
$$

and as a consequence

$$
\Theta | D, \Psi \sim WN \left(\mu + \frac{\sigma \lambda}{\sqrt{1+\lambda^2}} |D| - \frac{\sigma \lambda \sqrt{2}}{\sqrt{\pi(1+\lambda^2)}}, \frac{\sigma^2}{1+\lambda^2} \right). \qquad (7.12)
$$

TABLE 7.2

Wave data: mean CRPSs over 40 validation sets.

	Calm WS	Calm W	Storm WS	Storm W
Spatial	0.426	0.494	0.528	0.567
Temporal	0.520	0.628	0.446	0.476

Models are based on the wrapped skew normal (WS) and the wrapped normal (W)

The Wrapped Skew Gaussian Process

A natural way to construct a wrapped skew Gaussian process $\boldsymbol{\theta}(\mathbf{s}), \mathbf{s} \in \mathbb{R}^d$, is to start from a skew Gaussian process $Y(\mathbf{s})$ on the line and define, for each $\mathbf{s}$, $\boldsymbol{\theta}(\mathbf{s}) = Y(\mathbf{s}) \bmod 2\pi$, following the approach of [20]. To capture stationarity we use the skew Gaussian process proposed by [50]:

$$Y(\mathbf{s}) = \mu + \frac{\sigma\lambda}{\sqrt{1+\lambda^2}}|D(\mathbf{s})| + \frac{\sigma}{\sqrt{1+\lambda^2}}W(\mathbf{s}) - \frac{\sigma\lambda\sqrt{2}}{\sqrt{\pi(1+\lambda^2)}}. \qquad (7.13)$$

Here, $D(\mathbf{s})$ and $W(\mathbf{s})$ are independent zero mean Gaussian processes with isotropic parametric correlation functions, $\rho_d(h; \boldsymbol{\psi}_d)$ and $\rho_w(h; \boldsymbol{\psi}_w)$, respectively. Extension to a wrapped skew Gaussian space-time process is straightforward and is detailed in [33] while offering the following examples here.

7.2.5.1　Space-Time Analysis Using the Wrapped Skewed Gaussian Process

We fit the spatio-temporal wrapped skew process to two datasets. The first spans the period April 2010 between the 2nd at 00:00 and the 4th at 22:00, a calm period. The second spans the period April 2010 between the 5th at 00:00 and the 7th at 22:00, a storm period. We randomly select 220 spatial locations over the entire Adriatic area for both datasets, and we use 90% of the spatial locations and the first 48 time points to estimate models while the remaining locations and times are employed as the validation sets.

For each training set, we fitted a skew Gaussian process model and a wrapped Gaussian process model. We repeat the splitting procedure into training and validation sets 40 times and each time we compute the CRPS to compare the performance of the models. As prior distributions we use $\mu \sim U(0, 2\pi)$, $\gamma \sim U(-1, 1)$, $\psi_d \sim U(5^{-4}, 5^{-2})$, $\psi_w \sim U(10^{-3}, 10^{-1})$, where ψ_d and ψ_w are the decay parameters of the exponential covariance functions of $D(\mathbf{s})$ and $W(\mathbf{s})$, respectively, $\sigma^2 \sim U(0, 10)$ and a weak prior for λ, $\lambda \sim N(0, 100)$.

Again, employing the CRPS, for both validation sets under both sea states,

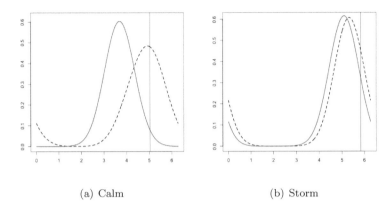

(a) Calm   (b) Storm

FIGURE 7.6
Examples of predictive distributions for one of the holdout sites in calm (a) and storm (b) sea state. The solid line is the predictive distribution under the wrapped Gaussian model while the dashed one is under the skew Gaussian model. The vertical line denotes the holdout value.

the wrapped skew Gaussian process shows a consequential gain in predictive ability compared with the standard wrapped Gaussian, see Table 7.2. Finally, Figure 7.6 shows predictive distributions for a holdout observation during a calm and during a storm state, respectively.

7.3 The Projected Gaussian Process

Here, we review the projected normal distribution and introduce the projected Gaussian spatial process and its properties. Then, we look at the distribution theory for fitting and kriging under the projected Gaussian process.

7.3.1 Univariate Projected Normal Distribution

Suppose a random vector $\mathbf{Y} = (Y_1, \ldots, Y_p)'$ follows a p-dimensional multivariate normal distribution, with mean $\boldsymbol{\mu}$ and covariance matrix $\boldsymbol{\Sigma}$ ($p \geq 2$). The corresponding random unit vector $\mathbf{U} = \mathbf{Y}/\|\mathbf{Y}\|$ is said to follow a *projected normal* distribution [31, 43] with the same parameters, denoted as $PN_p(\boldsymbol{\mu}, \boldsymbol{\Sigma})$. When $p = 2$ we have the circular projected normal distribution. Let $\phi_2(y_1, y_2; \boldsymbol{\mu}, \boldsymbol{\Sigma})$ be the density function of a bivariate normal distribution with mean vector $\boldsymbol{\mu} = (\mu_1, \mu_2)'$ and customary covariance matrix

$$\mathbf{\Sigma} = \begin{pmatrix} \sigma_1^2 & \rho\sigma_1\sigma_2 \\ \rho\sigma_1\sigma_2 & \sigma_2^2 \end{pmatrix}, \text{ (where } \rho \text{ is the correlation between the compo-}$$

nents). After transforming $\mathbf{Y}$ (equivalently $\mathbf{U}$) to an angular random variable Θ through $\Theta = \arctan^*(Y_2/Y_1) = \arctan^*(U_2/U_1),$[1] Θ can be shown [27, pp. 52] to have density

$$f(\theta|\boldsymbol{\mu}, \mathbf{\Sigma}) = \frac{\phi_2(\mu_1, \mu_2; \mathbf{0}, \mathbf{\Sigma})}{C(\theta)}$$

$$+ \frac{aD(\theta)\Phi_1\{D(\theta)\}\phi_1\left[a\{C(\theta)\}^{-\frac{1}{2}}(\mu_1\sin\theta - \mu_2\cos\theta)\right]}{C(\theta)} \quad (7.14)$$

where Φ_1 is the standard univariate normal, $a = \{\sigma_1\sigma_2\sqrt{1-\rho^2}\}^{-1}, C(\theta) = a^2(\sigma_2^2\cos^2\theta - \rho\sigma_1\sigma_2\sin 2\theta + \sigma_1^2\sin^2\theta)$, and $D(\theta) = a^2\{C(\theta)\}^{-\frac{1}{2}}\{\mu_1\sigma_2(\sigma_2\cos\theta - \rho\sigma_1\sin\theta) + \mu_2\sigma_1(\sigma_1\sin\theta - \rho\sigma_2\cos\theta)\}$.

The awkwardness of this distribution suggests that we introduce a latent length variable $R = \|\mathbf{Y}\|$ and work with the joint distribution of R and Θ, easily obtained through polar coordinate transformation from the joint distribution of Y_1 and Y_2. With $\mathbf{u} = (\cos\theta, \sin\theta)'$, it takes the form

$$f(r, \theta|\boldsymbol{\mu}, \mathbf{\Sigma}) = (2\pi)^{-1}|\mathbf{\Sigma}|^{-\frac{1}{2}}\exp\left(-\frac{(r\mathbf{u} - \boldsymbol{\mu})'\mathbf{\Sigma}^{-1}(r\mathbf{u} - \boldsymbol{\mu})}{2}\right) r. \quad (7.15)$$

The literature on the projected normal is small and has been confined to the special case $PN_2(\boldsymbol{\mu}, I)$ [16, 36, 37, 40], also known as the *displaced normal* [24]. Its corresponding density function is symmetric and unimodal, so $PN_2(\boldsymbol{\mu}, I)$ is comparable to other symmetric unimodal distributions such as the von Mises. Wang and Gelfand [46] study the projected normal family with a general covariance matrix $\mathbf{\Sigma}$ and refer to this richer class $PN(\boldsymbol{\mu}, \mathbf{\Sigma})$ as the *general projected normal* distribution. This version allows asymmetry and bimodality but is not identified; $\mathbf{U} = \mathbf{Y}/\|\mathbf{Y}\|$ is invariant to scale transformation. In $\mathbf{\Sigma}$, Wang and Gelfand [46] set $\sigma_1 = \tau$ and $\sigma_2 = 1$ to provide identifiability, resulting in a four-parameter $(\mu_1, \mu_2, \tau, \rho)$ distribution. They also demonstrate that the class of projected normal distributions is dense in the class of all densities obtained by projecting bivariate densities onto the unit circle.

7.3.2   Projected Gaussian Spatial Processes

By projecting a bivariate spatial process on $\mathbb{R}^2$ we can create a spatial stochastic process of random variables each taking values on a circle. Let $\mathbf{Y}(\mathbf{s}) = (Y_1(\mathbf{s}), Y_2(\mathbf{s}))'$ denote a 2×1 vector of random variables at location

---

[1]From [18, pp. 13], $\arctan^*(S/C)$ is formally defined as $\arctan(S/C)$ if $C > 0, S \geq 0$; $\pi/2$ if $C = 0, S > 0$; $\arctan(S/C) + \pi$ if $C < 0$; $\arctan(S/C) + 2\pi$ if $C \geq 0, S < 0$; undefined if $C = 0, S = 0$.

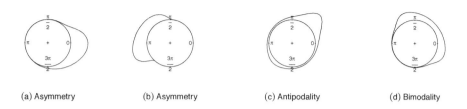

(a) Asymmetry   (b) Asymmetry   (c) Antipodality   (d) Bimodality

FIGURE 7.7
Density shapes for the general projected normal distribution.

$\mathbf{s}$, $\mathcal{D}$ be the domain of interest and $\{\mathbf{Y}(\mathbf{s}) : \mathbf{s} \in \mathcal{D}\}$ be a bivariate stochastic process. Letting $(\cos\Theta(\mathbf{s}), \sin\Theta(\mathbf{s}))' = (Y_1(\mathbf{s}), Y_2(\mathbf{s}))'/\|\mathbf{Y}(\mathbf{s})\|$, one obtains a circular process $\Theta(\mathbf{s})$. This projected process inherits properties of the inline bivariate process such as stationarity.

We let $\mathbf{Y}(\mathbf{s})$ be a bivariate Gaussian process with mean $\boldsymbol{\mu}(\mathbf{s})$ and cross-covariance function $C(\mathbf{s}, \mathbf{s}') = \text{cov}(\mathbf{Y}(\mathbf{s}), \mathbf{Y}(\mathbf{s}'))$. We define the induced circular process upon projection to be the *projected Gaussian process*. The multivariate distributional convenience of the Gaussian process enables straightforward distribution theory for the projected Gaussian process. Here, we consider a separable choice for the cross-covariance function, $C(\mathbf{s}, \mathbf{s}') = \varrho(\mathbf{s}, \mathbf{s}') \cdot T$, where ϱ is a valid correlation function and T is a 2×2 positive definite matrix. Since we are concerned only with the spatial dependence for the $\Theta(\mathbf{s})$ process, there is no need to introduce two correlation functions. In consideration of the identifiability constraint, T is set to be $\begin{pmatrix} \tau^2 & \rho\tau \\ \rho\tau & 1 \end{pmatrix}$.

The marginal distribution for the random variable at each location $\Theta(\mathbf{s})$ is a univariate projected normal. Formally, the joint distribution of $\Theta(\mathbf{s})$ and $\Theta(\mathbf{s}')$ is given by

$$f(\theta(\mathbf{s}), \theta(\mathbf{s}')) = \int_0^\infty \int_0^\infty f(r(\mathbf{s}), \theta(\mathbf{s}), r(\mathbf{s}'), \theta(\mathbf{s}')) \, dr(\mathbf{s}) dr(\mathbf{s}'). \quad (7.16)$$

The double integral in (7.16) has no explicit expression. Furthermore, we can not obtain a closed form for $f(\theta(\mathbf{s})|\theta(\mathbf{s}'))$.

Instead, to provide a feel for the scope of possible joint distributions, we examine three illustrative sets of parameters, reflecting marginals that are unimodal and symmetric, unimodal and asymmetric, bimodal, respectively. The joint density plots are shown in Figure 7.8; the rows reflect different cases of univariate marginals and the columns illustrate increasing levels of spatial dependence. Specifically, the parameters for the marginals are: $\mu_1 = -1, \mu_2 = 0, \tau = 1, \rho = 0$ for row 1, $\mu_1 = -1, \mu_2 = 0, \tau = 1, \rho = 0.4$ for row 2 and $\mu_1 = -0.32, \mu_2 = 0, \tau = 0.48, \rho = -0,62$ for row 3. For the columns, $\varrho(\mathbf{s}, \mathbf{s}')$ equals 0, 0.4, and 0.7 from left to right. The richness, departing from a symmetric case as with independent von Mises, wrapped normal, or $PN(\boldsymbol{\mu}, I)$, is evident. By column, $\varrho(\|\mathbf{s} - \mathbf{s}'\|)$ is the same, by row, $\boldsymbol{\mu}$ and T are the same. So, the nature of the joint distribution is more affected by changes in the latter than in the former.

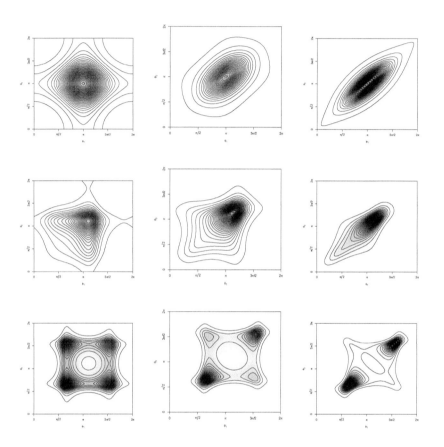

FIGURE 7.8

Bivariate joint distribution of $\Theta(\mathbf{s})$ and $\Theta(\mathbf{s}')$ of three different marginals (rows) and three different levels of spatial dependence (columns). Parameter values are given in the text.

The angular process $\{\Theta(\mathbf{s}), \mathbf{s} \in \mathcal{D}\}$ induces a multivariate circular distribution for $\boldsymbol{\Theta} = (\Theta(\mathbf{s}_1), \Theta(\mathbf{s}_2), \ldots, \Theta(\mathbf{s}_n))'$. First, we obtain the joint distribution of $\{\mathbf{Y}(\mathbf{s}_1), \ldots, \mathbf{Y}(\mathbf{s}_n)\}$ from the bivariate Gaussian process specification. For each location $\mathbf{s}_i$, $i = 1, \ldots, n$, we change variables to polar coordinates by letting $Y_1(\mathbf{s}_i) = R(\mathbf{s}_i) \cos \Theta(\mathbf{s}_i)$ and $Y_2(\mathbf{s}_i) = R(\mathbf{s}_i) \sin \Theta(\mathbf{s}_i)$. Integrating out $\mathbf{R} = (R(\mathbf{s}_1), \ldots, R(\mathbf{s}_n))'$ would yield the joint distribution of $\boldsymbol{\Theta} = (\Theta(\mathbf{s}_1), \ldots, \Theta(\mathbf{s}_n))'$. As noted above, we do not need to obtain this joint distribution. We introduce the latent vector of variables $\mathbf{R}$ and work with the joint distribution of $\boldsymbol{\Theta}$ and $\mathbf{R}$ which is available explicitly. Note the analogy with the introduction of latent winding numbers for the wrapped Gaussian process. Again, we see the power of employing latent variables.

The correlation behavior associated with the inline Gaussian process is inherited, in some fashion, by the projected Gaussian process. To understand the effect of the decay parameter ϕ in dictating the smoothness of the circular process, we obtained prior draws from the projected Gaussian process at 289 locations on a $[0,1] \times [0,1]$ grid for different values of ϕ and fixed values of $\boldsymbol{\mu}$ and T. In the Figure 7.9, ϕ increases (range decreases) from left to right; accordingly the corresponding circular process realization becomes less smooth.

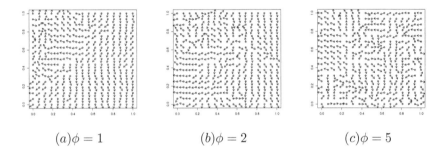

$(a)\phi = 1$ $\qquad\qquad$ $(b)\phi = 2$ $\qquad\qquad$ $(c)\phi = 5$

FIGURE 7.9
Simulated projected Gaussian process realizations for three values of ϕ.

7.3.3 Model Fitting and Inference

Suppose we have a projected Gaussian spatial process model, $\Theta(\mathbf{s})$, which is induced from a linear bivariate process $\mathbf{Y}(\mathbf{s})$ with mean $\boldsymbol{\mu}(\mathbf{s})$ and the separable cross-covariance $C(\mathbf{s}, \mathbf{s}') = \varrho(\mathbf{s}-\mathbf{s}'; \phi)\cdot T$. For simplicity, we only provide model fitting details with constant mean $\boldsymbol{\mu}(\mathbf{s}) \equiv \boldsymbol{\mu} = (\mu_1, \mu_2)'$ and an exponential correlation function in the cross-covariance denoted by $\varrho(\mathbf{s}-\mathbf{s}'; \phi) = e^{-\phi\|\mathbf{s}-\mathbf{s}'\|}$.

Again, $\mathbf{T} = \begin{pmatrix} \tau^2 & \rho\tau \\ \rho\tau & 1 \end{pmatrix}$. Therefore, we have five parameters: $\boldsymbol{\Psi} = (\mu_1, \mu_2, \tau,$ $\rho, \phi)$. For data $\boldsymbol{\theta} = (\theta(\mathbf{s}_1), \ldots, \theta(\mathbf{s}_n))'$, at each $\mathbf{s}_i$, we introduce $R(\mathbf{s}_i)$ and work in the space of $(\Theta(\mathbf{s}_i), R(\mathbf{s}_i))$, $i = 1, \ldots, n$. This data augmentation exploits the distributional convenience of the latent inline Gaussian process.

As a result, we have the corresponding "unobserved" latent linear variables $Y_1(\mathbf{s}_i) = R(\mathbf{s}_i)\cos\Theta(\mathbf{s}_i)$ and $Y_2(\mathbf{s}_i) = R(\mathbf{s}_i)\sin\Theta(\mathbf{s}_i)$. Denoting $\mathbf{Y}_1 = (Y_1(\mathbf{s}_1), \ldots, Y_1(\mathbf{s}_n))'$ and $\mathbf{Y}_2 = (Y_2(\mathbf{s}_1), \ldots, Y_2(\mathbf{s}_n))'$, $\mathbf{Y} = (\mathbf{Y}_1', \mathbf{Y}_2')'$ follows a multivariate normal with mean $(\mu_1 \mathbf{1}_{1\times n}, \mu_2 \mathbf{1}_{1\times n})'$ and covariance matrix $\tilde{\Sigma} = T \otimes \mathbf{H}(\phi)$, where $\{\mathbf{H}(\phi)\}_{j,k} = \varrho(\mathbf{s}_j - \mathbf{s}_k; \phi)$, $j, k = 1, \ldots, n$. From Section 7.3.2, the joint distribution of $\boldsymbol{\Theta}$ and $\mathbf{R}$ can be obtained by a change of variables from the joint distribution of $\mathbf{Y}_1$ and $\mathbf{Y}_2$. However, block updating of the latent $\mathbf{R}$'s still seems difficult. Instead, we utilize the properties of the inline GP to obtain the conditional distribution, $\mathbf{Y}(\mathbf{s}_i)|\mathbf{Y}(-\mathbf{s}_i), \boldsymbol{\Psi}$, thus inducing the conditional distribution $R(\mathbf{s}_i)|R(-\mathbf{s}_i), \boldsymbol{\Psi}$.

To complete a Bayesian formulation of our model, priors are needed for the model parameters. Conjugacy for $\boldsymbol{\mu}$ arises under a bivariate normal prior, e.g., $\boldsymbol{\mu} \sim N(\mathbf{0}, \lambda_0 I)$. For τ^2, we choose an inverse Gamma $IG(a_\tau, b_\tau)$ with mean $b_\tau/(a_\tau - 1) = 1$ while a uniform prior on $(-1, 1)$ is used for ρ. For the decay parameter ϕ of the exponential correlation function, we employ a uniform prior with support allowing ranges larger than the maximum distance over the region.

7.3.4 Kriging with the Projected Gaussian Processes

Given observations $\boldsymbol{\Theta} = (\Theta(\mathbf{s}_1), \ldots, \Theta(\mathbf{s}_n))'$, we seek to predict $\Theta(\mathbf{s}_0)$ at a new location $\mathbf{s}_0$. The conditional distribution for $\Theta(\mathbf{s}_0)|\boldsymbol{\Theta}$ can be expressed as

$$f(\theta(\mathbf{s}_0)|\boldsymbol{\theta}) = \int f(\theta(\mathbf{s}_0)|\mathbf{r}, \boldsymbol{\theta}) f(\mathbf{r}|\boldsymbol{\theta})\, d\mathbf{r}, \qquad (7.17)$$

where $\boldsymbol{\theta} = (\theta(\mathbf{s}_1), \ldots, \theta(\mathbf{s}_n))'$ and $\mathbf{r} = (r(\mathbf{s}_1), \ldots, r(\mathbf{s}_n))'$.

In order to derive $f(\theta(\mathbf{s}_0)|\boldsymbol{\Theta})$, we need to first obtain the integrand in (7.17). Let $\mathbf{Y}^* = (\mathbf{Y}'(\mathbf{s}_1), \ldots, \mathbf{Y}'(\mathbf{s}_n))'$ and $\mathbf{Y}(\mathbf{s}_0) = (Y_1(\mathbf{s}_0), Y_2(\mathbf{s}_0))'$. For the latent linear Gaussian process, the joint distribution for $\mathbf{Y}(\mathbf{s}_0)$ conditional on $\mathbf{Y}^*$ and parameters of the bivariate GP (denoted by $\boldsymbol{\Psi}$) is a bivariate normal distribution. Working with polar coordinates, we can obtain an analytical expression for $\Theta(\mathbf{s}_0), R(\mathbf{s}_0)|\mathbf{R}, \boldsymbol{\Theta}$. However, integration over $\mathbf{R}$ is intractable so we resort to Monte Carlo approximation. Thus, in the context of a Bayesian modeling framework, kriging can be implemented by drawing the appropriate latent variables and then utilizing the convenient distributional properties of the underlying inline bivariate Gaussian process.

We view the projected Gaussian spatial process model primarily as a tool for prediction. Interpretation of the parameters with regard to the distribution for the projected variable is difficult because of the complex interplay between the parameters. Moreover, these parameters also do not allow much interpretation on the linear scale since they are associated with latent variables. Therefore, we provide details of Bayesian kriging under the foregoing model specification. The joint distribution of $\mathbf{Y}(\mathbf{s}_0) = (Y_1(\mathbf{s}_0), Y_2(\mathbf{s}_0))'$ and $\mathbf{Y}^* = (Y_1(\mathbf{s}_1), Y_2(\mathbf{s}_1), \ldots, Y_1(\mathbf{s}_n), Y_2(\mathbf{s}_n))'$ is

$$\begin{pmatrix} \mathbf{Y}(\mathbf{s}_0) \\ \mathbf{Y}^* \end{pmatrix} \sim MVN\left(\begin{pmatrix} \boldsymbol{\mu}(\mathbf{s}_0) \\ \boldsymbol{\mu}^* \end{pmatrix}, \begin{pmatrix} 1 & \boldsymbol{\rho}'_{0,\mathbf{Y}}(\phi) \\ \boldsymbol{\rho}_{0,\mathbf{Y}}(\phi) & \mathbf{H}_{\mathbf{Y}}(\phi) \end{pmatrix} \otimes T \right), \qquad (7.18)$$

where $\boldsymbol{\mu}^* = (\boldsymbol{\mu}'(\mathbf{s}_1), \ldots, \boldsymbol{\mu}'(\mathbf{s}_n))'$, $\{\mathbf{H}_{\mathbf{Y}}(\phi)\}_{j,k} = \varrho(\mathbf{s}_j - \mathbf{s}_k; \phi)$ and $\{\boldsymbol{\rho}_{0,\mathbf{Y}}(\phi)\}_j = \varrho(\mathbf{s}_0 - \mathbf{s}_j; \phi)$ and MVN denotes the multivariate normal distribution.

Thus, the conditional distribution $\mathbf{Y}(\mathbf{s}_0)|\mathbf{Y}^*$ is a bivariate normal and the conditional distribution for $\Theta(\mathbf{s}_0)|\mathbf{Y}^*$ is a general projected normal. Therefore, we are able to draw samples from the predictive distribution in (7.17) using $\Theta(\mathbf{s}_0)|\mathbf{Y}^*$, equivalently, $\Theta(\mathbf{s}_0)|\Theta, \mathbf{R}$ as in expression (7.14).

With posterior draws of $\Theta(\mathbf{s}_0)$, we can infer about the mean direction and the mean resultant length of the predictive density at $\mathbf{s}_0$. By definition, the mean direction is $\omega = \arctan^*(\beta/\alpha)$ and the resultant length is $\gamma = \sqrt{(\alpha^2 + \beta^2)}$, where $\alpha = \mathrm{E}\cos\Theta$ and $\beta = \mathrm{E}\sin\Theta$. So, more explicitly, for each posterior draw, $\boldsymbol{\Psi}^{(l)}$ and $\mathbf{r}^{(l)}$, we draw a posterior sample $\mathbf{y}(\mathbf{s}_0)^{(l)}$ of $\mathbf{Y}(\mathbf{s}_0)$ from $N_2(E_{\mathbf{s}_0}^{(l)}, \Sigma_{\mathbf{s}_0}^{(l)})$, and then convert to $\theta(\mathbf{s}_0)^{(l)}$. Compute $\overline{C}(\mathbf{s}_0) = \Sigma_l \cos\theta(\mathbf{s}_0)^{(l)}/L$ and $\overline{S}(\mathbf{s}_0) = \Sigma_l \sin\theta(\mathbf{s}_0)^{(l)}/L$. Then, $\hat{\omega}(\mathbf{s}_0) = \arctan^*(\overline{S}(\mathbf{s}_0)/\overline{C}(\mathbf{s}_0))$ and $\hat{\gamma}(\mathbf{s}_0) = \sqrt{\overline{C}^2(\mathbf{s}_0) + \overline{S}^2(\mathbf{s}_0)}$.

7.3.4.1 An Example Using the Projected Gaussian Process

Again, we use data outputs from the deterministic wave model implemented by ISPRA (Istituto Superiore per la Protezione e la Ricerca Ambientale). These outputs are associated with deep waters and have a spatial resolution of 0.1 degree longitude, approximately on a grid with about 12.5×12.5 km cells. We selected a random set of 250 locations for illustration. We intentionally chose an irregular set of locations to provide an irregular set of interpoint distances for model fitting and kriging.

To illustrate a static spatial analysis, we utilize data from single time slices separately during a calm period and a stormy period in the Adriatic Sea choosing a single hour within each period. We randomly select 50 locations from the 250 for the purpose of validation. For the fitting, we again run the MCMC algorithm for 25,000 iterations, with a burn-in of 5000 and thinning by collecting every fifth sample. We compute kriging estimates using 4000 posterior samples. We adopt vague priors: $\boldsymbol{\mu} \sim N_2(\mathbf{0}, 4I)$, $\tau^2 \sim IG(2, 1)$ and $\rho \sim \mathrm{Unif}(-1, 1)$. We use a uniform prior, $\mathrm{Unif}(10^{-3}, 10^{-1})$ for ϕ, corresponding to a maximum and minimum range of 3000 km and 30 km respectively. Allowing the prior to have support over such large values of the range might seem inappropriate for the spatial scale of the process. However, the exceptionally strong spatial dependence for the directions, as exhibited in Figure 7.12, yields very slow estimated decay (see Figure 7.10) implying a range beyond the largest pairwise distance in our dataset.

In Figure 7.11 we present the posterior joint density for a pair $(\theta(\mathbf{s}), \theta(\mathbf{s}'))$ at two distances, 200 km and 1000 km, during the calm day. We see weaker dependence at the longer distance but it is still strong, consistent with the foregoing comment on long spatial ranges for the wave directions. Last, we provide model comparison between the general projected Gaussian model and the wrapped Gaussian model using average circular distance defined as $\frac{1}{m}\sum_{j=1}^{m}(1 - \cos(\hat{\theta}(\mathbf{s}_j^*) - \theta(\mathbf{s}_j^*)))$ where, in our notation, the $\hat{\theta}$'s are posterior predictive means under the model. For the calm time slice, the average circular distances are 0.0222 for the general projected Gaussian model and 0.3743

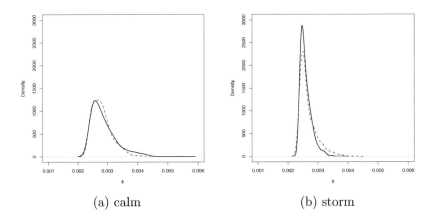

(a) calm   (b) storm

FIGURE 7.10
Posterior summaries of the decay parameter ϕ under the projected Gaussian
process model (solid line) and the reduced projected Gaussian process model
$T = I$ (dashed line) for two datasets.

for the wrapped Gaussian model; for the storm time slice, the average circular
distances are 0.0217 and 0.1516, respectively. The general projected Gaussian
model substantially outperforms the wrapped Gaussian model.

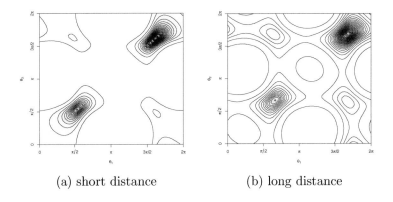

(a) short distance   (b) long distance

FIGURE 7.11
Posterior bivariate density plots for a pair of locations at short distance and
at long distance during a calm period.

7.3.5 The Space-Time Projected Gaussian Process

Wave directions are available over 24 time points, two hours apart; our goal is to model the spatial evolution of the wave directions over continuous time. Space-time modeling in the linear setting can be adapted for our projected Gaussian process setting; we specify a bivariate space-time Gaussian process model to induce a projected Gaussian space-time model. We imagine circular variables $\Theta(\mathbf{s}, t)$ over $(\mathbf{s}, t) \in \mathcal{D} \subset \mathbb{R}^2 \times \mathbb{R}^+$. For the bivariate space-time Gaussian process we again assume constant means, with a separable covariance structure,

$$C\left((\mathbf{s}, t), (\mathbf{s}', t')\right) = \varrho_{\mathbf{s}}(\mathbf{s} - \mathbf{s}'; \phi_{\mathbf{s}})\rho_t(t - t'; \phi_t) \cdot \mathbf{T}, \qquad (7.19)$$

where $\varrho_{\mathbf{s}}$ is the spatial correlation and ϱ_t is the temporal correlation. The parameters are now $\boldsymbol{\Psi} = (\boldsymbol{\mu}, T, \phi_{\mathbf{s}}, \phi_t)$, where $\phi_{\mathbf{s}}$ and ϕ_t are the decay parameters associated with their corresponding correlation functions and $\mathbf{T}$ is as in the previous section. Again, we only need one space-time covariance function.

Under this specification, we add one more parameter, ϕ_t, with a continuous prior, $\text{Unif}(3/(3t_{\max}), 3/(0.01t_{\max}))$. The model fitting is easily extended from that for the static spatial model and kriging proceeds similarly to that described above. We focus on the one-step ahead prediction; with values $t = t_1, t_2, ..., t_k$, we forecast to the time point t_{k+1} for the n observed locations.

We start from the joint distribution,

$$\left(\begin{array}{c} \mathbf{Y}_{t_{k+1}} \\ \mathbf{Y}_{t_1:t_k} \end{array} \right) \sim MVN \left(\left(\begin{array}{c} \boldsymbol{\mu}_{t_{k+1}} \\ \boldsymbol{\mu}_{t_1:t_k} \end{array} \right), \mathbf{H}_{k+1}(\phi_t) \otimes \mathbf{H}_n(\phi_{\mathbf{s}}) \otimes \mathbf{T} \right), \qquad (7.20)$$

where $\mathbf{Y}_{t_{k+1}} = (Y_1(s_1, t_{k+1}), Y_2(s_1, t_{k+1}), \ldots, Y_1(s_n, t_{k+1}), Y_2(s_n, t_{k+1}))'$, $\mathbf{Y}_{t_1:t_k} = (\mathbf{Y}'_{t_1}, \ldots, \mathbf{Y}'_{t_k})'$, $\mathbf{Y}_{t_j} = (Y_1(s_1, t_j), Y_2(s_1, t_j), \ldots, Y_1(s_n, t_j), Y_2(s_n, t_j))'$, $\mathbf{H}_{k+1}(\phi_t) = \{\varrho_t(t_j - t_q; \phi_t)\}$, $j, q = 1, \ldots, k+1$ and $\mathbf{H}_n(\phi_{\mathbf{s}}) = \{\varrho_{\mathbf{s}}(\mathbf{s}_i - \mathbf{s}_p; \phi_{\mathbf{s}})\}$, $i, p = 1, \ldots, n$. We obtain the conditional distribution $\mathbf{Y}_{t_{k+1}}|\mathbf{Y}_{t_1:t_k}, \boldsymbol{\Psi}$ and $\otimes$ is the Kronecker product. Then, posterior predictive samples of $\mathbf{Y}_{t_{k+1}}$ can be obtained using posterior samples of the parameters.

The case of a nonseparable covariance function is considered in [34]. Again, we employ a bivariate linear process but now we specify the dependence structure through a covariance function of the form $\text{Cov}(\mathbf{Y}(\mathbf{s}_i, t_j), \mathbf{Y}(\mathbf{s}_{i'}, t_{j'})) = \text{Cor}(\mathbf{h}_{i,i'}, u_{j,j'})\mathbf{T}$. An example of $\text{Cor}(\mathbf{h}_{i,i'}, u_{j,j'})$ is provided in (7.21) below.

7.3.6 A Separable Space-Time Wave Direction Data Example

To illustrate the separable case, we apply our model to data during a calm period and during a stormy period, each over 24 time points two hours apart. For the calm period, wave directions are outputs from April 2, 2010 and April 3, 2010, for the storm period, from April 5, 2010 and April 6, 2010. The observations are provided at regular discrete time points, that is, 00:00, 02:00 ,..., 22:00. We use the same collection of locations as in the static example. Again, our goal is to illustrate prediction at a future time point. Therefore, we hold out data for the last time point, $t = 24$, and fit the model using the data from the previous 23 time points. The same set of $n = 200$ locations was used (shown in Figure 7.12). We provide the predictions at $t = 24$ for each of these locations. In Figure 7.12 we graphically compare the observations at $t = 24$ and the posterior mean directions at each of the 200 locations, with (a) for the calm period and (b) for the storm period. Arrow length captures the relative concentration of the posterior estimate.

Calm period, 22:00 on April 3rd, 2010

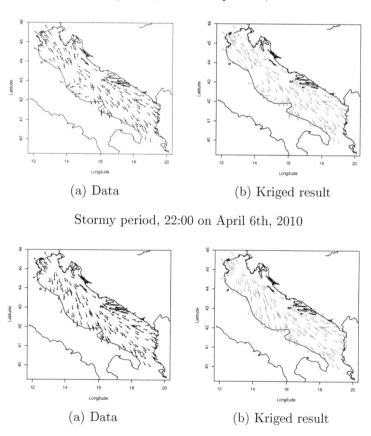

Stormy period, 22:00 on April 6th, 2010

FIGURE 7.12
Comparison of hold out data and model fitting results for the PN separable space-time model.

7.4   Space-Time Comparison of the WN and PN Models

Here, we fit and compare nonseparable space-time models for both the WN and PN models. We choose a nonseparable covariance function from the class proposed by Gneiting in [13]. This is a stationary function with variance σ^2 and correlation function in the form:

$$\text{Cor}(Y(\mathbf{s},t), Y(\mathbf{s}',t')) \equiv \rho(\mathbf{h}, u) = \frac{1}{(a|u|^{2\alpha}+1)^\tau} \exp\left(-\frac{c\|\mathbf{h}\|^{2\gamma}}{(a|u|^{2\alpha}+1)^{\beta\gamma}}\right),\tag{7.21}$$

where $(\mathbf{h}, u) \in \mathbb{R}^d \times \mathbb{R}$, $\mathbf{h} = \mathbf{s} - \mathbf{s}'$ and $u = t - t'$. Here, a and c are non-negative scaling parameters for time and space, respectively. The smoothness parameters α and γ take values in $(0, 1]$, and the space-time interaction parameter β is in $[0, 1]$. Attractively, as β tends to 0, we tend to separability in space and time.

To complete the model specification we need to supply prior distributions. We adopt the following choices. Since a and c are positive, a and $c \sim \Gamma(\cdot, \cdot)$. Since α, β, and γ are bounded between 0 and 1, we adopt a beta distributions $(B(\cdot, \cdot))$. Priors for the variances and the mean direction are given the usual normal-inverse gamma forms, i.e., $\sigma_Y^2, \phi_Y^2 \sim Inv\Gamma(\cdot, \cdot)$ and $\mu_y \sim WrapN(\cdot, \cdot)$. We use the same priors for the covariance function parameters for both the WN and PN models. Specifically: $a \sim \Gamma(1.5, 1)$, $c \sim \Gamma(1.5, 1)$, $\alpha \sim B(2, 2.5)$, $\beta \sim B(1.1, 2)$, $\gamma \sim B(2, 2.5)$, $\sigma_Y^2 \sim Inv\Gamma(2, 2)$, $\phi_Y^2 \sim Inv\Gamma(1, 0.25)$, $\mu_Y \sim WrapN(\pi, 10)$, $\mu_{Z_1} \sim N(0, 10)$, $\mu_{Z_2} \sim N(0, 10)$, $\rho_Z \sim N(0, 5)I(-1, 1)$, $\sigma_Z^2 \sim Inv\Gamma(2, 2)$ and $\phi_Z^2 \sim Inv\Gamma(1, 0.25)$. All priors are weakly informative. Also, the prior for β is centered near 0.1, i.e., close to the separable model. Decay parameter priors in space and time are related to the minimum and maximum distances in space and time, chosen to ensure that the probability mass is concentrated over such intervals.

We again selected the datasets shown in Figure 7.2 which report the three sea states, *calm, transition*, and *storm*. Here, we seek to learn about the spatio-temporal structure of the data relying only on the specification of the correlation function. We fitted the model using 100 spatial points $\times$ 10 time points six hours apart (1000 observations in total) in order to have a dataset including all sea states. Notice that spatial distances are evaluated in kilometres. Then, we developed four validation datasets, each with 350 spatial points and 1 time point. Specifically, we have one dataset for each sea state plus one for a one-step forward prediction. Finally, we used the model fitted over the 1000 points to predict each validation dataset. Three of the datasets are inside the time window used for model estimation, one in calm sea, one in transition and one during a storm. The fourth validation set is at 12:00 on May 7, 2010, 6 hours after the last time used for model fitting. The observed circular process in each of these four time windows can be seen in Figure 7.2. Again for each time window and model we computed the mean CRPS and APE, see Table

TABLE 7.3
Wave data: CRPS and APE for the WN and PN models computed on each
validation dataset.

		WN	PN
Average	CRPS	0.655	0.629
	APE	0.437	0.421
Calm	CRPS	1.450	1.398
	APE	0.995	0.973
Transition	CRPS	0.082	0.074
	APE	0.033	0.028
Storm	CRPS	0.063	0.042
	APE	0.026	0.009
One-step prediction	CRPS	1.024	1.001
	APE	0.693	0.674

TABLE 7.4
Wave data: mean point estimate (PE) and 95% credible interval (CI) for the
correlation parameters for the WN and PN models

		WN	PN
a	PE	0.076	0.009
	(CI)	(0.019,0.200)	(0.005,0.019)
c	PE	3.2×10^{-4}	1.4×10^{-4}
	(CI)	$(1.3 \times 10^{-4}, 7.1 \times 10^{-4})$	$(7.0 \times 10^{-4}, 2.9 \times 10^{-4})$
α	PE	0.495	0.693
	(CI)	(0.288,0.744)	(0.562,0.819)
β	PE	0.592	0.430
	(CI)	(0.158,0.915)	(0.101,0.774)
γ	(PE)	0.797	0.872
	(CI)	(0.697 0.897)	(0.779,0.939)

7.3. Furthermore, we computed the mean CRPS and APE over the 4 time
windows.

Both models show the same behavior in terms of CRPS and APE: they
produce poorer estimates in a calm state, the variance being larger than in
other states. However, estimation is very accurate during a storm or a transition for both models. The PN always performs better that the WN. The
largest difference between the APE values of the two models (0.022) is observed during the calm sea time window.

In Table 7.4 we give credible intervals and posterior mean estimates for the
value of the parameters of the correlation function. For both models nonseparable correlation structure is strongly supported. The point estimates of the
spatial (c) and temporal (a) decay are smaller in the PN model. Notice that

data are bimodal whenever the wave directions look like those in Figure 7.2 (c) and (d), i.e., when over a large region at a given time a storm is rotating or two different weather systems are meeting. Then, scalar statistics, such as the overall mean direction or the overall concentration, may not be informative enough regarding this behavior.

To analyze the local behavior of model fitting, in Figure 7.13 we report CRPS surfaces, evaluated in calm, transition and storm for both models. Here, we simulate 2000 values from the estimated model over all locations in order to compute CRPS. Then, we smooth the surfaces using multilevel B-splines as implemented in the `mba.surf` function [11, 26]. Again we find the same behavior: larger CRPS values when the phenomenon variability is larger. Further, we can identify areas that are estimated, better or poorer, under each model.

7.5   Joint Modeling of Wave Height and Direction

Here, we extend our interest to include wave heights along with outgoing wave directions and provide a framework for joint modeling of these two measurements. More generally, we offer a modeling approach for joint spatial and spatio-temporal analysis of an angular and a linear variable. Figure 7.14 displays both the outgoing wave directions and wave heights over a region of the Adriatic Sea, for a subset of available locations, at a particular time point during a storm. The wave heights are displayed in the image plot, as an additional layer over the arrow plots of the directions. Perhaps not surprising, there is strong visual evidence of spatial dependence for both heights and directions, although with quite different patterns. So, it is natural to build a joint model to accommodate the underlying association between them. Modeling linear and circular variables in a marine context has been considered in a likelihood framework by [25] and [8] (cross-reference to F. Lagona's Chapter 3).

Here, the contribution is to develop a fully model-based approach to capture joint structured spatial dependence for modeling linear data and directional data at different spatial locations. We employ a projected Gaussian process for directions and, given direction, a linear Gaussian process for heights. We show that Bayesian model fitting under such specification is straightforward using a suitable data augmented Markov chain Monte Carlo (MCMC) algorithm. This joint modeling framework allows natural extension to space-time data and can directly incorporate space-time covariate information, enabling both spatial interpolation and temporal forecasting.

In this regard, we build our joint model through a conditional times marginal specification where the marginal specification is a space or space-time directional data process and then, conditionally, we specify a space or space-time linear process. The only other spatial work involving a circular-

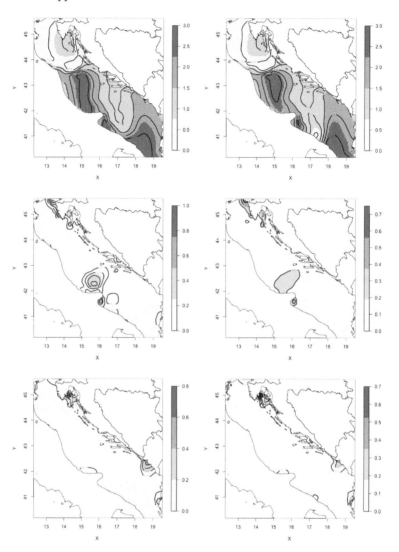

FIGURE 7.13
Wave data: CRPS surfaces for the space-time nonseparable WN (first column) and PN (second column) models, under calm (first row), transition (second row), and storm (third row) states. Scales differ across states

linear regression model we are aware of is the paper of [35]. They model wind angle and wind speed given wind angle. They model wind angle as a wrapped normal model, introducing conditionally autoregressive (CAR) spatial random effects. They model wind speed on the log scale, conditional on wind angle with a cosine link, again adding CAR spatial random effects. For our

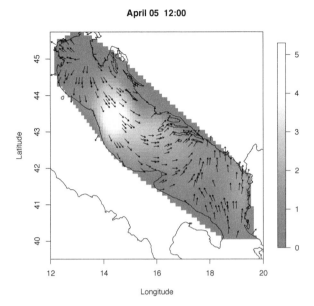

April 05 12:00

FIGURE 7.14
Plot of wave heights (meters) and wave directions for 200 locations at 12:00 on April 5, 2010.

setting with wave directions, a projected Gaussian process seems physically more appropriate than a wrapping model.

In the literature to date, the discussion usually focuses on *conditional* modeling. For instance, if the angular variable Θ depends on linear variables, it is referred to as a *Linear-Circular regression* model, a regression specification that uses the linear covariates to explain a directional response. Such modeling typically adopts the von Mises distribution with a suitable link function, e.g., the arctan function, that maps each regressor to $(-\pi, \pi)$ [12, 19]. If the linear variable X depends on independent angular variables, Θ's, we have a regression model with a linear response and angular covariates, known as a *Circular-Linear regression* model. Again, a link function is employed [29].

We begin with a single linear variable and a single directional variable which, illustratively, we refer to as the wave height and wave direction in the sequel. We build a joint parametric model for the wave height H and the wave direction Θ, recalling the latent variable R, in the form

$$f(H, \Theta | \boldsymbol{\Psi}_h, \boldsymbol{\Psi}_\theta) = \int f(H, \Theta, R | \boldsymbol{\Psi}_h, \boldsymbol{\Psi}_\theta) \, dR$$

$$= \int f(H | \Theta, R, \boldsymbol{\Psi}_h) f(\Theta, R | \boldsymbol{\Psi}_\theta) \, dR, \qquad (7.22)$$

where $\boldsymbol{\Psi}_h$ and $\boldsymbol{\Psi}_\theta$ are sets of parameters associated with the conditional model for height and the marginal model for direction, respectively, and are elaborated below. The model for $f(\Theta, R|\boldsymbol{\Psi}_\theta)$ is the projected normal Gaussian process from the previous section so needs no further detail.

Here, the conditional density, $f(H|\Theta, R, \boldsymbol{\Psi}_h)$ arises as a univariate Bayesian spatial regression (a customary *geostatistical* model) for the wave height $H(\mathbf{s})|\Theta(\mathbf{s}), R(\mathbf{s}), \boldsymbol{\Psi}_h$, with $\Theta(\mathbf{s})$ and $R(\mathbf{s})$ included in the mean through a link function $g(\cdot)$,

$$H(\mathbf{s}) = g(\Theta(\mathbf{s}), R(\mathbf{s})) + w(\mathbf{s}) + \epsilon(\mathbf{s}).$$

As usual, the residual is partitioned into the spatial effect term $w(\mathbf{s})$ and the non-spatial error term $\epsilon(\mathbf{s})$, where $w(\mathbf{s})$ is assumed to follow a zero mean stationary Gaussian process with covariance function $C_h(\mathbf{s} - \mathbf{s}')$ and $\epsilon(\mathbf{s})$'s are uncorrelated pure errors. Under this conditional specification, we are actually implementing a regression model with both circular (Θ) and linear (R) covariates. A natural choice for the link function $g(\cdot)$ will revert to the linear regression on $Y_1(\mathbf{s})$ and $Y_2(\mathbf{s})$ from the "unobserved" inline Gaussian process $\mathbf{Y}(\mathbf{s})$. Explicitly, we have

$$H(\mathbf{s}) = \beta_0 + \beta_1 R(\mathbf{s}) \cos \Theta(\mathbf{s}) + \beta_2 R(\mathbf{s}) \sin \Theta(\mathbf{s}) + w(\mathbf{s}) + \epsilon(\mathbf{s}) \qquad (7.23)$$
$$= \beta_0 + \beta_1 Y_1(\mathbf{s}) + \beta_2 Y_2(\mathbf{s}) + w(\mathbf{s}) + \epsilon(\mathbf{s}), \qquad (7.24)$$

where the spatial random effect $w(\mathbf{s})$ follows a zero-centered GP with covariance function $C_h = \sigma_h^2 \varrho_h(\mathbf{s} - \mathbf{s}'; \phi_h)$ and the error term $\epsilon(\mathbf{s}) \overset{\text{iid}}{\sim} N(0, \tau_h^2)$.

Prior specification, model fitting, and kriging follow similarly to that of the previous sections. Full details and an example are supplied in [48]. We note that, altogether, $(H(\mathbf{s}), Y_1(\mathbf{s}), Y_2(\mathbf{s}))'$ specifies a trivariate Gaussian process whose mean structure and cross-covariance structure, under the above specification, are easily calculated.

We can envision an underlying process for heights and directions in continuous space and time; fitting such a model would enable both interpolation and forecasting at future time points. So now, denote the wave height and wave direction measurements at location $\mathbf{s}$ and time t as $H(\mathbf{s}, t)$ and $\Theta(\mathbf{s}, t)$. Again, this joint space-time model is provided conditionally. For the marginal distribution of the circular variables $\Theta(\mathbf{s}, t)|\boldsymbol{\Psi}_\theta$, we propose a spatio-temporal projected Gaussian process with a constant mean $\boldsymbol{\mu} = (\mu_1, \mu_2)'$ and, for illustration, a separable cross-covariance form (as in Section 7.3.5),

$$C_\theta((\mathbf{s}, t), (\mathbf{s}', t')) = \varrho_{\theta,\mathbf{s}}(\mathbf{s} - \mathbf{s}'; \phi_{\theta,\mathbf{s}}) \varrho_{\theta,t}(t - t'; \phi_{\theta,t}) \cdot T, \qquad (7.25)$$

where $\varrho_{\theta,\mathbf{s}}$ is the spatial correlation and $\varrho_{\theta,t}$ is the temporal correlation. This directional data model has parameters $\boldsymbol{\Psi}_\theta = (\boldsymbol{\mu}, T, \phi_{\theta,\mathbf{s}}, \phi_{\theta,t})$, $\phi_{\theta,\mathbf{s}}$ and $\phi_{\theta,t}$ and again, $T = \begin{pmatrix} \tau_\theta^2 & \rho\tau_\theta \\ \rho\tau_\theta & 1 \end{pmatrix}$.

At the conditional level, we introduce

$$H(\mathbf{s},t) = \beta_0 + \beta_1 Y_1(\mathbf{s},t) + \beta_2 Y_2(\mathbf{s},t) + w(\mathbf{s},t) + \epsilon(\mathbf{s},t). \tag{7.26}$$

In fact, we might simplify $w(\mathbf{s},t)$ to $w(\mathbf{s})$ so that the spatial random effect $w(\mathbf{s})$ follows a zero-centered GP with covariance function $C_h = \sigma_h^2 \varrho_h(\mathbf{s} - \mathbf{s}'; \phi_h)$. Regardless, the independent and identically distributed random error terms $\epsilon(\mathbf{s},t) \sim N(0, \tau_h^2)$. Now the parameters associated with the model for the heights are $\boldsymbol{\Psi}_h = \{\boldsymbol{\beta}, \phi_h, \sigma_h^2, \tau_h^2\}$. Again, prior specification, model fitting, and kriging follow similarly to that of the previous section (see [48]).

7.6　Concluding Remarks

Analyzing directional data presents many novel challenges. In this chapter we have taken up the issue of exploring directional data observed in space and time. Our motivation is wave direction data. The key points we have focused on here are as follows:

- Two classes of models are attractive for analyzing such data, a wrapped process and a projected process. In some applications, as in the case where directions are induced by time, e.g., time of day, it may be most appropriate to wrap linear time to a circle, e.g., to clock time, creating, upon rescaling to $(0, 2\pi]$, circular data. Examples include times of crimes with spatial locations or times of hospital admissions with addresses of patients. In other applications, it may be most natural to project a linear process onto a circle as perhaps with wind or wave directions.

- In particular, starting from linear variables and wrapping or projecting offers modeling convenience. That is, spatial and space-time data require multivariate specifications capturing space or space-time dependence. So, wrapping or projecting linear multivariate Gaussian variables provides attractive multivariate circular distributions. Arguably, this is easier and more flexible than building say multivariate von Mises distributions.

- Furthermore, since our need is for modelling realizations of a stochastic process over space and time, it is attractive to start with linear Gaussian processes and let the resulting circular processes inherit the dependence structure from the linear process. The resulting wrapped Gaussian process enables straightforward interpretation of parameters while the resulting projected Gaussian process, though less interpretable, is a bit more flexible.

- We have presented these models as hierarchical specifications, fitted within a Bayesian framework. This enables full posterior inference for both parameters and predictions or forecasts, avoiding possibly inappropriate asymptotics. We have shown that fitting can be done straightforwardly through the inclusion of a suitable process of latent variables, i.e., winding numbers for the wrapped Gaussian process and lengths for the projected Gaussian process. We have also demonstrated that spatial prediction – kriging – is easily done as a post-model fitting activity.

- The flexibility of the hierarchical specification typically leads to exploration of many models. So, we have proposed model comparison criteria, Average Predictive Error, and Continuous Rank Probability Scores, to accomplish such selection.

- Either of our proposed space-time directional data models can be readily used in applications. Again, our working example involved wave directions and wave heights.

Bibliography

[1] A. Azzalini. A class of distributions which includes the normal ones. *Scandinavian Journal of Statistics*, 12:171–178, 1985.

[2] A. Azzalini. The skew-normal distribution and related multivariate families. *Scandinavian Journal of Statistics*, 32(2):159–188, 2005.

[3] A. Azzalini and A. Capitanio. Statistical applications of the multivariate skew normal distribution. *Journal of the Royal Statistical Society. Series B (Statistical Methodology)*, 61(3):pp. 579–602, 1999.

[4] A. Azzalini and A. Dalla Valle. The multivariate skew-normal distribution. *Biometrika*, 83(4):715–726, 1996.

[5] Y. Baba. Statistics of angular data : wrapped normal distribution model (in japanese). In *Proceedings of the Institute of Statistical Mathematics*, volume 28, pages 179–195, 1981.

[6] S. Banerjee, A. E. Gelfand, and B. P. Carlin. *Hierarchical Modeling and Analysis for Spatial Data*. Chapman & Hall/CRC, New York, second edition, December 2014.

[7] L. Bao, T. Gneiting, E. P. Grimit, P. Guttorp, and A. E. Raftery. Bias correction and Bayesian model averaging for ensemble forecasts of surface wind direction. *Monthly Weather Review*, 138(5):1811–1821, 2015/02/09 2009.

[8] J. Bulla, F. Lagona, A. Maruotti, and M. Picone. A multivariate hidden Markov model for the identification of sea regimes from incomplete skewed and circular time series. *Journal of Agricultural, Biological, and Environmental Statistics*, 17(4):544–567, 2012.

[9] S. Coles. Inference for circular distributions and processes. *Statistics and Computing*, 8(2):105–113, 1998.

[10] C. Engel and E. Ebert. Performance of hourly operational consensus forecasts (ocfs) in the Australian region. *Weather and Forecasting*, 22(6):1345–1359, 2015/02/09 2007.

[11] A. O. Finley and S. Banerjee. *MBA: Multilevel B-spline Approximation*, 2014. R package version 0.0-8.

[12] N. I. Fisher and A. J. Lee. Regression models for an angular response. *Biometrics*, 48(3):665–677, 1992.

[13] T. Gneiting. Nonseparable, Stationary Covariance Functions for Space-Time Data. *Journal of the American Statistical Association*, 97(458):590–600, 2002.

[14] E. P. Grimit, T. Gneiting, V. J. Berrocal, and N. A. Johnson. The continuous ranked probability score for circular variables and its application to mesoscale forecast ensemble verification. *Quarterly Journal of the Royal Meteorological Society*, 132(621C):2925–2942, 2006.

[15] D. Harrison and G. K. Kanji. The development of analysis of variance for circular data. *Journal of Applied Statistics*, 15:197–224, 1988.

[16] D. Hernandez-Stumpfhauser, F. J. Breidt, and J. D. Opsomer. Hierarchical Bayesian small area estimation for circular data. *Canadian Journal of Statistics*, 44(4):416–430, 2016.

[17] H. Holzmann, A. Munk, M. Suster, and W. Zucchini. Hidden Markov models for circular and linear-circular time series. *Environmental and Ecological Statistics*, 13(3):325–347, 2006.

[18] S. Rao Jammalamadaka and A. SenGupta. *Topics in Circular Statistics*. World Scientific, Singapore, 2001.

[19] R. Johnson and T. Wehrly. Measures and models for angular correlation and angular-linear correlation. *Journal of the Royal Statistical Society Series B*, 39:222–229, 1977.

[20] G. Jona Lasinio, A. E. Gelfand, and M. Jona Lasinio. Spatial analysis of wave direction data using wrapped Gaussian processes. *Annals of Applied Statistics*, 6(4):1478–1498, 2012.

[21] G. Jona Lasinio, A. Orasi, F. Divino, and P. Conti. Statistical contributions to the analysis of environmental risks along the coastline. In *Società Italiana di Statistica - rischio e previsione. Venezia, 6-8 giugno*, pages 255–262. Società Italiana di Statistica, CLEUP, ISBN/ISSN: 978-88-6129-093-8., 2007.

[22] E. Kalnay. *Athmospheric Modeling, Data Assimilation and Predictability.* Cambridge University Press, Cambridge, 2002.

[23] S. Kato, K. Shimizu, and G. S. Shieh. A circular-circular regression model. *Statistica Sinica*, 18:633–645, 2008.

[24] D. G. Kendall. Pole-seeking brownian motion and bird navigation. *Journal of the Royal Statistical Society. Series B (Statistical Methodology)*, 36(3):365–417, 1974.

[25] F. Lagona, M. Picone, A. Maruotti, and S. Cosoli. A hidden Markov approach to the analysis of space-time environmental data with linear and circular components. *Stochastic Environmental Research and Risk Assessment*, 29(2):397–409, 2015.

[26] S. Lee, G. Wolberg, and S. Y. Shin. Scattered data interpolation with multilevel b-splines. *IEEE Transactions on Visualization and Computer Graphics*, 3(3):228–244, 1997.

[27] K. V. Mardia. *Statistics of Directional Data.* Academic Press, London and New York, 1972.

[28] K. V. Mardia. Statistics of directional data. *Journal of the Royal Statistical Society Series B*, 37:349–393, 1975.

[29] K. V. Mardia. Linear-Circular Correlation Coefficients and Rhythmometry. *Biometrika*, 63(2), 1976.

[30] K. V. Mardia, G. Hughes, C. C. Taylor, and H. Singh. A multivariate von Mises distribution with applications to bioinformatics. *Canadian Journal of Statistics*, 36(1):99–109, 2008.

[31] K. V. Mardia and P. E. Jupp. *Directional Statistics.* John Wiley and Sons, Chichster, 1999.

[32] K.V. Mardia. Characterizations of directional distributions. In G. P. Patil, S. Kotz, and J.K. Ord, editors, *A Modern Course on Statistical Distributions in Scientific Work*, volume 17 of *NATO Advanced Study Institutes Series*, pages 365–385. Springer Netherlands, 1975.

[33] G. Mastrantonio, A. E. Gelfand, and G. Jona Lasinio. The wrapped skew Gaussian process for analyzing spatio-temporal data. *Stochastic Environmental Research and Risk Assessment*, 30(8):2231–2242, 2016.

[34] G. Mastrantonio, G. Jona Lasinio, and A. E. Gelfand. Spatio-temporal circular models with non-separable covariance structure. *TEST*, 25:331–350, 2016.

[35] D. Modlin, M. Fuentes, and B. Reich. Circular conditional autoregressive modeling of vector fields. *Environmetrics*, 23(1):46–53, 2012.

[36] G. Nuñez-Antonio and E. Gutiérrez-Peña. A Bayesian analysis of directional data using the projected normal distribution. *Journal of Applied Statistics*, 32(10):995–1001, 2005.

[37] G. Nuñez-Antonio, E. Gutierrez-Peña, and G. Escarela. A Bayesian regression model for circular data based on the projected normal distribution. *Statistical Modeling*, 11(3):185–201, 2011.

[38] A. Pewsey. The wrapped skew-normal distribution on the circle. *Communications in Statistics - Theory and Methods*, 29(11):2459–2472, 2000.

[39] A. Pewsey. Modelling asymmetrically distributed circular data using the wrapped skew-normal distribution. *Environmental and Ecological Statistics*, 13(3):257–269, 2006.

[40] B. Presnell, S. P. Morrison, and R. C. Littell. Projected multivariate linear models for directional data. *Journal of the American Statistical Association*, 93(443):1068–1077, 1998.

[41] E. H. Sánchez and B. Scarpa. A wrapped flexible generalized skew-normal model for a bimodal circular distribution of wind direction. *Chilean Journal of Statistics*, 3(2):129–141, 2012.

[42] H. Singh, V. Hnizdo, and E. Demchuk. Probabilistic model for two dependent circular variables. *Biometrika*, 89(3):719–723, 2002.

[43] C. G. Small. *The Statistical Theory of Shape*. Springer New York, 1996.

[44] A. Speranza, C. Accadia, M. Casaioli, S. Mariani, G. Monacelli, R. Inghilesi, N. Tartaglione, P. M. Ruti, A. Carillo, A. Bargagli, G. Pisacane, F. Valentinotti, and A. Lavagnini. Poseidon: An integrated system for analysis and forecast of hydrological, meteorological and surface marine fields in the Mediterranean area. *Nuovo Cimento*, 27(C):329–345, 2004.

[45] A. Speranza, C. Accadia, S. Mariani, M. Casaioli, N. Tartaglione, G. Monacelli, P. M. Ruti, and A. Lavagnini. Simm: An integrated forecasting system for the Mediterranean area. *Meteorological Applications*, 14(4):337–350, 2007.

[46] F. Wang and A. E. Gelfand. Directional data analysis under the general projected normal distribution. *Statistical Methodology*, 10(1):113–127, 2013.

[47] F. Wang and A. E. Gelfand. Modeling space and space-time directional data using projected Gaussian processes. *Journal of the American Statistical Association*, 109(508):1565–1580, 2014.

[48] F. Wang, A. E. Gelfand, and G. Jona Lasinio. Joint spatio-temporal analysis of a linear and a directional variable: space-time modeling of wave heights and wave directions in the Adriatic Sea. *Statistica Sinica*, 25(1):25–39, 2015.

[49] D. S. Wilks. Comparison of ensemble-mos methods in the Lorenz '96 setting. *Meteorological Applications*, 13:243–256, 2006.

[50] H. Zhang and A. El-Shaarawi. On spatial skew-Gaussian processes and applications. *Environmetrics*, 21(1):33–47, 2010.

8

Cylindrical Distributions and Their Applications to Biological Data

Toshihiro Abe[1] **and Ichiro Ken Shimatani**[2]

[1]*Nanzan University, Nagoya, Japan*
[2]*The Institute of Statistical Mathematics, Tokyo, Japan*

CONTENTS

8.1	Introduction	164
8.2	Example: Commonly Observed Patterns for Cylindrical Data ..	165
8.3	A Brief Review of the Univariate Probability Distribution	165
	8.3.1 Probability Distributions on $[0, \infty)$	165
	8.3.2 Circular Distributions	167
	8.3.3 Sine-Skewed Perturbation to the Symmetric Circular Distributions	168
8.4	Cylindrical Distributions	168
	8.4.1 The Johnson–Wehrly Distribution	168
	8.4.2 The Weibull–von Mises Distribution	169
	8.4.3 Gamma–von Mises Distribution	170
	8.4.4 Generalized Gamma–von Mises Distribution	171
	8.4.5 Sine-Skewed Weibull–von-Mises Distribution	172
	8.4.6 Parameter Estimation	173
8.5	Application 1: Quantification of the Speed/Turning Angle Patterns of a Flying Bird	173
8.6	Application of Cylindrical Distributions 2: How Trees Are Expanding Crowns	174
	8.6.1 Crown Asymmetry in Boreal Forests	176
	8.6.2 Crown Asymmetry Model	177
	8.6.3 Results of the Cylindrical Models	180
8.7	Concluding Remarks	184
	Acknowledgments	184
	Bibliography	185

8.1 Introduction

Correlation/covariance is a fundamental concept, and multivariate analyses often begin by investigating their presence in given multivariate data. In particular, if an objective variable is not a scalar but a vector with correlations, to construct a statistical model, we need multivariate probability distributions that can flexibly capture the correlations between the objective variables. For example, the multivariate Gaussian distribution is used in the Gaussian process and the geostatistical process (Cressie & Wikle, 2011) to express stochastic uncertainty in temporally or spatially correlated objective variables.

If the objective variable is circular, we need multivariate versions of circular probability distributions, and in the two-dimensional case, there are two types. The first is when two objective variables are circular; then, the probability distribution is defined on the torus. The second type is when one is circular and the other is linear; then, the probability distribution should be defined on $S^1 \times \mathbb{R}$, or if linear variables take only non-negative values, the domain is $S^1 \times [0, \infty)$. In both cases, the probability distribution is defined on a "cylinder" and such distributions are called "cylindrical distributions."

The latter type of data are commonly seen, for example, in velocity data, which consist of a speed and a direction. Often, these show specific correlations. For example, the directions are concentrated on a specific direction when speeds are high; on the contrary, if slow, the directions are diverse. Similar patterns can be seen in moving objects; turning angles are close to zero if speeds are high, while the object may turn to any direction if slow (see Figure 8.1). In such cases, correlations are present between the linear and circular variables in the form that the variances over the circular variables become high when speeds are slow.

In this chapter, by focusing on cylindrical distributions on $S^1 \times [0, \infty)$, we introduce those distributions that are mathematically tractable, namely the probability density function (pdf) can be expressed by elementary functions or, at most, well-known special functions, and provide examples of their applications. In Section 8.2, we show examples of velocity data. After a brief review of probability distributions on $[0, \infty)$ and on the circle in Section 8.3, Section 8.4 introduces cylindrical distributions. In Section 8.5, we apply the cylindrical distributions to the data given in Section 8.2. Section 8.6 demonstrates an application of the cylindrical distributions, using a statistical model of forest tree data in Finland, to quantify the factors that affect asymmetric crown expansion. Finally, we offer some remarks in Section 8.7.

8.2   Example: Commonly Observed Patterns for Cylindrical Data

For the trajectory of a moving object, speeds and turning angles per unit time form a cylindrical distribution. Figure 8.1a is a part of a GPS trajectory of a seabird (streaked shearwater) flying on the Pacific Ocean near the northeast coast of Japan (Shimatani et al., 2012). In Goto et al. (2017), a circular model for inferring wind velocity and animal heading (the direction in which an animal is oriented) from seabird's trajectories was proposed. For details and recent studies about this seabird species, see Goto et al. (2017) and references therein. Figure 8.1b shows scatterplots of the speeds and turning angles of the three selected parts shown in Figure 8.1a (throughout the paper, we set east as direction 0 (radian), north as $\pi/2$, west as π, and south as $-\pi/2$).

We can see that the turning angles are concentrated around zero, speeds have a positive mode, and variances in turning angles increase when speeds are low. At the same time, we can see some variations in the three scatterplots; for example, the left panel in Figure 8.1b looks more concentrated around 0 at high speeds than do the other panels, whereas the right panel might have larger variances at low speeds.

Visually, when a bird is flying straightforward (the left panel), the variances in turning angles look small, while if a bird frequently changes direction and fluctuates around a specific area (the middle and right panels), the variances seem to be larger, and the latter two are hardly distinguishable for (a) and (b). The straightforward trajectories are often seen when an animal wants to move to some far area, while the latter is a typical foraging pattern.

The degree of concentration for foraging is an important factor for animal foraging strategy studies, and their quantification is one of the important initial statistical tasks by simply calculating the passage of time (the time duration until an animal left the specific area) or some summary statistics. The application of a cylindrical distribution gives us a new quantification, which is demonstrated in Section 8.5.

8.3   A Brief Review of the Univariate Probability Distribution

8.3.1   Probability Distributions on $[0, \infty)$

For explanatory purposes, we review the fundamental distributions on $[0, \infty)$. The first distribution is the exponential distribution, denoted by $\text{Exp}(\beta)$, whose pdf is given by

$$f_E(l; \beta) = \beta \exp(-\beta l), \quad l \geq 0,$$

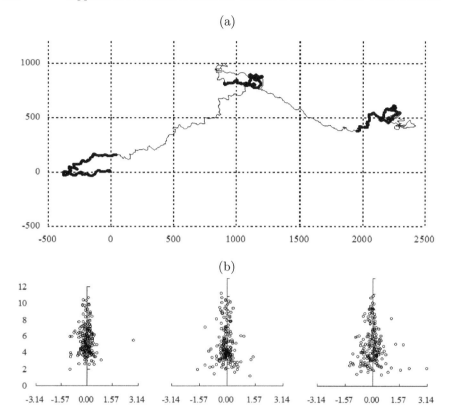

FIGURE 8.1

Examples of cylindrical data on $[-\pi, \pi) \times [0, 12]$: velocity. (a) The trajectory of a flying seabird obtained by GPS logger. (b) The scatterplots between turning angles (the horizontal axes) and speeds (the vertical axes, m/0.5 sec) for the three parts indicated by the bold lines in (a).

where $\beta > 0$ is the scale parameter.

Next, the Gamma distribution, $\mathrm{Gam}(\alpha, \beta)$, has the following density function:

$$f_G(l; \alpha, \beta) = \frac{\beta^\alpha}{\Gamma(\alpha)} l^{\alpha-1} \exp(-\beta l), \quad l \geq 0,$$

where $\alpha > 0$ and $\beta > 0$ are the shape and scale parameters, respectively and $\Gamma(\cdot)$ denotes the Gamma function.

The density of the Weibull distribution, $\mathrm{Wei}(\alpha, \beta)$, is given by

$$f_W(l; \alpha, \beta) = \alpha \beta^\alpha l^{\alpha-1} \exp\left\{-(\beta l)^\alpha\right\}, \quad l \geq 0,$$

where $\alpha > 0$ and $\beta > 0$ are the shape and scale parameters, respectively.

Finally, as the extension of the above distributions, the Generalized Gamma distribution, $GG(\alpha, \beta, \gamma)$ (Stacy, 1962), has the following density function:

$$f_{GG}(l; \alpha, \beta, \gamma) = \frac{\gamma\beta^\alpha}{\Gamma(\alpha/\gamma)} l^{\alpha-1} \exp\{-(\beta l)^\gamma\}, \quad l \geq 0,$$

where $\beta > 0$ is the scale parameter and $\alpha > 0$ and $\gamma > 0$ are the shape parameters.

8.3.2   Circular Distributions

For the circular part of the cylindrical distributions, we also review some circular distributions on $[-\pi, \pi)$. As a perturbation of the circular uniform distribution, the cardioid distribution is defined by the following pdf:

$$f_{CA}(\theta; \mu, \rho) = \frac{1}{2\pi}\{1 + 2\rho\cos(\theta - \mu)\}, \quad -\pi \leq \theta < \pi,$$

where $\mu \in [-\pi, \pi)$ is the location parameter and $|\rho| < 1/2$ controls the concentration.

Next, the density of the von Mises distribution, $VM(\mu, \kappa)$, is given by

$$f_{VM}(\theta; \mu, \kappa) = \frac{1}{2\pi I_0(\kappa)} \exp\left\{\kappa\cos(\theta - \mu)\right\}, \quad -\pi \leq \theta < \pi,$$

where $\mu \in [-\pi, \pi)$ is the location parameter, $\kappa \in [0, \infty)$ is the concentration parameter, and I_p is the modified Bessel function of the first kind of order p, defined as

$$I_p(\kappa) = \frac{1}{2\pi}\int_0^{2\pi} \cos(p\theta)\exp(\kappa\cos\theta)d\theta = \sum_{r=0}^{\infty} \frac{1}{\Gamma(p+r+1)r!}\left(\frac{\kappa}{2}\right)^{2r+p}.$$

The density of the wrapped Cauchy distribution $WC(\mu, \rho)$ is given by

$$f_{WC}(\theta; \mu, \rho) = \frac{1}{2\pi}\frac{1 - \rho^2}{1 + \rho^2 - 2\rho\cos(\theta - \mu)}, \quad -\pi \leq \theta < \pi,$$

where $\mu \in [-\pi, \pi)$ is the location parameter and $\rho \in [0, 1)$ is the concentration parameter.

The following density function is essentially the same density as Jones and Pewsey (2005):

$$f_{JP}(\theta; \mu, \kappa, \alpha) = \frac{\{1 - \tanh(\kappa)\cos(\theta - \mu)\}^{-\alpha}}{2\pi\cosh^\alpha(\kappa)P_{\alpha-1}(\cosh(\kappa))}, \quad -\pi \leq \theta < \pi, \quad (8.1)$$

where $P_{\alpha-1}(\cdot)$ is the associated Legendre function of the first kind of degree $\alpha - 1$ and order 0. Its integral expression is given by

$$P_{\alpha-1}(z) = \frac{1}{\pi}\int_0^\pi \left(z + \sqrt{z^2 - 1}\cos x\right)^{\alpha-1} dx.$$

Depending on the values of α, the distribution with density (8.1) reduces to the von Mises when $\alpha \to 0$, cardioid when $\alpha = -1$, wrapped Cauchy when $\alpha = 1$, and a few other distributions. Note that the mode of the distribution occurs at $\mu - \pi$.

For more details about these distributions, see Chapter 2 of Ley and Verdebout (2017).

8.3.3   Sine-Skewed Perturbation to the Symmetric Circular Distributions

Abe and Pewsey (2011) proposed a universal method to construct a skewed distribution from a symmetric distribution. For a base circular density $f_0(\theta; \phi)$ symmetric at $\theta = 0$ (ϕ is a set of parameters), the sine-skewed circular distribution has the pdf:

$$f_0^{SS}(\theta; \phi, \lambda) = (1 + \lambda \sin \theta) f_0(\theta; \phi), \quad \theta \in [-\pi, \pi),$$

where the parameter $|\lambda| \le 1$ controls skewness. If $\lambda = 0$, the density reduces to the original base symmetric density.

8.4   Cylindrical Distributions

Let (Θ, L) be a pair of circular and linear random variables, respectively. When we observe cylindrical data as in Figure 8.2, if the linear components increase, the degree of concentration around a certain direction also tends to increase. This section introduces models that satisfy these conditions. The basic strategy is to combine the distributions in Sections 8.1 and 8.2.

8.4.1   The Johnson–Wehrly Distribution

Johnson and Wehrly (1978) proposed a cylindrical distribution whose joint pdf is given by

$$f_{JW}(\theta, l; \mu, \nu, \kappa) = \frac{\sqrt{\nu^2 - \kappa^2}}{2\pi} \exp\{-\nu l + \kappa l \cos(\theta - \mu)\},$$
$$(\theta, l) \in [-\pi, \pi) \times [0, \infty), \tag{8.2}$$

where $0 \le \kappa < \nu$ and $-\pi \le \mu < \pi$. The distribution is obtained by maximizing the (Shannon) entropy with $E(L)$, $E(L \cos \Theta)$ and $E(L \sin \Theta)$ constant (see Mardia, 1975, p. 352). The functional form of the density indicates that as the length part $(= l)$ increases, the concentration around the location $(= \mu)$ also increases. Clearly, the mode of the density always occurs at $l = 0$. The marginal distribution of Θ is the wrapped Cauchy distribution

$\mathrm{WC}\left(\mu, \kappa / \left(\nu + \sqrt{\nu^2 - \kappa^2}\right)\right)$. The marginal distribution of L has the following density:

$$f_{JW}^L(l) = \sqrt{\nu^2 - \kappa^2}\, I_0(\kappa l) \exp(-\nu l), \quad l > 0.$$

The conditional distribution $\Theta | (L = l)$ is the von Mises distribution $\mathrm{VM}(\mu, \kappa l)$. The conditional distribution $L | (\Theta = \theta)$ is the exponential distribution $\mathrm{Exp}(\{\nu - \kappa \cos(\theta - \mu)\} l)$. Panels (b) and (d) in Figure 8.2 are examples of the two-dimensional contour plots of this pdf.

8.4.2   The Weibull–von Mises Distribution

As a more flexible symmetric cylindrical distribution, we show a special case of the density in Abe and Ley (2017), whose joint pdf is given by

$$f_{WM}(\theta, l; \mu, \kappa, \alpha, \beta) = \frac{\alpha \beta^\alpha}{2\pi \cosh(\kappa)} l^{\alpha-1} \exp\left[-(\beta l)^\alpha \left\{1 - \tanh(\kappa) \cos(\theta - \mu)\right\}\right], \quad (8.3)$$

where $(\theta, l) \in [-\pi, \pi) \times [0, \infty)$, $\alpha > 0$ and $\beta > 0$ are the linear scale and shape parameters, and $-\pi \leq \mu < \pi$ and $\kappa \geq 0$ are the circular location and concentration parameters, respectively. We call this cylindrical distribution the Weibull–von Mises distribution. The density (8.3) is basically proportional to the product of the Weibull and von Mises (replacing κ with κl) distributions,

$$f_{WM}(\theta, l) \propto l^{\alpha-1} \exp\left[-(\beta l)^\alpha\right] \times \exp\left\{(\beta l)^\alpha \tanh(\kappa) \cos(\theta - \mu)\right\},$$

divided by the normalizing constant. Again, as the length part ($= l$) increases, the concentration around the location ($= \mu$) also increases. Here, the mode of the density does not always occur at $l = 0$. The marginal density of the circular component Θ is given by

$$
\begin{aligned}
f_{WM}^C(\theta) &= \frac{1}{2\pi \cosh(\kappa)} \int_0^\infty \alpha \beta^\alpha l^{\alpha-1} \exp\left[-(\beta l)^\alpha \left\{1 - \tanh(\kappa) \cos(\theta - \mu)\right\}\right] dl \\
&= \frac{1}{2\pi \cosh(\kappa)} \frac{1}{1 - \tanh(\kappa) \cos(\theta - \mu)} \\
&= \frac{1 - \tanh^2(\kappa/2)}{2\pi} \frac{1}{1 + \tanh^2(\kappa/2) - 2\tanh(\kappa/2) \cos(\theta - \mu)},
\end{aligned}
$$

which is $\mathrm{WC}(\mu, \tanh(\kappa/2))$. The marginal density of the linear component L in turn corresponds to

$$
\begin{aligned}
f_{WM}^L(l) &= \frac{1}{2\pi \cosh(\kappa)} \alpha \beta^\alpha l^{\alpha-1} \int_{-\pi}^{\pi} \exp\left[-(\beta l)^\alpha \left\{1 - \tanh(\kappa) \cos(\theta - \mu)\right\}\right] d\theta \\
&= \frac{1}{2\pi \cosh(\kappa)} \alpha \beta^\alpha l^{\alpha-1} \exp\{-(\beta l)^\alpha\} \int_{-\pi}^{\pi} \exp\left\{(\beta l)^\alpha \tanh(\kappa) \cos(\theta - \mu)\right\} d\theta \\
&= \frac{I_0(l^\alpha \beta^\alpha \tanh(\kappa))}{\cosh(\kappa)} \alpha \beta^\alpha l^{\alpha-1} \exp\{-(\beta l)^\alpha\}.
\end{aligned}
$$

This is an extended version of the marginal density obtained in Johnson and Wehrly (1978). It simplifies to the Weibull distribution when $\kappa = 0$. The conditional densities from (8.3) are now readily given by

$$f_{WM}(\theta|l) = \frac{f_{WM}(\theta, l)}{f_{WM}^L(l)} = \frac{1}{2\pi I_0(l^\alpha \beta^\alpha \tanh(\kappa))} \exp\{(\beta l)^\alpha \tanh(\kappa) \cos(\theta - \mu)\}$$

$$(8.4)$$

and

$$f_{WM}(l|\theta) = \frac{f_{WM}(\theta, l)}{f_{WM}^C(\theta)}$$

$$= \alpha \left[\beta \{1 - \tanh(\kappa) \cos(\theta - \mu)\}^{1/\alpha} \right]^\alpha l^{\alpha - 1}$$

$$\exp\left[-\left\{ \beta (1 - \tanh(\kappa) \cos(\theta - \mu))^{1/\alpha} l \right\}^\alpha \right]. \qquad (8.5)$$

The density (8.4) is the von Mises distribution with concentration $(\beta l)^\alpha \tanh(\kappa)$, whereas (8.5) is the Weibull distribution with shape parameter α and scale parameter $\beta \{1 - \tanh(\kappa) \cos(\theta - \mu)\}^{1/\alpha}$. The mode of the density (8.5) occurs at

$$l = \frac{(1 - 1/\alpha)^{1/\alpha}}{\beta \{1 - \tanh(\kappa) \cos(\theta - \mu)\}^{1/\alpha}}$$

if $\alpha \geq 1$.

By decomposing the density function $f(\theta, l)$ into $f(l|\theta) f(\theta)$, we can easily generate a random number by first generating $\Theta \sim f(\theta)$ and then $L|(\Theta = \theta) \sim f(l|\theta)$. The algorithm goes as follows:

Step 1: Generate a random variable Θ following a $WC(\mu, \tanh(\kappa/2))$.

Step 2: Generate a random variable L from the Weibull distribution with shape parameter α and scale parameter $\beta \{1 - \tanh(\kappa) \cos(\Theta - \mu)\}^{1/\alpha}$.

8.4.3   Gamma–von Mises Distribution

As in Abe and Ley (2017), another cylindrical distribution is obtained by replacing the Weibull distribution with the Gamma distribution. The pdf is given by

$$f_{GM}(\theta, l; \mu, \kappa, \alpha, \beta) = C^{-1} l^{\alpha - 1} \exp[-\beta l \{1 - \tanh(\kappa) \cos(\theta - \mu)\}] \quad (8.6)$$

where $(\theta, l) \in [-\pi, \pi) \times [0, \infty)$, $\alpha, \beta, \gamma, \kappa > 0$, and $-\pi \leq \mu < \pi$. The normalizing constant C is given by

$$\int_{-\pi}^{\pi} \int_0^\infty l^{\alpha - 1} \exp[-\beta l \{1 - \tanh(\kappa) \cos(\theta - \mu)\}] dl d\theta$$

$$= \frac{\Gamma(\alpha)}{\beta^\alpha} \int_{-\pi}^{\pi} \frac{1}{(1 - \tanh(\kappa) \cos \theta)^\alpha} d\theta$$

$$= \frac{2\pi \Gamma(\alpha)(\cosh(\kappa))^\alpha P_{\alpha-1}(\cosh(\kappa))}{\beta^\alpha}.$$

We call this distribution the Gamma–von Mises distribution. The marginal density of the circular component is the distribution given in Section 8.3.2:

$$
\begin{aligned}
f_{GM}^C(\theta) &= C^{-1} \int_0^\infty l^{\alpha-1} \exp[-\beta l \{1 - \tanh(\kappa)\cos(\theta - \mu)\}] dl \\
&= \frac{\{1 - \tanh(\kappa)\cos(\theta - \mu)\}^{-\alpha}}{2\pi \cosh^\alpha(\kappa) P_{\alpha-1}(\cosh(\kappa))}.
\end{aligned}
$$

The conditional density given the angle is the Gamma distribution $\Gamma(\alpha, \beta\{1 - \tanh(\kappa)\cos(\theta - \mu)\})$:

$$
f_{GM}(l|\theta) = \frac{\beta^\alpha \{1 - \tanh(\kappa)\cos(\theta - \mu)\}^\alpha}{\Gamma(\alpha)} l^{\alpha-1} \exp[-\beta l \{1 - \tanh(\kappa)\cos(\theta - \mu)\}].
$$

The mode of this distribution occurs at $l = (\alpha-1)/[\beta\{1 - \tanh(\kappa)\cos(\theta - \mu)\}]$ if $\alpha \geq 1$.

The conditional density given the length is the von Mises distribution $VM(\mu, \beta l \tanh(\kappa))$:

$$
f_{GM}(\theta|l) = \frac{\exp\{\beta l \tanh(\kappa)\cos(\theta - \mu)\}}{2\pi I_0(\beta l \tanh(\kappa))}.
$$

Figure 8.2 provides examples of the two-dimensional contour plots of this pdf.

8.4.4   Generalized Gamma–von Mises Distribution

Another extension is obtained by using the generalized Gamma distribution, and the pdf is then given by

$$
f_{GG}(\theta, l; \alpha, \beta, \gamma) = C^{-1} l^{\alpha-1} \exp\left[-(\beta l)^\gamma \{1 - \tanh(\kappa)\cos(\theta - \mu)\}\right],
$$

where $(\theta, l) \in [-\pi, \pi) \times [0, \infty)$, $\alpha, \beta, \gamma, \kappa > 0$ and $-\pi \leq \mu < \pi$, and the normalizing constant is given by

$$
\begin{aligned}
C &= \int_{-\pi}^\pi \int_0^\infty l^{\alpha-1} \exp[-(\beta l)^\gamma \{1 - \tanh(\kappa)\cos(\theta - \mu)\}] dl d\theta \\
&= \frac{\Gamma(\alpha/\gamma)}{\gamma \beta^\alpha} \int_{-\pi}^\pi \frac{1}{(1 - \tanh(\kappa)\cos\theta)^{\alpha/\gamma}} d\theta \\
&= \frac{2\pi \Gamma(\alpha/\gamma)(\cosh(\kappa))^{\alpha/\gamma} P_{\alpha/\gamma-1}(\cosh(\kappa))}{\gamma \beta^\alpha}.
\end{aligned}
$$

The Weibull–von Mises distribution is a special case of the distribution and obtained by setting $\gamma = \alpha$. Another special case is the Gamma–von Mises distribution obtained as $\gamma = 1$.

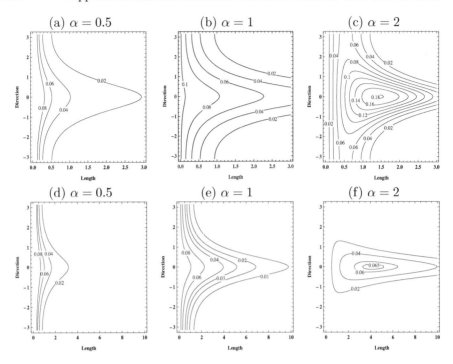

FIGURE 8.2

Contour plots of the Weibull–von Mises (a-c) and Gamma–von Mises (d-f) densities over $(0,3) \times [-\pi, \pi]$ and $(0,10) \times [-\pi, \pi]$, respectively, for $\mu = 0$, $\kappa = 1$ and $\beta = 1$. (b) and (e) are the same as the density of the Johnson–Wehrly distribution.

8.4.5   Sine-Skewed Weibull–von-Mises Distribution

These cylindrical distributions can be easily extended to each sine-skewed version (for the circular part) by using the method described in Section 8.3.3. For example, the sine-skewed Weibull–von Mises distribution is given by

$$f_{SWM}(\theta, l; \mu, \alpha, \beta, \lambda) = \frac{\alpha \beta^{\alpha}}{2\pi \cosh(\kappa)} \{1 + \lambda \sin(\theta - \mu)\} l^{\alpha - 1}$$
$$\exp\left[-(\beta l)^{\alpha} \{1 - \tanh(\kappa) \cos(\theta - \mu)\}\right].$$

When the concentration parameter κ vanishes, this distribution reduces to

$$\frac{1}{2\pi} \{1 + \lambda \sin(\theta - \mu)\} \times \alpha \beta^{\alpha} l^{\alpha - 1} \exp\{-(\beta l)^{\alpha}\},$$

which is a product of the Weibull and cardioid distributions.

We can describe a simple random number generation algorithm in the same way as the Weibull–von Mises just by inserting the second step:

Step 1: Generate a random variable Θ_1 following a (symmetric) wrapped Cauchy law with location μ and concentration $\tanh(\kappa/2)$.

Step 2: Generate a random variable independently $U \sim Unif[0, 1]$ and define Θ as
$$\begin{cases} \Theta_1 & \text{if } U < \{1 + \lambda \sin(\Theta_1 - \mu)\}/2 \\ -\Theta_1 & \text{if } U \geq \{1 + \lambda \sin(\Theta_1 - \mu)\}/2; \end{cases}$$

Θ then follows the sine-skewed wrapped Cauchy distribution.

Step 3: Generate X from a Weibull with shape parameter $\beta\{1 - \tanh(\kappa) \cos(\Theta - \mu)\}^{1/\alpha}$.

8.4.6 Parameter Estimation

Let $(\theta_1, l_1), \ldots, (\theta_n, l_n)$ be a sample of n independent and identically distributed couples of circular and linear observations drawn from a cylindrical distribution. Then, the likelihood function can be expressed, in the cases of the basic three symmetric distributions, namely the Johnson–Wehrly, Weibull–von Mises, and Gamma–von Mises distributions, respectively, as

$$L_{JW}(\mu, \kappa, \nu) = \prod_{i=1}^{n} f_{JW}(l_i, \theta_i; \mu, \kappa, \nu), \tag{8.7}$$

$$L_{WM}(\mu, \kappa, \alpha, \beta) = \prod_{i=1}^{n} f_{WM}(l_i, \theta_i; \mu, \kappa, \alpha, \beta), \tag{8.8}$$

$$L_{GM}(\mu, \kappa, \alpha, \beta) = \prod_{i=1}^{n} f_{GM}(l_i, \theta_i; \mu, \kappa, \alpha, \beta). \tag{8.9}$$

The likelihood functions of the other distributions can be written in the same way. In general, it is difficult to give closed-form expressions for the maximum likelihood estimates (MLEs); hence, numerical methods should be used to find the solutions.

We used the function **NMaximize** in *Mathematica* or **Maximize** in *Mathcad*, and encountered no problems in the optimization procedure.

8.5 Application 1: Quantification of the Speed/Turning Angle Patterns of a Flying Bird

For the cylindrical data in Figure 8.1 and another trajectory of the same bird with a similar time duration, we maximized Equations (8.8) and (8.9). Table 8.1 summarizes the maximum log-likelihood (MLL), MLEs, and modes

TABLE 8.1
The MLL for the Weibull– and Gamma–von Mises cylindrical distributions, MLEs of the selected model (shown by the bold letters), and modes of the speeds derived from the selected model.

Patterns	MLE				Mode	MLL	
	μ	κ	α	β		Weibull	Gamma
(a)	−0.02	0.53	11.22	3.76	5.30	−496.2	**−445.8**
(b)	−0.04	0.77	6.08	3.43	4.19	−523.1	**−473.1**
(c)	0.01	0.63	5.85	2.25	4.87	−580.0	**−554.3**
(d)	−0.02	2.53	3.57	0.53	5.80	**−281.2**	−289.6

calculated from the model with a larger log-likelihood (shown by the bold letters in Table 8.1). The left column in Figure 8.3 illustrates the trajectories and the right column summarizes the results when the cylindrical distributions were fitted to the cylindrical data.

The Gamma–von Mises was selected for (a)–(c), while the Weibull was selected for (d). The cylindrical distributions successfully evaluated each flying pattern, and we obtained a quantitative evaluation of the trajectories.

Visually, (b) and (c) are both just surrounding in each specific area, but the quantification by the cylindrical model suggested that (c) is more wandering because of smaller κ and more fat-tailed contours (Figures 8.3b and 8.3c). For the two seemingly straightforward trajectories, the selection of the different cylindrical models may mean that these two have qualitatively distinct characteristics. Because the function in κ differs between the Weibull– and Gamma–von Mises distributions, we cannot directly compare the κ values to evaluate the degree of the variances in turning angles.

Based on the expressions for the mode of the conditional distributions described in Sections 8.4.2 and 8.4.3, Table 8.1 also presents the modes of the speed for each pattern.

If the cylindrical data are influenced by certain factors, they should be evaluated by using a regression model in which the parameters of the distribution are expressed by the functions of those factors. The next section describes an example of such a situation.

8.6 Application of Cylindrical Distributions 2: How Trees Are Expanding Crowns

In this section, we demonstrate an application of cylindrical distributions by using the forest data in Aakala et al. (2016). For details of the measurements, data, ecological background, and implications, see their paper.

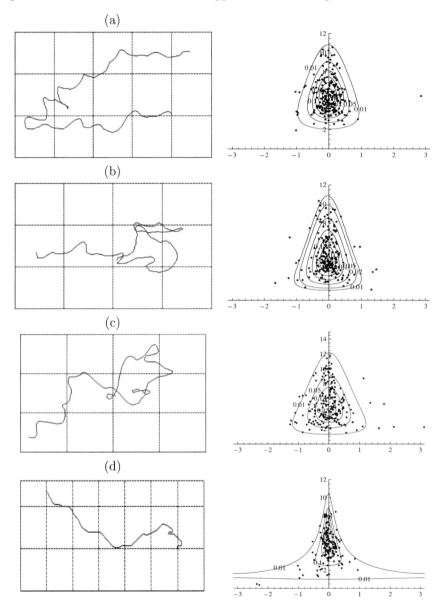

FIGURE 8.3

Examples of the application of the cylindrical distributions to velocity data. (a), (b), and (c) are drawn by the data shown in the left, middle, and right panels in Figure 8.1, respectively. (d) is another trajectory of the same seabird. The left column: trajectories of a seabird obtained by GPS logger. The mesh unit is 100 m. The right column: The scatterplot between the turning angles (horizontal axes) and speeds (vertical axes, m/0.5 sec) on the contour plots drawn by the MLEs of each selected model.

8.6.1   Crown Asymmetry in Boreal Forests

Boreal forests in northern Finland are characterized by the dominance of conifers, low density, and strong solar radiation from southern directions during the summer because of the low sun angles. Consequently, trees tend to expand crowns toward the south, not just above each trunk, causing crown asymmetry. On the contrary, because trees compete for light with neighboring trees, tree crowns tend to expand away from nearby trees as a strategy to reduce competition. This also causes crown asymmetry. In the case of directional solar radiation and competition with nearby trees, the question then becomes how trees respond to these conditions and what type of crown asymmetry appears. If a competitor stands to the south, in which direction does the tree expand its crown: the south for solar radiation, the north to escape competition, or other directions? If a competitor stands to the east, does the crown expand toward the west, south, or south-west?

In Aakala et al. (2016), sampling plots were located in the field, and all trees with a diameter greater than 10 cm at a height of 1.3 m were measured. These measurements included their trunk positions, sizes (diameter, height), species, and tree canopy projections (i.e., which part on the forest floor is covered by which tree(s)). Based on these measurement, tree trunk position maps and canopy projection maps were drawn. The centroid of each canopy projection was computed, and based on these data, crown asymmetry was defined as follows.

Let x_i be the x-y coordinate of the trunk of tree i (hereafter, this is called the "focal tree") and u_i be the centroid of the canopy projection. Crown asymmetry is expressed by the vector $\overrightarrow{x_i u_i}$ (Figure 8.4). We denote the length by $l_i = \|\overrightarrow{x_i u_i}\|$ and the direction by $\theta_i = \arg \overrightarrow{x_i u_i}$, and we call these "intensity" and "orientation," respectively.

In boreal forests in Finland, previous studies have suggested that neighboring trees affect the growth of the focal tree by up to about 5 m. Here, trees j satisfying $\|x_i - x_j\| < 5$ m are defined as the "competitors" of focal tree i. Among the 1721 trees in the sampling plots, only those trees in the inside 5 m from the plot edges were used as focal trees. Consequently, 661 were focal trees, while 64 had no competitors and 137 had a single competitor. The remainder (460) had two or more competitors (the maximum was 10 competitors).

Figure 8.5a illustrates the scatterplot of the directions from the competitor to the focal tree (given by $\arg \overrightarrow{x_i x_j}$) and θ_i for the focal trees with a single competitor. This figure shows positive correlations, and focal trees with high intensity (the black symbols) tend to be close to the diagonal, while trees with low intensity (white symbols) seem to be distributed more diversely. Figure 8.5b is the scatterplot of the orientations and intensities for focal trees with no competitors. Orientations are weakly concentrated around the south and trees with high intensities seem to be more concentrated around the south. These findings support the above statements.

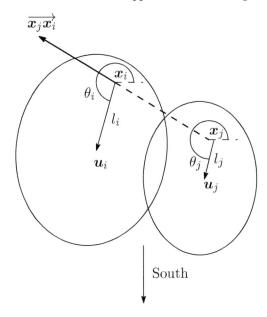

FIGURE 8.4
The illustration of the notations for crown asymmetry. i is the focal tree and j indicates its competitor. $\boldsymbol{x}_i$ and $\boldsymbol{x}_j$ are the trunk positions. $\boldsymbol{u}_i$ and $\boldsymbol{u}_j$ are the centroids of the canopy projections. The arrows from $\boldsymbol{x}_i$ to $\boldsymbol{u}_i$ and from $\boldsymbol{x}_j$ to $\boldsymbol{u}_j$ indicate the crown asymmetry vectors. θ_i and θ_j are their orientations and l_i and l_j are their intensities.

However, Figure 8.5a does not reflect the combined effects of a competitor and solar radiation. If a focal tree had two or more competitors, focal trees had several undesirable directions, and we cannot visually check these tendencies. More importantly, the tendencies are unclear for trees with low intensities. The easiest way in which to clarify these patterns would be to delete such trees (e.g., set a threshold value L and delete tree i if $l_i < L$), although this results in many problems (e.g., the threshold is arbitrary and the sample size decreases). Moreover, if this increasing variance reflects some essential feature of boreal forests, we would fail to find it.

Statistical modeling may solve these problems, and we may be able to quantity the relative importance of solar radiation and competitors for boreal forests. We therefore constructed a statistical model by using cylindrical distributions that have greater variance in certain directions when the intensity is low.

8.6.2 Crown Asymmetry Model

Based on exploratory analyses such as Figure 8.5 together with field observations and previous studies, we hypothesized that the preferred direction was

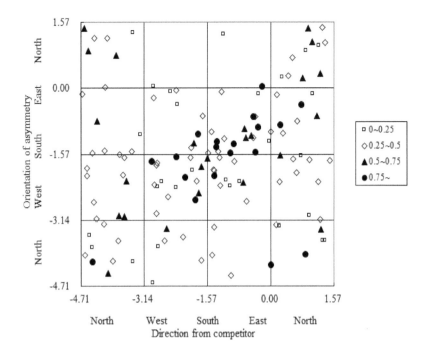

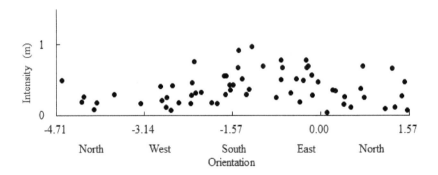

FIGURE 8.5

(a) The scatterplot between θ_i and l_i for trees with a single competitor. The symbols are classified into four classes depending on intensities. (b) The frequencies of θ_i for trees with no competitors.

determined by solar radiation and competitors. To construct statistical models for crown asymmetry, we set the preferred (south) and the opposites to unpreferred (away from competitors) directions as the explanatory variables and the crown asymmetry vector as the objective variable; in other words,

(l_i, θ_i) were assumed to be random samples from a symmetric cylindrical distribution. This modeling approach simply extends a general formulation of the linear regression (i.e., it expresses the location parameter (instead of the expected value) by using a linear function of (the weighted average over) the preferred and unpreferred directions. Note that because the explanatory variables are circular, the weighted average means the argument of the weighted sum over the corresponding vectors ($\vec{S} = (0, -1)$ and $\overrightarrow{x_j x_i}$). Hence, the preferred direction of focal tree i, μ_i, is expressed by

$$\mu_i = \arg\left(w_0 \vec{S} + \sum_{j=1}^{n_i} w_j \overrightarrow{x_j x_i} \right),$$

where n_i is the number of competitors and the summation covers all the competitors. For the weights of the competitors (w_j), because nearby larger trees should have a stronger influence, we used the following competition index from Rouvinen and Kuuluvainen (1997):

$$CI^{ji} = \frac{d_j}{\|\overrightarrow{x_j x_i}\|},$$

where d_j indicates the diameter of tree j. This index means that the degree of competition is inversely proportional to the distance to the competitor $(\|\overrightarrow{x_j x_i}\|)$ and proportional to the size of the competitor (d_j). For the weight of solar radiation (w_0), we set this as a free parameter denoted by a. Then, the most preferable orientation of asymmetry for focal tree i is written as

$$\mu_i(a) = \arg\left(a\vec{S} + \sum_{j=1}^{n_i} CI^{ji} \overrightarrow{x_j x_i} \right). \tag{8.10}$$

The parameter a quantifies the relative importance of solar radiation and competitors with a given size and distance.

By substituting $\mu = \mu_i(a)$ into the symmetric cylindrical distributions with density (8.2), (8.3), or (8.6) and letting $(\theta_1, l_1), \ldots, (\theta_n, l_n)$ be a sample of n independent and identically distributed couples of circular and linear observations drawn from each distribution, we obtain the statistical model that has a cylindrical variable as an objective variable. Then, the likelihood functions (8.7), (8.8), and (8.9), respectively become

$$L_{JW}(a, \kappa, \nu) = \prod_{i=1}^{n} f_{JW}(l_i, \theta_i; \mu_i(a), \kappa, \nu),$$

$$L_{WM}(a, \kappa, \alpha, \beta) = \prod_{i=1}^{n} f_{WM}(l_i, \theta_i; \mu_i(a), \kappa, \alpha, \beta),$$

$$L_{GM}(a, \kappa, \alpha, \beta) = \prod_{i=1}^{n} f_{GM}(l_i, \theta_i; \mu_i(a), \kappa, \alpha, \beta).$$

Because the cylindrical distributions have different numbers of parameters, we evaluated the models by using the Akaike information criterion (AIC; Konishi & Kitagawa, 2008).

8.6.3   Results of the Cylindrical Models

We obtained the MLEs and AIC values as summarized in Table 8.2. Gamma–von Mises exhibited the best AIC value.

TABLE 8.2
The MLEs, MLL, and AIC values for the three cylindrical distribution models.

	MLE				MLL	AIC
	a	κ	α	β		
Johnson–Wehrly	5.95	0.63		3.11	-1249.1	2504.2
Weibull–von Mises	5.94	0.68	1.35	2.65	-1196.7	2401.4
Gamma–von Mises	5.95	0.43	1.96	5.43	-1163.5	2335.0

Figure 8.6 illustrates the model fit by overlapping the three-dimensional graphs of the cylindrical distributions using the MLEs on the frequencies of observations. The directional axis indicates the difference between the expectations $\mu_i(\hat{a}) = \arg(\hat{a}\overrightarrow{\boldsymbol{S}} + \sum_{j=1}^{n_i} CI^{ji}\overrightarrow{\boldsymbol{x}_j\boldsymbol{x}_i})$ and observed orientation of tree i (θ_i). The Johnson–Wehrly distribution overestimated the frequencies for low intensities and this was simply because this distribution always takes the maximum at $l = 0$ (Figure 8.6a), while the real data had a positive mode. Compared with the Weibull–von Mises distribution, the Gamma–von Mises distribution fit better primarily since it captured the frequencies around the high mode.

The MLE $\hat{a}$ allowed us to quantify the relative contribution of solar radiation and competitors. $\hat{a} = 5.95$ corresponds to the competitor with the competition index $CI_{ji} = d_j/\|x_i x_j\| = 5.95$, for example, diameter $d_j = 5.95$ cm if the tree stands with $\|x_i x_j\| = 1$ m distance, 17.9 cm if 3 m, and about 30 cm if 5 m. In other words, the attraction of solar radiation is as much as the negative effects of these competitors. Such evaluation and insight would not be obtained without a quantification by the cylindrical models.

Aakala et al. (2016) applied only the Johnson–Wehrly distribution simply because neither the Weibull–von Mises nor the Gamma–von Mises had been developed. Although the fitting was insufficient for the intensities (the mode of observed intensities was not at zero), the estimate of a was surprisingly close to that when we applied the new cylindrical distributions. The results here confirm the quantitative evaluation in Aakala et al. (2016) as well as the applicability of the cylindrical distributions.

Figure 8.7 illustrates the evaluation of the best fitted model by using the Gamma–von Mises. We simulated random numbers (θ_i, l_i) for each focal tree 200 times and summarized them to frequencies of the marginal distributions. Then, 95% confidence envelopes were drawn by computing the top and bot-

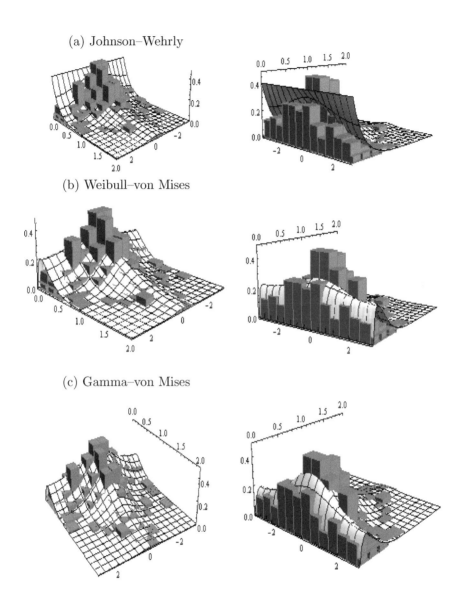

FIGURE 8.6
Three-dimensional histogram of crown asymmetry in the boreal tree data together with the three-dimensional densities of the maximum likelihood fits for the (a) Johnson–Wehrly, (b) Weibull–von Mises, and (c) Gamma–von Mises distributions.

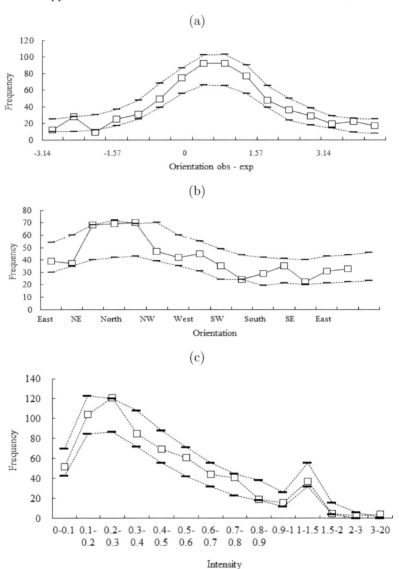

FIGURE 8.7

The model fit comparisons between the marginal densities of the selected cylindrical model (Gamma–von Mises) and the marginal frequencies of the observations; (a) orientation: the horizontal axis indicates the observations (expectations); (b) orientation: the horizontal axis indicates the directions; (c) intensity: the vertical axes are frequencies. The white squares indicate the observed frequencies. The horizontal bars indicate the confidence envelopes drawn by 100 simulations.

tom 2.5% percentiles. The observations are inside of the confidence envelopes, which also demonstrates the improvement over the Johnson–Wehrly distributions used in Aakala et al. (2016).

Figure 8.8 illustrates the correlations between the expectations and observations separately shown depending on intensities. When intensities are low, the observations are widely distributed around the expectations, whereas when they are high, observations are concentrated around the expectations.

Aakala et al. (2016) applied and compared several models, including the cases when (i) the solar radiation component in (8.10) was deleted, (ii) the

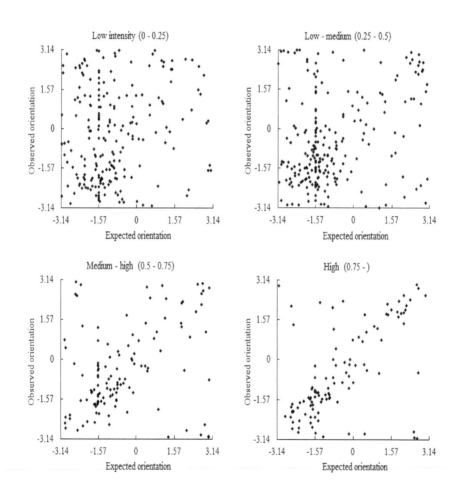

FIGURE 8.8
The correlations between the expected orientations ($= \mu_i(\hat{a})$) and observed ones ($= \theta_i$) separated by four intensity classes from small (the top left) to large (the bottom right).

competitor component in (8.10) was deleted, and (iii) only taller trees were used as competitors. The model using (10) gives a smaller AIC value than others, and Aakala et al. (2016) discussed boreal forest ecological implications induced from these results.

8.7   Concluding Remarks

Cylindrical distributions provide a basic statistical tool for analyzing cylindrical data in which the linear and circular variables are correlated. By applying cylindrical probability distributions, we quantify cylindrical data such as velocity. If an objective variable is a correlated pair of linear and circular variables, models whose objective variable is on the cylinder (e.g., the model in Section 8.6) provide predictions, the coefficient of an explanatory variable may measure the relative importance of the covariates, and model selection evaluates the covariates likely to be affected by the objective variables.

So far, we have cylindrical distributions that have increasing variances in the circular component when the linear component decreases to zero. The simplest way in which to construct a distribution is to begin with a product of a linear and a circular distribution inserting some correlations, if the normalizing constant is represented by, at most, easily computable special functions. By applying these distributions to correlated circular and linear data, we can expect to find characteristics that were not visually clear, and each distribution may evaluate the data differently.

Currently, various cylindrical data exist globally, while applications of cylindrical distributions are rare. However, rigorous analysis of such data will require applications of cylindrical distributions, and such demands together with interdisciplinary collaborative studies promote the development of circular statistics.

Acknowledgments

We thank Tuomas Aakala, Timo Kuuluvainen, and Yasuhiro Kubota for the collaborative forest ecological studies in Section 8.6, Ken Yoda, Katsufumi Sato, and Nobuhiro Katsumata for the collaborative bio-logging zoological studies in Sections 8.2 and 8.5, and the editors for their helpful advice and comments in this revision. Toshihiro Abe was supported in part by JSPS KAKENHI Grant Number 15K17593 and Nanzan University of Pache Research Subsidy I-A-2 for the 2017 academic year.

Bibliography

[1] Aakala, T., Shimatani, K., Abe, T., Kubota, Y. & Kuuluvainen, T. (2016). Crown asymmetry in high latitude forests: disentangling the directional effects of tree competition and solar radiation. *Oikos*, **125**, 1035–1043.

[2] Abe, T. & Ley, C. (2017). A tractable, parsimonious and flexible model for cylindrical data, with applications. *Econometrics and Statistics*, **4**, 91–104.

[3] Abe, T. & Pewsey, A. (2011). Sine-skewed circular distributions. *Statistical Papers*, **52**, 683–707.

[4] Cressie, N. & Wikle, C.K. (2011). *Statistics for Spatio-Temporal Data*. John Wiley & Sons, Hoboken.

[5] Goto, Y., Yoda, K. & Sato, K. (2017). Asymmetry hidden in birds' tracks reveals wind, heading, and orientation ability over the ocean. *Science Advance*, **3**, no. 9, e1700097. DOI: 10.1126/sciadv.1700097.

[6] Johnson, R.A. & Wehrly, T.E. (1978). Some angular-linear distributions and related regression models. *Journal of the American Statistical Association*, **73**, 602–606.

[7] Jones, M.C. & Pewsey, A. (2005). A family of symmetric distributions on the circle. *Journal of the American Statistical Association*, **100**, 1422–1428.

[8] Konishi, S. & Kitagawa, G. (2008). *Information Criteria and Statistical Modeling*. Springer, New York.

[9] Mardia, K.V. (1975). Statistics of Directional Data. *Journal of the Royal Statistical Society: Series B*, **37**, 349–393.

[10] Ley, C. & Verdebout, T. (2017). *Modern Directional Statistics*. CRC Press, Boca Raton, FL.

[11] Rouvinen, S. & Kuuluvainen, T. (1997). Structure and asymmetry of tree crowns in relation to local competition in a natural mature scots pine forest. *Canadian Journal of Forest Research*, **27**, 890–902.

[12] Shimatani, I. K., Yoda, K., Katsumata, N. & Sato, K. (2012). Toward the quantification of a conceptual framework for movement ecology using circular statistical modeling. *PLoS One*, 7(11), e50309.

[13] Stacy, E.W. (1962). A generalization of the Gamma distribution. *The Annals of Mathematical Statistics*, **33**, 1187–1192.

9

Directional Statistics for Wildfires

Jose Ameijeiras–Alonso

Universidade de Santiago de Compostela

Rosa M. Crujeiras

Universidade de Santiago de Compostela

Alberto Rodríguez Casal

Universidade de Santiago de Compostela

CONTENTS

9.1	Introduction to Wildfire modeling	187
9.2	Fires' Seasonality	188
	9.2.1 Landscape Scale	189
	9.2.2 Global Scale	195
9.3	Fires' Orientation	197
	9.3.1 Main Spread on the Orientation of Fires	199
	9.3.2 Orientation–Size Joint Distribution	200
	9.3.3 Orientation–Size Regression Modeling	205
9.4	Open Problems	206
	Bibliography	207

9.1 Introduction to Wildfire modeling

As pointed out by [6], fires are a part of the Earth system process, which combines biogeochemical cycles, vegetation dynamics, and human activities. Fires' regimes play an important role in this system, along with their causes and consequences, and can be characterized in terms of frequency, intensity, seasonality, extent and type, considering different spatial and also temporal scales [3]. The main drivers of fires' regimes are climate, fuel, and land use management practices [25] and, although their interactions are complex, their characterization presents important implications *at local and global scales* of wildfire patterns [20].

At a global scale, the climatological conditions have an important effect in fires' seasonality [11] and vegetation fires should be characterized by an annual cycle presenting an alternation of a fire–free and a fire–active season [9]. Therefore, in general, this pattern should lead to a single annual season of fires in the different regions of the world [16]. However, human activity has also influenced and modified wildfire regimes, altering the natural (climatological) fires' seasonality, as fires are employed for many purposes related to land use practices [19]. Fires produced as a consequence of land use show also preferential timings, usually out of the climatological fires' seasons creating human–altered fires' regimes [23]. Then, focusing on an annual period, the identification of more fires' seasons than the ones expected due to the climatological reasons (usually a single season) reveal the anthropogenic fingerprints on the fires' regimes, helping to understand where the fire is used as a land management tool [6].

At a local scale, the characterization of fires has also important management implications [26]. For instance, landscape fuel reduction treatments will only be successful if placed aligned with the correct orientation to slow down or stop the fires' spread [33]. Then, to determine if in a specific region there is any preferential orientation is important [5], and its relation with other variables [13, 14], such as burnt area, wind speed, or wind direction.

Finally, as mentioned before, fires' regimes have consequences on the Earth system process. The effect that wildfires have over the vegetation regrowth also affects the wind direction response. An example can be found in the study of how the wind response across Flea Creek Valley (Australia) had been altered by the post-fire vegetation regrowth [32].

Some examples, at global and local scales, of directional statistics applied to wildfire modeling will be provided in the following sections. Two case studies will be taken into consideration. In the first one, the focus will be placed on fires' seasonality modeling, a periodic phenomenon is considered from local and global perspectives. In the second example, fires' orientation distribution, marginally or jointly with other related variables, will be analyzed.

9.2   Fires' Seasonality

As mentioned in Section 9.1, vegetation fires are characterized by a strong annual periodicity (driven mainly by climatological factors), with fires usually appearing in the same period of the year. Climate controls most of the fires' cycle where a season of fires is followed by some months with almost no fires. For instance, Giglio et al. [16] showed that in most of the regions in the northern part of the world (in latitudes over the Tropic of Cancer), the principal peak of fires occurs during the Northern Hemisphere summer (July, August, and September), which is the suitable season (in terms of climatological and weather conditions) for fires. However, as it has already been mentioned, land use management practices also influence fires' seasonality. Agriculture burning for preparing fields (for harvest work) or for clearing the crop residues (after harvesting or in order to avoid future burnings) are causes of fires occurring out of the expected season. Then, the existence of several fires' seasons indicates whether human activity has changed the fires' seasonality by using fires as a land management tool.

The previous problem was studied considering the circular variable of time (along the year) of fires' occurrence in [6] and [2]. The studied dataset includes the date (in a daily resolution) of all the fires detected around the world by the *MODerate resolution Imaging Spectroradiometer* (MODIS), launched into Earth orbit by the *National Aeronautics and Space Administration* (NASA) on board of the Terra (*EOS AM*) and the Aqua (*EOS PM*) satellites, from July 10, 2002 to July 9, 2012. When considering the fires' time occurrence at a global scale, as pointed out by [6], the objective is to understand the role of fires in the Earth system, studying where it is actively employed as a land management tool. For doing so, [6] and [2] divided the world in regions of size 0.5° (latitude–longitude, in the geographic coordinate system). After this division of the globe, the regions having enough fires' incidence for extracting seasonal conclusions about fires were considered with the aim of studying their fires' seasonality. From a global perspective, it should be noticed that regions with low fires' incidence (fewer than 10 fires in 7 or more years) will be discarded.

At a local scale, there exist different regions in the world where modeling the temporal behavior of wildfires is of special interest [see, for example, [37], where the county of Los Angeles is studied], as this can help in getting a better understanding of the fires' behavior and the human activity in the region. As an illustrative example for showing different statistical techniques, this section will be focused on one particular region, Galicia (in the north–western part of Spain), where fires' seasonality will be also analyzed. For this region, MODIS data for August 7 (2006) is included in Figure 9.1 (left).

In what follows, how to take into account the fires' seasonality is illustrated with parametric and nonparametric tools at local and global scales, employing the circular random variable of the time occurrence of wildfires.

9.2.1   Landscape Scale

As already mentioned, at a local scale, the different statistical techniques will be illustrated in the region delimited by the Galician borders from 2002 until 2012. The different fires detected in Galicia along this 10–year period, in their corresponding locations, are represented in Figure 9.1 (right). In this plot, one can see that even during (the Northern Hemisphere) winter, far away from the climatological peak of fires (which is produced in summer), there is a large amount of fires. Then, a natural question is if land use management practices have influenced the fires' seasonality in Galicia. Here, considering the circular

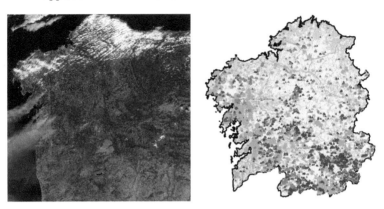

FIGURE 9.1
Left: fires detected (in red) by MODIS on board of the Terra satellite in the north–western part of the Iberian Peninsula, from [28]. Right: fires detected in Galicia (Spain) from July 10, 2002 to July 9, 2012 (represented with triangles), the fires are colored according to when they were detected and divided in the four Northern Hemisphere calendar seasons: spring (green), summer (yellow), autumn (brown), and winter (blue).

distribution of the fires' occurrence time, a complete characterization of the random variable is not needed as, in this case, the question of interest (are the humans altering the fires' seasonality?) could be solved by determining the number of fires' seasons. For doing so, to determine the number of modes of the distribution is enough driving fires' occurrence times as a continuous random variable on the circle. These modes are identified as the values at which the probability density function reaches a local maximum. Nevertheless, a complete characterization of the random variable studying the occurrence time of fires can be helpful to answer different questions such as, when the principal peaks of fires are produced [20], the delay of the agricultural fires with respect to the climatological ones [23] or the mass associated to each mode for a better understanding of the importance of the different human activities [19].

Classical as well as modern circular statistical tools can be used for trying to elucidate if human activity has influenced fires' seasonality. The analysis of the rose diagram represented in Figure 9.2 (left) may be seen as an initial exploratory approach. Then, observing this rose diagram (where the area of each sector is proportional to the corresponding group frequency), it seems that a principal peak is produced between August and the beginning of September (climatological season of fires) and a secondary peak is produced around the end of March. In any case, the conclusions about the number of fires' seasons will depend on where each sector begins and the number of groups specified.

Benali et al. [6], who tackled the problem from a global perspective, initially considered a regionwise analysis. In particular, they compared which of the following models fits in a better way the underlying distribution of the data: a von Mises (Model 1) or a mixture of two von Mises (Model 2). The density functions of those models are

$$\text{Model 1}: \quad f(\theta; \mu, \kappa) = \frac{1}{2\pi I_0(\kappa)} \exp(\kappa \cos(\theta - \mu)),$$

$$\text{Model 2}: \quad f(\theta; \mu_1, \mu_2, \kappa_1, \kappa_2) = \frac{p}{2\pi I_0(\kappa_1)} \exp(\kappa_1 \cos(\theta - \mu_1))$$
$$+ \frac{1-p}{2\pi I_0(\kappa_2)} \exp(\kappa_2 \cos(\theta - \mu_2)),$$

$$(9.1)$$

with $\theta, \mu, \mu_1, \mu_2 \in [0, 2\pi)$ (or, depending on the characterization of the circular random variables, any other support of length 2π); $\kappa, \kappa_1, \kappa_2 \geq 0$ and $p \in (0,1)$ [or $p \in (0.05, 0.95)$ in the case of 6]. As usual, I_0 denotes the modified Bessel function of the first kind and order zero. In general, the modified Bessel function of first kind and order r is defined, in κ, as:

$$I_r(\kappa) = \frac{\left(\frac{\kappa}{2}\right)^r}{\pi^{1/2}\Gamma\left(r + \frac{1}{2}\right)} \int_{-1}^{1} (1 - t^2)^{r - \frac{1}{2}} \exp(\kappa t) dt.$$

In Figure 9.2 (right), the two models are fitted and represented for the sample with the fires detected in Galicia. The parameters in (9.1) were estimated via maximum likelihood in Model 1 and using the Expectation Maximization (EM) algorithm, introduced in [4] and implemented by [17], for carrying out maximum likelihood estimation in Model 2. For selecting which model should be used, one can use some kind of information criterion, such as the Akaike Information Criterion (AIC). The AIC, introduced by [1], takes into account the model complexity by means of the number of parameters in the model (d) and the model fit by means of the maximum value of the likelihood function ($\hat{\Lambda}$), with the following expression:

$$\text{AIC} = -2\log(\hat{\Lambda}) + 2d.$$

According to this criterion, a model is better than another one if it has a smaller AIC value. In this example, it can be concluded that Model 2 ($\text{AIC}_2 = 18726$) fits in a better way the data than Model 1 ($\text{AIC}_1 = 21592$). Then, according to this parametric fit, it seems that humans are altering the fires' seasonality in Galicia, creating a new fires' season in the middle of March.

Since the human–altered seasons of fires can be caused by several factors, the underlying structure of the wildfires may be very complex (presenting not only multimodal patterns but also asymmetry). This complex structure can be observed in the counts of human fires in Figure 5 of [19] and a simple parametric fit may lead to a misspecification of the model. Even more, depending

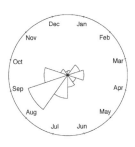 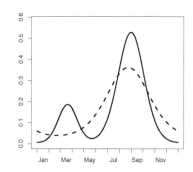

FIGURE 9.2
Sample with the fires detected in Galicia (Spain) from July 10, 2002 to July 9, 2012. Left: raw circular plot and rose diagram. Right: representation of the fitted von Mises density function (dashed line) and the fitted mixture of two von Mises densities (solid line).

on the estimated values of $\mu_1, \mu_2, \kappa_1, \kappa_2$ and p, the mixture of two von Mises model can be unimodal indicating the absence of human fires for both models in (9.1). For constructing more complicated but flexible models, reflecting the complex structure of fires, two approaches can be followed. On the one hand, taking a parametric route, the number of components in the mixture could be increased (at the cost of increasing also the number of parameters). As a second approach, a nonparametric density estimator could be constructed. In this case, given a random sample of angles $(\Theta_1, ..., \Theta_n)$, the kernel density estimation (KDE) can be used for estimating the probability density function f. This KDE (see, for example, [21] Chapter 3) is defined as

$$\hat{f}_\nu(\theta) = \frac{1}{n} \sum_{i=1}^{n} K\left(\theta; \Theta_i, \nu\right), \tag{9.2}$$

where K_ν is the circular kernel with concentration parameter ν. An example of circular kernel is the von Mises distribution (Model 1 in 9.1). Taking this kernel, the KDE can be seen as a mixture of n von Mises centered in each sample data and with the same weights $(1/n)$ and common concentration $\kappa = \nu$. Depending on this concentration parameter, different conclusions about the number of modes can be derived where, in general, small (large) values of ν provide estimations with less (more) modes. For the example in Galicia, as shown in Figure 9.3, when the concentration parameter is "small" ($\nu = 2$) the density estimation is unimodal with a mode in the middle of August, while larger values of ν lead to multimodal estimations with modes also in the middle of March ($\nu = 5, 10, 20$) and in the middle of October ($\nu = 20$).

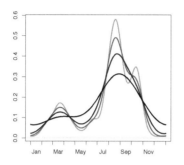 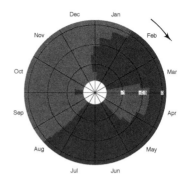

FIGURE 9.3

Sample with the fires detected in Galicia (Spain) from July 10, 2002 to July 9, 2012. Left: KDE with von Mises kernel and concentration parameters (from dark to light gray): 2, 5, 10, and 20. Right: CircSiZer map, given a concentration parameter (as indicated in the radius in $-\log_{10}$ scale) and an angle, blue color indicates locations where the curve is significantly increasing, red color shows where it is significantly decreasing, and magenta indicates where its derivative is not significantly different from zero; thus, for a given concentration and reading the CircSiZer in the clockwise sense of rotation, a blue–red pattern indicates where there is a significant peak.

An important issue in the use of kernel methods is whether observed features in the smoothed curve (e.g., the KDE), such as modes, are artifacts of the sampling variability. Based on the idea of exploring the significant modes for different values of ν, the CircSiZer, introduced by [30] using the von Mises kernel and by [18] employing the wrapped normal kernel (for more details, see [21], Section 4.3), reveals if the smoothed curve is significantly increasing/decreasing and displays this information in a color map. For doing so, a confidence interval is constructed for the scale–space versions of $\hat{f}_{\nu}(\theta)$, that is, in different locations θ and for different concentration parameters ν. There are several ways of constructing these confidence intervals (see [30]). In the CircSiZer plotted in Figure 9.3 (right), the so–called "bootstrap–t" approach, implemented by [31], was employed. The confidence intervals for the derivative of the KDE for a given location (a specific angle of the annulus) and a concentration parameter (determined by the radius of the annulus in the plot, in $-\log_{10}$ scale) are used, in the CircSiZer map, to indicate, in blue color, where the smoothed curve is significantly increasing, red color shows where it is significantly decreasing and magenta indicates where its derivative is not significantly different from zero. Then, given a value of $-\log_{10}(\nu)$ (which corresponds to a certain ring) and following the sense of rotation given by the arrow in the upper–right part, modes can be detected as blue–red patterns.

In case there is not enough data to make statements about significance, a gray area would appear in the CircSiZer. In the largest rings (far away from the center), oversmoothed estimations with large values of $-\log_{10}(\nu)$ (small values of ν) are obtained. Hence, the conclusions about the number of modes depend on the concentration parameter. This is the case for the fires detected in Galicia where, depending on the observed ring, the estimated number of modes, in the CircSiZer map represented in Figure 9.3 (right), is different. In this case, the CircSiZer map detects one mode in the middle of August for all the concentration parameters (all the rings show a blue–red pattern) and this would correspond with the climatological season. For the "medium and small" rings two modes (the second one in the middle of March) are detected by the CircSiZer and a third mode (mid–October) appears when taking "small" values of ν. These other modes would be related with the human activity.

Observing the CircSiZer (Figure 9.3, right), it seems reasonable to think that humans are altering the fires' seasonality in Galicia, but this is just an exploratory tool and does not provide a formal answer to the question about the number of fires' seasons. As mentioned before, the conclusions based on the CircSiZer depend on ν and require an expert eye for interpretation. A way to avoid this problem (without imposing any parametric model) is testing the number of modes in a nonparametric way. It will be seen later that this testing approach allows for a systematic application on a global scale. In this local problem with the Galicia data example, it is of interest to test if the fires detected show just one mode or if more than one mode are significant. Formally, if j is the number of modes of the random variable, the testing problem can be formulated as

$$H_0 : j = 1 \quad \text{vs.} \quad H_a : j > 1. \tag{9.3}$$

For solving this testing problem, two proposals can be found in the circular nonparametric literature. The first one, based on the U^2 of [36], was proposed by [12]. The second one, using the excess mass statistic of [27], was introduced in [2]. The test statistic used in [12] is defined as

$$U^2 = n \int_0^{2\pi} \left[F_n(\theta) - F_0(\theta) - \int_0^{2\pi} (F_n(\phi) - F_0(\phi)) dF_0(\phi) \right]^2 dF_0(\theta). \tag{9.4}$$

It measures the distance between the empirical distribution function F_n and a given distribution function F_0. For testing k–modality (that is $H_0 : j = k$), this F_0 is replaced by a kernel distribution estimation with k modes. To approximate the distribution of the test statistic (9.4) under the null hypothesis, a smoothed bootstrap procedure is proposed. To perform the bootstrap procedure the new resamples are also generated from a smoothed distribution with k modes. Ameijeiras-Alonso [2] showed through a simulation study that the proposal of [12] is poorly calibrated in practice even for "large" sample sizes. For that reason, Ameijeiras-Alonso [2] provided a new proposal based on the excess mass of [27], adapted to the circular setting. Given a sample

$(\Theta_1, \ldots, \Theta_n)$, the excess mass statistic is defined as

$$\Delta_{n,k+1} = \max_\lambda \{D_{n,k+1}(\lambda)\}, \tag{9.5}$$

where $D_{n,k+1}(\lambda) = E_{n,k+1}(\mathbb{P}_n, \lambda) - E_{n,k}(\mathbb{P}_n, \lambda)$ and where $E_{n,k}(\mathbb{P}_n, \lambda)$ is called the empirical excess mass function for k modes and it is defined as

$$E_{n,k}(\mathbb{P}_n, \lambda) = \sup_{C_1, \ldots, C_k} \left\{ \sum_{l=1}^{k} (\mathbb{P}_n(C_l) - \lambda ||C_l||) \right\}, \tag{9.6}$$

where the supremum is taken over all families $\{C_l : l = 1, \cdots, k\}$ of pairwise disjoint connected arcs on the circle. In the definition (9.6), $\mathbb{P}_n$ denotes the empirical probability, $\mathbb{P}_n(C_l) = (1/n) \sum_{i=1}^{n} \mathcal{I}(\Theta_i \in C_l)$, $\mathcal{I}$ is the indicator function and $||C_l||$ is the measure of C_l. An example of the theoretical excess mass function, for two modes and a given value of λ, is shown in Figure 9.4 for illustrative purposes. For testing k modes, the test statistic (9.5) measures the difference in mass probability between considering that the density has $(k+1)$ or k modes. "Large" values of this test statistic will indicate that at least one more mode is present, i.e., that the null hypothesis is false. To approximate the distribution of this test statistic, in [2], a bootstrap procedure, generating the resamples from a modification of the kernel density estimation with k modes, was proposed. It was also showed that this proposal has a correct behavior even for relatively "small" sample sizes. Using this method for testing if there is just one season of fires in Galicia, the null hypothesis in (9.3) is rejected even for a significance level of 1% (with $B = 1000$ bootstrap replicates), supporting evidence against the null hypothesis of just one (climatological) season of fires and therefore confirming the interference of human activity on fires' seasonality.

9.2.2   Global Scale

As already mentioned, the problem of determining the number of seasons of fires at a global scale was considered by [6], from a parametric perspective, and by [2], using nonparametric techniques both for estimation and testing. It should be noted that an important advantage of the aforementioned testing tools is that their availability enables a systematic application on different regions toward a global analysis. For that reason, with the aim of determining the number of modes in thousands of regions of the globe, in [2] also the testing approach was considered for studying the entire world (divided in regions of size $0.5°$). When doing this study, the issues related with performing a multiple testing and the spatial dependence (in terms of the fires' behavior) between the regions of study must be considered. For taking into account these two issues, an adaptation of the [7] method was employed in [2]. The method allows to correct the *False Discovered Rate* (FDR) accounting for the spatial dependence of the data just considering the dependence structure between the regions, which is obtained from land use information. Once this is done, using

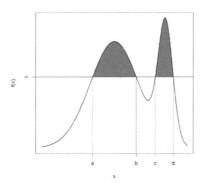

FIGURE 9.4

The excess mass is the probability mass exceeding a given level λ. In this case, it is equal to the gray area and it is obtained when taking the union of two arcs $[a, b]$ and $[c, d]$.

$B = 5000$ bootstrap replicates and correcting the p–values, considering both an underlying spatial dependence structure and the multiple testing scenario, it is obtained that almost all the world presents a multimodal pattern and the nonrejected regions are scattered. This suggests the prevalence of fires' use as a land management tool at a global scale.

In [6], a parametric approach was developed for studying the number of fires' seasons. For each region of the world and in each year, they proposed to select one of the two distributions described in (9.1) using the model efficiency index [29], or neither of them when the "goodness of fit is poor." As pointed out before, even selecting the two von Mises distribution (Model 2 in 9.1), the estimation output could be unimodal. Being aware of this problem, Benali et al. [6] propose to identify just one season of fires each time that the distance between the estimated location parameters in the two fitted von Mises ($\hat{\mu}_1$ and $\hat{\mu}_2$) is less than 3 months and a half. Then, the same study was done using the 10 years data and the regions cataloged as bimodal were divided into three categories: predominantly, frequently, and sporadically bimodal. This last classification was done according to the previous study where the seasonality is studied every year separately. Also, some of the regions cataloged as other (with a "poor" goodness of fit) were reclassified when a clear pattern was found in the year by year study. After doing this classification, seven categories (shown in Figure 9.5) were obtained: no fire, low fires' incidence, other, unimodal, predominantly bimodal (when 7 out of 10 years were bimodal), frequently bimodal (when 4 to 6 out of 10 years were bimodal) and sporadically bimodal (when up to 3 out of 10 years were bimodal). With this parametric fit, about 25% of the regions with relevant fires' activity present a bimodal pattern. These regions are concentrated in the 11 areas delimited

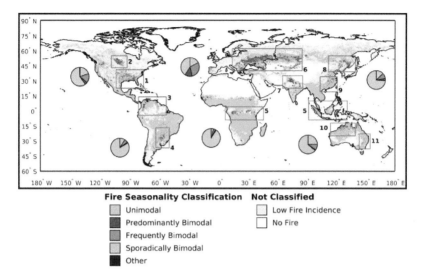

FIGURE 9.5
In colors, fires' seasonality classification according to [6]. The pie charts indicate the percentage of regions cataloged in each pattern (unimodal or the different kinds of bimodality) in each of the six continents: Africa, Asia, Australia, Europe, and North and South America. The rectangles delimit the principal areas with regions cataloged as bimodal.

with rectangles in Figure 9.5. The main conclusions, from the map obtained by [6], are that just in Europe humans are clearly altering the fires' seasonality, with more than 50% of the regions cataloged as bimodal. In addition, it should be noted that Africa has the lowest percentage of bimodal fires' activity, with less than 10% of bimodal regions.

Just some final comments should be made about the parametric approach. During the last few years, more flexible parametric models have been introduced in the circular literature (see, for example, [21] Section 2.2). Then, an interesting challenge is to find a parametric flexible distribution that allows modeling the fires' seasonality taking into account the spatial dependence of fires and also other covariates that influence the fires' behavior such as land use, in order to understand in a better way how humans are altering the fires' seasonality.

9.3 Fires' Orientation

When studying wildfires' behavior at a local scale, understanding how fire spreads is important to prevent and avoid fire growth. In this sense, fires' orientation, and also its relation with other variables such as terrain orientation and slope, fires' size or wind characteristics, are important issues that should be taken into account to explain fires' propagation. The dataset that will be used for illustrating how fires' orientation can be modelled contains information about the occurrence of wildfires in Portugal (with a summarized version of data grouped by 102 watersheds) and was initially analyzed by [5] with the aim of studying the spread of fires. This dataset was obtained from the *Instituto Superior de Agronomia National* fires' atlas, based on late summer/fall Landsat imagery. The period of study ranged from 1975 to 2005 in [5], while in [13] the fires until 1985 were removed due to lack of accuracy. A total of 34,345 wildfires were considered [26,870 in the study of 13, 14].

For such an analysis, fires' orientations must be defined and computed, in the plane or in the sphere. Specifically, following the proposal by [5], fires' perimeters (see Figure 9.6, obtained from Figure 9.4 of [5]) are considered as a collection of points on the plane. The fires' orientation is determined with a principal component analysis following the approach proposed by Luo [22, pp. 131–136]. On the plane, the orientation of a fire is taken as the minimum angle (in the clockwise sense of rotation) that the first principal component of the perimeter points forms with respect to the meridian passing through the perimeter center of the fire. In this case, the orientations constitute axial data, and for introducing circular models the solution in [5] and in García-Portugués et al. [13, 14] was taking twice the observed orientation. This first approach on the plane considers, for constructing the fires' perimeter, values of latitude and longitude. In a second step, altitude can be also added, obtaining with the same procedure an orientation on the three-dimensional sphere [13]. Analyzing if this orientation and the burnt area (linear variable) are related may help to understand the behavior of fires in Portugal [13], in that case, this relationship can be modelled via joint (multidimensional cylindrical) distributions [14] or with a regression model.

Before entering the example in detail, it should be mentioned that there are also other directional variables that may alter the fires' size or orientation. In particular, modeling the effects that wind direction has on fires' size [37, 38] or its relation with the spread orientation of wildfires (see [34] for an introduction to this topic) provides other directional sources of data. In the first case, in Los Angeles county, the wind direction [38] and the spatial location of wildfires [37], among other meteorological variables, are considered as dependent variables for studying in a regression model the response of fires' size. Up to the authors' knowledge, the second case still remains unsolved in the directional field, although some initial studies were already developed for the South–East of Australia in [35], where the implications that wind orientation have on the fires' behavior and bushfire risk management are analysed. This relationship between the wind and fires' orientation may be studied considering a multidimensional torus (directional–directional joint distribution or regression).

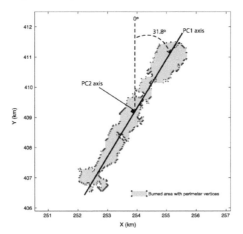

FIGURE 9.6
Example of how the fires' orientation is obtained from its perimeter in [5].

9.3.1 Main Spread on the Orientation of Fires

Consider fires' orientations in the plane (circular data), obtained as described in the previous section. First, in [5], different circular tests (Kuiper, Watson, and Rayleigh) are applied for determining if the underlying distribution of the fires, in each watershed, is uniform (i.e., the value of its density function is constant for all the orientation parameters), follows a von Mises (model 1 in 9.1) or if there is a preferential orientation and with which angle it corresponds. The obtained results are showed in Figure 9.7 (left, from Figure 9.6 of [5]). When neither the uniform nor the von Mises are suitable models for the data, it should be figured out if there is a unique mode and estimate its location nonparametrically.

Since the mean slopes of fires present marked differences between the different regions (see Figure 9.6, center, obtained from Figure 9.1 of [13]), in [13], for studying the relationship of orientation and size, the authors also take the slope of the fire considering the first principal component of the fires' perimeter in the three-dimensional space. In general, if Ω_q is a q–dimensional sphere $\Omega_q = \{\boldsymbol{\theta} \in [0, \pi)^{q-1} \times [0, 2\pi)\}$, the random variable $\boldsymbol{\Theta}$ will be directional when supp$(\boldsymbol{\Theta}) \subseteq \Omega_q$. In this general case, parametric approaches for modeling this random variable include, among others [21, Section 2.3], the von Mises–Fisher distribution, with probability density function:

$$f_q(\boldsymbol{\theta}; \boldsymbol{\mu}, \kappa) = C_q(\kappa) \exp\left(\kappa \boldsymbol{\theta}_l^T \boldsymbol{\mu}_l\right) \tag{9.7}$$

with $\boldsymbol{\theta} \in \Omega_q$, location parameter $\boldsymbol{\mu} \in \Omega_q$, concentration parameter $\kappa \geq 0$, where the subindex l refers to the Cartesian coordinates, that is, if $\boldsymbol{\phi} \in$

$[0, \pi)^{q-1} \times [0, 2\pi)$, the elements of ϕ_l will be equal to

$$
\begin{aligned}
\phi_{l1} &= \cos(\phi_1), \\
\phi_{l(q+1)} &= \prod_{i=1}^{q} \sin(\phi_i), \\
\phi_{lj} &= \cos(\phi_j) \prod_{i=1}^{j-1} \sin(\phi_i), \text{ if } q > 1, \text{ with } j = 2, \ldots, q;
\end{aligned}
$$

and the normalization constant C_q is a function defined in κ as:

$$
C_q(\kappa) = \frac{\kappa^{\frac{q-1}{2}}}{(2\pi)^{\frac{q+1}{2}} I_{(q-1)/2}(\kappa)}.
$$

As mentioned before, if orientations on the circle are considered, in some occasions, none of the previously mentioned distributions are valid or flexible enough for modeling the data at hand. As a next step, [13] proceeded from a nonparametric perspective with a kernel density estimation. Given a directional sample in the general sphere, this estimation can be obtained in a similar way as in (9.2) just considering a spherical kernel (see [21] Section 3.2.1). In particular, taking (9.7) as a kernel, the density estimation can be seen as a mixture of n von Mises–Fisher densities centered in each sample data, with the same weights $(1/n)$ and with the same concentration parameter.

9.3.2　Orientation–Size Joint Distribution

Once it is shown how to model the orientation of the fires (parametrically and nonparametrically, on a circular and on a spherical domain), another important question is to determine if there is any relationship between fires' orientation and their size. With descriptive purposes and for a first visual comparison with fires' orientation, the area burnt in each watershed is plotted in Figure 9.7 (right). The previous problem can be approached studying the relationship between both components (fires' orientation and size) by a joint distribution estimation with directional and linear random variables, (Θ, Z), with support in the multidimensional cylinder supp$(\Theta, Z) \subseteq \Omega_q \times \mathbb{R}$. For a parametric approach in this setting, some references can be found in Ley and Verdebout [21, Section 2.4]. However, since no more information is provided, a nonparametric estimation can be an alternative. Given a random sample $\{(\Theta_1, Z_1), \ldots, (\Theta_n, Z_n)\}$, the cylindrical (directional and linear components) kernel density estimator at a point (θ, z) is defined (see, for example, [21], Section 3.2.2) as follows

$$
\hat{f}_{\nu,h}(\theta, z) = \frac{1}{n} \sum_{i=1}^{n} KL(\theta, z; \Theta_i, Z_i, \nu, h), \tag{9.8}
$$

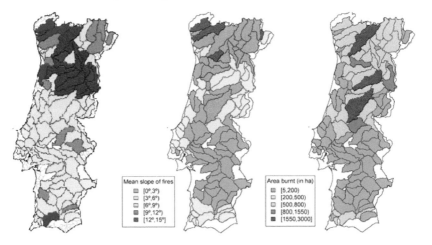

FIGURE 9.7
Map of Portugal divided in 102 watersheds. Left: map from [5], where the watersheds with no preferential orientation are presented in gray color and the ones with preferential orientations are in other colors; according to their mean fire perimeter orientation: around 0° (±22.5°; North/South spread) in blue, around 45° (North–East/South–West) in red, around 90° (East/West) in green and around 135° (South–East/North–West) in yellow. Center: mean fires' slope on each watershed, where the slope is measured in degrees, from plain (0°) to vertical slope (90°). Right: total burnt area in each watershed.

where KL is a directional–linear kernel, $h > 0$ the linear bandwidth parameter and $\nu > 0$ the directional concentration parameter. In particular, when considering the product kernel $KL(\cdot, \cdot; \Theta_i, Z_i, \nu, h) = K(\cdot; \Theta_i, \nu) \times L(\cdot; Z_i, h)$, K is the von Mises–Fisher density (9.7) and L the standard normal density, this density estimation can be seen as a mixture of n components, with the same weights $(1/n)$, of the density product of a von Mises–Fisher density centered in each Θ_i, with concentration parameter ν, and a normal density centered in each Z_i and common standard deviation h. In [13], the kernel density estimation was used for testing the null hypothesis of independence between both components (orientation and size), which can be stated in terms of density functions as:

$$H_0 : f_{(\Theta, Z)}(\theta, z) = f_\Theta(\theta) f_Z(z), \quad \forall (\theta, z) \in \Omega_q \times \mathbb{R},$$

against the alternative

$$H_a : f_{(\Theta, Z)}(\theta, z) \neq f_\Theta(\theta) f_Z(z), \quad \text{for any } (\theta, z) \in \Omega_q \times \mathbb{R},$$

where f_Θ and f_Z denote the directional and the linear marginals. The test statistic considered is the $\mathcal{L}_2$ distance in $\Omega_q \times \mathbb{R}$ between the nonparametric estimation of the joint distribution, $\hat{f}_{\nu, h}$, and its estimation under the null, that is, the product of the marginal directional and linear kernel estimators, denoted by $\hat{f}_\nu$ and $\hat{f}_h$, respectively. Specifically:

$$T_n = \int_{\Omega_q \times \mathbb{R}} \left(\hat{f}_{\nu,h}(\boldsymbol{\theta}, z) - \hat{f}_\nu(\boldsymbol{\theta}) \hat{f}_h(z) \right)^2 w_q(d\boldsymbol{\theta}) dz,$$

where w_q is the Lebesgue measure of Ω_q. The null hypothesis of independence between both components is rejected when this test statistic is too "large." Two considerations must be emphasized about this test statistic. First, it depends on the concentration–bandwidth parameter (ν, h) and second the "slow" rate of convergence of the test statistic distribution to the Gaussian limit under independence hampers the use of an asymptotic test in practice. Since there is no proper bandwidth for testing, two approaches can be followed. The first one could be to consider a data–driven bandwidth, García-Portugués et al. [13] suggest a Likelihood Cross Validation (LCV) or the so–called Bootstrap Likelihood Cross Validation concentration–bandwidth (see [13] for more details). Another alternative is to run the test for a collection of bandwidths and analyze its trace (its behavior along the selected range). Regarding the slow rate of convergence to the asymptotic distribution, this problem can be overcome by resampling methods. For instance, García-Portugués et al. [13] propose a bootstrap method where the test statistic is computed from the σ–permuted resample, that is from the random sample $\{(\boldsymbol{\Theta}_1, Z_{\sigma(1)}), \ldots, (\boldsymbol{\Theta}_n, Z_{\sigma(n)})\}$, where σ is a random permutation of n elements. The results reported in [13] showed that this test has a good behavior in terms of calibration and power, even for "small" sample sizes. Then, on the aggregated data by watersheds, this test can be used for testing the relation between orientation and size (burnt area), taking in both cases *mean* values as representatives of the fires in each region. When considering the detailed sample (not aggregated by watershed), then the null hypothesis is clearly rejected (with $B = 1000$ bootstrap replicates), even with a significance level $\alpha = 0.01$, indicating that there is a relation between orientation and size, both when the orientations are considered on the circle or on the sphere.

When studying the fires independently in each watershed, the maps in Figure 9.8 (from left to right, first for circular and third for spherical orientations) are obtained. From these maps, it can be observed that the independence between both variables is only supported for some watersheds and, again, a multiple testing problem appears. In this example, García-Portugués et al. [13] argued that it seems that there is no spatial dependence and the multiple testing problem is corrected, in their case, with the FDR procedure of [8]. The corrected p–values are shown in Figure 9.8 (from left to right, second for circular and forth for spherical orientations). From the results of [13], for a significance level of $\alpha = 0.05$, just in one watershed the null hypothesis is rejected. If also the altitude coordinate is considered, in 15 watersheds (18 for $\alpha = 0.10$) independence between orientation and size is rejected.

Once the null hypothesis of independence between fires' orientation and size is rejected, a parametric cylindrical distribution model could be consid-

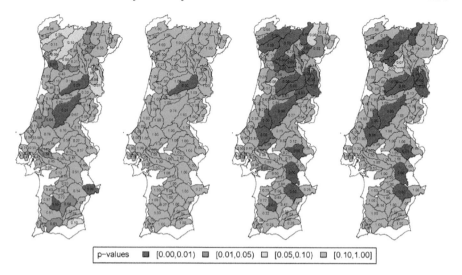

FIGURE 9.8

P–values obtained after applying the independence test between the orientation and the size of the fire in each watershed, from [13]. From left to right: considering the circular orientation (first, without correcting and second, using the FDR procedure) and the spherical orientation (third, without correcting and fourth, using the FDR procedure).

ered, but which model should be appropriate? In [14], a test for assessing a specific parametric form was proposed and applied to the summary of the wildfires detected in Portugal (average fires' orientation and size in each watershed). Using this cylindrical data (considering the circular orientation and the linear size), it was studied if the distribution introduced by [24] is a suitable model. Given a value on a cylinder (θ, z), with $\theta \in [0, 2\pi)$ and $z \in \mathbb{R}$, the density function of [24] is defined as

$$f(\theta, z; \mu_1, \mu_2, \kappa, \sigma, \rho_1, \rho_2) = \frac{1}{2\pi I_0(\kappa)} \exp(\kappa \cos(\theta - \mu_1)) \frac{1}{\sqrt{2\pi\sigma_c^2}} \exp\left(-\frac{(x - \mu_c)^2}{2\sigma_c^2}\right),$$

where $\mu_1 \in [0, 2\pi)$, $\kappa > 0$ and

$$\mu_c = \mu_2 + \sqrt{\kappa}\sigma(\rho_1(\cos(\theta) - \cos(\mu_1)) + \rho_2(\sin(\theta) - \sin(\mu_1))),$$

$$\sigma_c^2 = \sigma^2(1 - \rho^2), \text{ where } \rho = \sqrt{\rho_1^2 + \rho_2^2},$$

with $0 \le \rho \le 1$, $\sigma > 0$ and $\mu_2 \in \mathbb{R}$. For illustration purposes, in Figure 9.9, the parametric fit is represented of the 102 circular–linear sample points using the [24] distribution and also the kernel density estimation (9.8) with the von Mises–normal product kernel.

Given the random cylindrical sample $\{(\boldsymbol{\Theta}_1, Z_1), \ldots, (\boldsymbol{\Theta}_n, Z_n)\}$ the problem of interest, in this case, is testing if the distribution of the data belongs to a

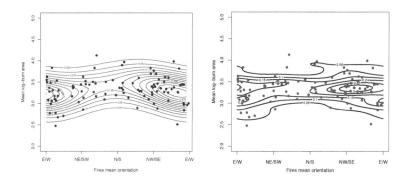

FIGURE 9.9
Represented with points on the scatter plot: sample with the mean orientation
and size of the wildfires in each of the 102 watersheds in Portugal. Left: parametric fit using the [24] distribution, obtained from [14]. Right: nonparametric
kernel density estimation (9.8) with the product von Mises–normal kernel.

family of parametric densities indexed by the parameter ψ, $\mathcal{F}_\Psi = \{f_\psi : \psi \in \Psi\}$, that is,

$$H_0 \;:\; f(\boldsymbol{\theta}, z) = f_{\psi_0}(\boldsymbol{\theta}, z) \text{ for all } (\boldsymbol{\theta}, z) \in \Omega_q \times \mathbb{R} \text{ vs.}$$
$$H_1 \;:\; f(\boldsymbol{\theta}, z) \neq f_{\psi_0}(\boldsymbol{\theta}, z) \text{ for some } (\boldsymbol{\theta}, z) \in \Omega_q \times \mathbb{R}$$

with ψ_0 known (simple null hypothesis) or unknown (composite null hypothesis). As mentioned before, in the wildfires detected in Portugal, the family
tested by [14] was the one introduced by [24] with unknown parameters, using
as a test statistic:

$$R_n = \int_{\Omega_q \times \mathbb{R}} (\hat{f}_{\nu,h}(\boldsymbol{\theta}, z) - KL_{\nu,h} f_{\hat{\psi}}(\boldsymbol{\theta}, z))^2 dz \omega_q(d\boldsymbol{\theta}), \qquad (9.9)$$

where $\hat{\psi}$ denotes a $\sqrt{n}$–consistent estimator of the unknown parameter ψ_0 (in
the simple null hypothesis, this test statistic is constructed in the same way,
just replacing the estimator by the known parameter). In the expression (9.9),
$KL_{\nu,h} f_{\hat{\psi}}(\boldsymbol{\theta}, z)$ is the expected value of $f_{\nu,h}(\boldsymbol{\theta}, z)$ under H_0, that is,

$$KL_{\nu,h} f_{\hat{\psi}}(\boldsymbol{\theta}, z) = \int_{\Omega_q \times \mathbb{R}} KL\left(\boldsymbol{\theta}, z; \mathbf{y}, t, \nu, h\right) f_{\hat{\psi}}(\mathbf{y}, t) dt \omega_q(d\boldsymbol{\theta})(d\mathbf{y}). \qquad (9.10)$$

The test statistic (9.9) depends on the concentration–bandwidth parameter
(ν, h) and it exhibits the same problems of the independence test previously
presented. Hence, its calibration in practice is carried out using a parametric
bootstrap procedure. The bootstrap statistic is defined as in (9.9) replacing
the kernel density estimation and the estimator of ψ_0 by those ones calculated

from the bootstrap resamples generated by the density $f_{\hat{\psi}}$. For the practical application, a LCV concentration–bandwidth is taken and $B = 1000$ bootstrap replicates are generated for calibration. The obtained p–value is equal to 0.156, showing that there are no evidences against the null hypothesis of the data following the [24] model.

9.3.3   Orientation–Size Regression Modeling

Although the joint distribution modeling approach is interesting for providing some insight on the joint behavior of the variables, a regression approach could be also useful. In this case, the influence of fires' orientation on fires' size could be modelled by means of directional–linear regression:

$$Z = m(\boldsymbol{\Theta}) + \sigma(\boldsymbol{\Theta})\epsilon, \tag{9.11}$$

where $m(\boldsymbol{\theta}) = \mathbb{E}(Z|\boldsymbol{\Theta} = \boldsymbol{\theta})$ is the conditional mean, $\sigma(\boldsymbol{\theta}) = \mathrm{Var}(Z|\boldsymbol{\Theta} = \boldsymbol{\theta})$, the conditional variance and the errors ϵ are random variables with zero mean and unit variance. From a nonparametric approach, the local directional–linear regression estimator of [10] (and later modified by [15]) can be employed for estimating m in a nonparametric way. Given the random sample $\{(\boldsymbol{\Theta}_1, Z_1), \ldots, (\boldsymbol{\Theta}_n, Z_n)\}$, where $(\boldsymbol{\Theta}_i, Z_i) \in \Omega_q \times \mathbb{R}$, the authors propose to obtain an estimator $\hat{m}_{\hat{\beta},\nu}$ for the regression function m, using a local constant or a local linear approximation, namely $\tilde{m}_{\beta}$ (with $\beta = \beta_0 \in \mathbb{R}$ for the local constant estimator and $\beta \in \mathbb{R}^{q+1}$ for the local linear). The estimators are derived by solving a weighted least squared problem in each case:

$$\text{Constant}\ :\ \hat{\beta}_0 = \min_{\beta_0 \in \mathbb{R}} \sum_{i=1}^{n} (Z_i - \tilde{m}_{\beta_0})^2 \, K(x; \boldsymbol{\Theta}_i, \nu),$$

$$\text{Linear}\ :\ \hat{\beta} = \min_{\beta \in \mathbb{R}^{q+1}} \sum_{i=1}^{n} (Z_i - \tilde{m}_{\beta})^2 \, K(x; \boldsymbol{\Theta}_i, \nu),$$

where weights are taken according to K, a directional kernel.

In fact, when a nonparametric regression fit is available, this estimator could be used to explore in more detail how the fires' orientation influences the fires' size. In Figure 9.10, the 102 data points with the average fires' orientation and size in the Portuguese watersheds are represented for illustration purposes. From the scatter plot and the nonparametric estimation, it can be observed that the regression relationship between the fires' orientation and size cannot be explained with a non-effect model (this conclusion can be also tested with the [15] proposal). In addition, using the CircSiZer map of [30], where the construction of this exploratory tool for circular–linear regression is similar to the one in the density case (see Section 9.2.1), one could analyze if there are fires' orientations linked to the largest fires. From this plot, it can be

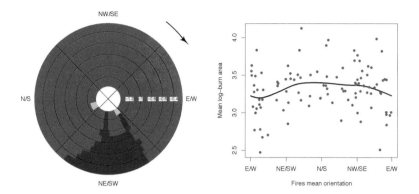

FIGURE 9.10

Sample with the mean orientation and size of the wildfires in each of the 102 watersheds in Portugal. Left: CircSiZer map for the orientation–size regression model, given a concentration parameter (as indicated in the radius in $-\log_{10}$ scale) and an angle, blue color indicates locations where the regression curve is significantly increasing, red color shows where it is significantly decreasing and magenta indicates where its derivative is not significantly different from zero; gray color shows where there is not enough data to make statements about significance. Right: scatter plot of the data and local circular–linear regression estimator of [10].

concluded that the largest burnt areas in Portugal are produced when the principal spread of the fires is in the North/South or North–West/South–East orientations. This exercise could be reproduced in a local level, such a knowledge would be helpful for a better planning of resources deployment during fires' suppression activities.

9.4   Open Problems

Different applications of directional statistics in the study of wildfires have been discussed. Interesting examples are obtained when modeling the time of fires' occurrence as a circular random variable considering an annual cycle. Circular and cylindrical techniques are also required for analyzing the fires' orientation and its relation with other variables, such as their size. The results presented allow to obtain relevant insight into fires' behavior, both globally and locally. The illustration was focused on the Iberian Peninsula (south–western part of Europe), but as shown along the chapter, there are other locations where understanding the behavior of fires and its relation with other variables is an important issue. Yet, there are several open problems that can be approached from a directional point of view.

As previously pointed out, in those regions where it has a special relevance, a complete characterization of the time occurrence of fires is still necessary for a better understanding of wildfires' seasonality and how humans are altering the fires' annual cycle. This can also help to understand in a better way human activity in different parts of the world. For doing so, different covariates can be considered and, in particular, as shown by [2] and [37], the spatial correlation structure of fires will play an important role. This spatial dependence at local scale can be studied in the plane, but when considering a global scale, the Earth must be considered as a spherical object and more sophisticated covariance models are required.

There is also further interesting work to do in order to understand how fires' spread and grow. In particular, having a better knowledge of which directional variables are creating larger fires and the influence that they have on different fires' characteristics are important issues. Among others, the terrain orientation, wind direction and speed can be used as circular or spherical responses for fires' orientation or size prediction in different regions of the world. In the fires' suppression activities, a better understanding of these relations will allow those concerned with fires' management practices to estimate how a fire will evolve under certain conditions.

Bibliography

[1] H. Akaike. A new look at the statistical model identification. *IEEE Transactions on Automatic Control*, 19:716–723, 1974.

[2] J. Ameijeiras-Alonso. *Assessing Simplifying Hypotheses in Density Estimation*. PhD thesis, Universidade de Santiago de Compostela, 2017. Available at https://www.educacion.gob.es/teseo/mostrarRef.do?ref= 1573365.

[3] S. Archibald, C. E. Lehmann, J. L. Gómez-Dans, and R. A. Bradstock. Defining pyromes and global syndromes of fire regimes. *Proceedings of the National Academy of Sciences*, 110:6442–6447, 2013.

[4] A. Banerjee, I. S. Dhillon, J. Ghosh, and S. Sra. Clustering on the unit hypersphere using von Mises–Fisher distributions. *Journal of Machine Learning Research*, 6:1345–1382, 2005.

[5] A. M. G. Barros, J. Pereira, and U. J. Lund. Identifying geographical patterns of wildfire orientation: A watershed–based analysis. *Forest Ecology and Management*, 264:98–107, 2012. ISSN 0378–1127. doi: https://doi.org/10.1016/j.foreco.2011.09.027.

[6] A. Benali, B. Mota, N. Carvalhais, D. Oom, L. M. Miller, M. L. Campagnolo, and J. Pereira. Bimodal fire regimes unveil a global–scale anthropogenic fingerprint. *Global Ecology and Biogeography*, 26:799–811, 2017.

[7] Y. Benjamini and R. Heller. False discovery rates for spatial signals. *Journal of the American Statistical Association*, 102:1272–1281, 2007.

[8] Y. Benjamini and D. Yekutieli. The control of the false discovery rate in multiple testing under dependency. *Annals of Statistics*, 29:1165–1188, 2001.

[9] W. J. Bond and J. E. Keeley. Fire as a global herbivore: the ecology and evolution of flammable ecosystems. *Trends in Ecology & Evolution*, 20: 387–394, 2005.

[10] M. Di Marzio, A. Panzera, and C. C. Taylor. Local polynomial regression for circular predictors. *Statistics & Probability Letters*, 79:2066–2075, 2009.

[11] E. Dwyer, J. Pereira, J.-M. Grégoire, and C. C. DaCamara. Characterization of the spatio–temporal patterns of global fire activity using satellite imagery for the period April 1992 to March 1993. *Journal of Biogeography*, 27:57–69, 2000.

[12] N. I. Fisher and J. S. Marron. Mode testing via the excess mass estimate. *Biometrika*, 88:419–517, 2001.

[13] E. García-Portugués, A. M. G. Barros, R. M. Crujeiras, W. González-Manteiga, and J. Pereira. A test for directional–linear independence, with applications to wildfire orientation and size. *Stochastic Environmental Research and Risk Assessment*, 28:1261–1275, 2014.

[14] E. García-Portugués, R. M. Crujeiras, and W. González-Manteiga. Central limit theorems for directional and linear random variables with applications. *Statistica Sinica*, 25:1207–1229, 2015.

[15] E. García-Portugués, I. Van Keilegom, R. M. Crujeiras, and W. González-Manteiga. Testing parametric models in linear–directional regression. *Scandinavian Journal of Statistics*, 43:1178–1191, 2016.

[16] L. Giglio, I. Csiszar, and C. O. Justice. Global distribution and seasonality of active fires as observed with the Terra and Aqua Moderate Resolution Imaging Spectroradiometer (MODIS) sensors. *Journal of Geophysical Research: Biogeosciences*, 111:201601–201612, 2006.

[17] K. Hornik and B. Grün. movMF: An R package for fitting mixtures of von Mises–Fisher distributions. *Journal of Statistical Software*, 58:1–31, 2014. doi: 10.18637/jss.v058.i10.

[18] S. Huckemann, K.-R. Kim, A. Munk, F. Rehfeldt, M. Sommerfeld, J. Weickert, and C. Wollnik. The circular SiZer, inferred persistence of shape parameters and application to early stem cell differentiation. *Bernoulli*, 22:2113–2142, 11 2016. doi: 10.3150/15-BEJ722.

[19] S. Korontzi, J. McCarty, T. Loboda, S. Kumar, and C. O. Justice. Global distribution of agricultural fires in croplands from 3 years of moderate resolution imaging spectroradiometer (MODIS) data. *Global Biogeochemical Cycles*, 20:202101–202115, 2006.

[20] Y. Le Page, D. Oom, J. Silva, P. Jönsson, and J. Pereira. Seasonality of vegetation fires as modified by human action: observing the deviation from eco–climatic fire regimes. *Global Ecology and Biogeography*, 19:575–588, 2010.

[21] C. Ley and T. Verdebout. *Modern Directional Statistics*. Chapman & Hall/CRC Press, Boca Raton, FL, 2017.

[22] D. Luo. *Pattern Recognition and Image Processing*. Horwood Publishing Limited, Cambridge, United Kingdom, 1998.

[23] B. I. Magi, S. Rabin, E. Shevliakova, and S. Pacala. Separating agricultural and non-agricultural fire seasonality at regional scales. *Biogeosciences*, 9:3003–3012, 2012.

[24] K. V. Mardia and T. W. Sutton. A model for cylindrical variables with applications. *Journal of the Royal Statistical Society. Series B (Methodological)*, 40:229–233, 1978.

[25] J. R. Marlon, P. J. Bartlein, C. Carcaillet, D. G. Gavin, S. P. Harrison, P. E. Higuera, F. Joos, M. Power, and I. Prentice. Climate and human influences on global biomass burning over the past two millennia. *Nature Geoscience*, 1:697, 2008.

[26] F. Moreira, O. Viedma, M. Arianoutsou, T. Curt, N. Koutsias, E. Rigolot, A. Barbati, P. Corona, P. Vaz, G. Xanthopoulos, F. Mouillot, and E. Bilgilij. Landscape–wildfire interactions in southern Europe: implications for landscape management. *Journal of Environmental Management*, 92: 2389–2402, 2011.

[27] D. W. Müller and G. Sawitzki. Excess mass estimates and tests for multimodality. *Journal of the American Statistical Association*, 86:738–746, 1991.

[28] NASA Visible Earth. Fires in Spain and Portugal, from Jeff Schmaltz, MODISRapid Response Team, Goddard Space Flight Center. https://visibleearth.nasa.gov/view.php?id=17136, 2006. [Online; accessed September 14, 2017].

[29] J. E. Nash and J. V. Sutcliffe. River flow forecasting through conceptual models part I – A discussion of principles. *Journal of Hydrology*, 10: 282–290, 1970.

[30] M. Oliveira, R. M. Crujeiras, and A. Rodríguez-Casal. CircSizer: an exploratory tool for circular data. *Environmental and Ecological Statistics*, 21:143–159, 2014.

[31] M. Oliveira, R. M. Crujeiras, and A. Rodríguez-Casal. NPCirc: An R package for nonparametric circular methods. *Journal of Statistical Software*, 61:1–26, 2014.

[32] R. Quill, J. J. Sharples, and L. A. Sidhu. Effects of post–fire vegetation regrowth on wind fields over complex terrain. In T. Weber, M. J. McPhee, and R. S. Anderssen, editors, *MODSIM2015, 21st International Congress on modeling and Simulation. modeling and Simulation Society of Australia and New Zealand*, pages 277–283, December 2015.

[33] D. A. Schmidt, A. H. Taylor, and C. N. Skinner. The influence of fuels treatment and landscape arrangement on simulated fire behavior, Southern Cascade range, California. *Forest Ecology and Management*, 255: 3170–3184, 2008.

[34] J. J. Sharples. Review of formal methodologies for wind–slope correction of wildfire rate of spread. *International Journal of Wildland Fire*, 17: 179–193, 2008.

[35] J. J. Sharples, R. H. D. McRae, and R. O. Weber. Wind characteristics over complex terrain with implications for bushfire risk management. *Environmental modeling & Software*, 25:1099–1120, 2010.

[36] G. S. Watson. Goodness–of–fit tests on a circle. *Biometrika*, 48:109–114, 1961.

[37] H. Xu and F. P. Schoenberg. Point process modeling of wildfire hazard in Los Angeles County, California. *The Annals of Applied Statistics*, 5: 684–704, 2011.

[38] H. Xu, K. Nichols, and F. P. Schoenberg. Kernel regression of directional data with application to wind and wildfire data in Los Angeles County, California. *Forest Science*, 57:343–352, 2011.

10

Bayesian Analysis of Circular Data in Social and Behavioral Sciences

Irene Klugkist

Utrecht University and Twente University, the Netherlands

Jolien Cremers

Utrecht University, The Netherlands

Kees Mulder

Utrecht University, The Netherlands

CONTENTS

10.1	Introduction		212
10.2	Introducing Two Approaches Conceptually		213
	10.2.1	Intrinsic	214
	10.2.2	Embedding	215
10.3	Bayesian Modeling		216
	10.3.1	Intrinsic	217
	10.3.2	Embedding	218
10.4	The Development of Spatial Cognition		218
	10.4.1	The Data	218
	10.4.2	Bayesian Inference	219
	10.4.3	Inequality Constrained Hypotheses	225
10.5	Basic Human Values in the European Social Survey		227
	10.5.1	The Data	228
	10.5.2	The Model	229
	10.5.3	Variable Selection	230
	10.5.4	Bayesian Inference	231
	10.5.5	Comparison of Approaches	234
10.6	Discussion		236
	Notes		237
	Bibliography		237

10.1   Introduction

Circular data can arise from different measurement scales like the compass and the clock. Although such data may arise more often in fields like biology and environmental sciences, also in the social and behavioral sciences researchers are occasionally confronted with circular measurements. Examples of circular data representing directions in degrees or radians are, for instance, found in human movement psychology and cognitive psychology. Examples are studies using a moving room experiment [36], experiments on bimanual coordination [23], and studies on spatial cognition and learning [28]. The outcome of interest in such studies often is an angular deviation, that is, the observed response is compared to the correct response. The first illustration in this chapter uses data of the study by Bullens et al. [5] which is also an example from the field of spatial cognition.

Another instrument providing circular data is the clock. There are ample studies with a specific interest in the periodicity of events happening. As an example, we mention studies in criminology, where the time of day or day in the year that certain crimes occur are of interest [1, 4]. Another example is a study investigating cyclic trends in the occurrence of suicide [30].

In the social and behavioral sciences there is an additional instrument providing circular data, called the circumplex model. Examples of circumplex instruments in educational sciences are Leary's circumplex scale [21] and the interpersonal circumplex [39]. An example from psychology is Russel's mood scale [32]. In this chapter, in the second illustration, we will use data from a sociological circumplex instrument, that is, the Basic Human Values (BHV) scale by Schwartz [33, 34].

For the analysis of circular data, three main approaches are distinguished: the intrinsic, embedding, and wrapping approach. In this chapter we will analyze two example data sets with the intrinsic and the embedding approach and discuss and compare the results. Since all analyses will be executed in the Bayesian framework, we will also discuss several aspects of the Bayesian modeling, like prior specification and Markov chain Monte Carlo (MCMC) sampling. We will omit many of the technical details but provide references for additional reading.

The main contribution of this chapter is the presentation of two worked examples, one from psychology and one from sociology. We will show results of several analyses, both using the intrinsic and the embedding approach, and discuss aspects that are relevant for users of circular statistics, e.g., model assumptions, convergence checks, and interpretation of results. We will also use the examples to illustrate some of the advantages of using a Bayesian approach.

In Section 10.2, we will introduce the intrinsic and embedding approach in a simple and conceptual manner. In Section 10.3, we will provide the model

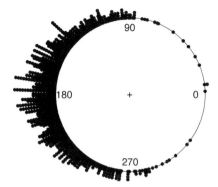

FIGURE 10.1
An example of data distributed on the circle in a symmetric and unimodal shape.

and its notations and some more details on the Bayesian estimation approach. Then, the first illustration is presented in Section 10.4. It deals with circular data representing angular errors in a spatial learning experiment, where the main focus lies on comparing three independent mean directions. We will also use this example to introduce a test for the evaluation of inequality constrained hypotheses using the Bayes factor. Section 10.5 will provide a multiple regression example, where the outcome variable of interest is the score on the BHV circumplex instrument. First, we will use the Deviance Information Criterion (DIC; [35]) to select relevant predictor variables from four candidates. Then, for the best fitting model, we will provide estimates for the relations between predictor(s) and outcome with elaborate discussions of the interpretation of these estimates. The chapter is concluded with a discussion of results in Section 10.6.

10.2   Introducing Two Approaches Conceptually

We will limit the discussion to data on a circle where the distribution looks very much like a normal distribution, that is, it is symmetric and unimodal. An example is provided in Figure 10.1, where the dots represent observations on the circle. Note that they are spread around the entire circle, but that there is a clear single mode and the distribution around the mode is rather symmetric. In Figure 10.1, the value zero is (conventionally though arbitrarily) located on the right-hand side of the circle and scores are in degrees, i.e., from $0 - 360°$. Equivalently, one could use radians, i.e., from $0 - 2\pi$, as for instance in Figure 10.2.

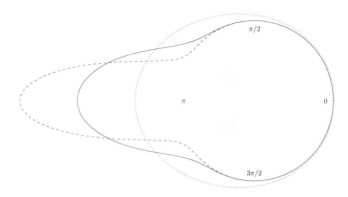

FIGURE 10.2
Three von Mises distributions, all with mean direction π but with different dispersion measures, namely $\kappa = 2$ (dotted), $\kappa = 10$ (solid), and $\kappa = 25$ (dashed).

Within both the intrinsic and the embedding approach the "normality" assumption leads to a natural choice for modeling the data. Both approaches also have, to a more or lesser extent, the flexibility to model data that is distributed differently on the circle, as for instance skewed, or bimodal. This will however not be considered in this chapter.

10.2.1 Intrinsic

The intrinsic approach models a distribution directly on the circle. The circular analogue of a normal distribution is the von Mises (vM) distribution, which has two parameters, one for the mean direction (denoted μ) and one for the concentration (denoted κ). In Figure 10.2, three $vM(\mu, \kappa)$ distributions are plotted. They all have mean direction $\mu = \pi$ (i.e., 180°), with different dispersion levels $\kappa = 2$ (dotted), $\kappa = 10$ (solid), and $\kappa = 25$ (dashed).

For angular data $\theta \in [0, 2\pi)$, the vM probability density function (pdf) is

$$f(\theta \mid \mu, \kappa) = \frac{1}{2\pi I_0(\kappa)} \exp\{\kappa \cos(\theta - \mu)\}, \tag{10.1}$$

where $I_0(\cdot)$ represents the modified Bessel function of the first kind and order zero.

A sample of observations $\boldsymbol{\theta} = \theta_1, \ldots, \theta_n$ is assumed to come from a $vM(\mu, \kappa)$ distribution and then used for inference on the mean direction μ and the level of concentration κ. In a Bayesian framework this implies that a prior distribution must be specified for the model parameters (μ, κ) which is then updated with the information in the sample $\boldsymbol{\theta}$ to provide the posterior distribution, that summarizes our knowledge about the vM parameters.

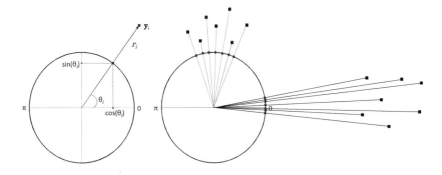

FIGURE 10.3
Projecting $\boldsymbol{y}$ on the circle (left), and the relation between distance of the bivariate data to the circle and the circular spread (right).

Summaries of the posterior distribution provide all desired quantities for each parameter, like the posterior (circular) mean, median, or mode (the last is comparable to the frequentist maximum likelihood estimate), a posterior (circular) standard deviation (comparable to the standard error), and a credible interval (comparable to the frequentist confidence interval). Furthermore, by using an MCMC approach, that is, iteratively sampling values for each parameter from the posterior distribution, it is straightforward to obtain estimates for functions of parameters.

10.2.2 Embedding

The embedding approach is based on the idea that observations on the circle can be considered projections of data in the (linear) bivariate space. Assuming a bivariate normal distribution, $MVN(\boldsymbol{\mu}, \boldsymbol{\Sigma})$, for the unobserved bivariate data leads to the projected normal (PN) distribution for the observed circular data.

Figure 10.3 provides an illustration. In the plot on the left, one can see how an observation $\boldsymbol{y}_i = (y_i^I, y_i^{II})^t$ in bivariate (linear) space (the square) can be projected on the circle (the dot). Two notations can be used to locate the circular observation, the angle θ_i or its decomposition in the sine and cosine components denoted $\boldsymbol{u}_i = (\cos\theta_i, \sin\theta_i)^t$. The length of $\boldsymbol{y}_i$, that is, $\|\boldsymbol{y}_i\|$, is denoted r_i and the relation between $\boldsymbol{y}_i$ and $\boldsymbol{u}_i$ is given as

$$\boldsymbol{u}_i = \frac{\boldsymbol{y}_i}{r_i}. \tag{10.2}$$

Note that projecting any set of data points in bivariate linear space to the circle is straightforward. However, in the analysis of circular data, the situation is the other way around. We have observed data on the circle but we lack knowledge about their "originating" positions in bivariate space since the r_i are not observed. There is no unique solution, i.e., without further constraints we have an identification problem. The assumption that is often made is the restriction of the variance-covariance matrix of the bivariate data to be an identity matrix ($\boldsymbol{\Sigma} = \boldsymbol{I}$). Given this assumption and the observed spread of the data on the circle, the distances r_i and subsequently the bivariate mean $\boldsymbol{\mu}$ of the unobserved linear data can be estimated.

The plot on the right in Figure 10.3 shows the relation between the spread of data on the circle and the distance of the data points in bivariate space by providing two differently diffused circular data sets. The one with mean direction around zero has less variation than the one in direction $\pi/2$ ($90°$). Because the variance-covariance matrix is fixed ($\boldsymbol{\Sigma} = \boldsymbol{I}$), the bivariate linear data related to the more spread circular data must be located closer to the circle compared to the other data set that leads to less variation on the circle.

The probability density function of the projected normal distribution is

$$PN(\theta \mid \boldsymbol{\mu}, \boldsymbol{I}) = \frac{1}{2\pi} \exp\left(-\frac{1}{2}||\boldsymbol{\mu}||^2\right)\left[1 + \frac{\boldsymbol{u}^t\boldsymbol{\mu}\ \Phi(\boldsymbol{u}^t\boldsymbol{\mu})}{\phi(\boldsymbol{u}^t\boldsymbol{\mu})}\right], \qquad (10.3)$$

where $\boldsymbol{u}_i = (\cos\theta_i, \sin\theta_i)^t$, $\boldsymbol{\mu} = (\mu^I, \mu^{II})^t$, and $\Phi(\cdot)$ and $\phi(\cdot)$ denote the cumulative distribution function and the probability density function of the standard normal distribution.

A sample of observations $\boldsymbol{\theta} = \theta_1, \ldots, \theta_n$ is assumed to come from a PN distribution and used for inference on the mean $\boldsymbol{\mu}$ in bivariate space. Note that this is not yet the parameter that we are interested in and that reparametrization is needed to obtain circular effects. This will be elaborated in the illustrations, especially in the regression model presented in Section 10.5.

10.3   Bayesian Modeling

To illustrate the approaches, we will use two statistical models that are common in psychological and sociological research. In the first illustration we will compare the mean directions of K independent groups (i.e., the circular analogue of the analysis of variance model), and in the second illustration we will investigate a regression model with a circular outcome and P linear predictors. For the general model that contains both types of predictors we will provide the choice of prior distributions and an overview of the steps of the MCMC algorithm, omitting all technical details that were previously presented elsewhere.

10.3.1   Intrinsic

In Section 10.2.1, the $vM(\theta \mid \mu, \kappa)$ model for a circular variable θ was provided. In the regression model, for observation i $(i = 1, \ldots, n)$, the mean vector is $\mu_i = \beta_0 + \boldsymbol{\delta}^t \boldsymbol{d}_i + g(\boldsymbol{\beta}^t \boldsymbol{x}_i)$ where $\boldsymbol{d}_i$ is a column vector of observed values for J dummy variables indicating group membership and $\boldsymbol{x}_i$ is a column vector of observed values for P continuous linear predictors, that are all centered. Further, $\beta_0 \in [-\pi, \pi)$ is an offset parameter which serves as a circular intercept, $\boldsymbol{\delta} \in [-\pi, \pi)^J$ is a column vector of circular group difference parameters, $g(\cdot) : \mathbb{R} \to (-\pi, \pi)$ is a twice differentiable link function, and $\boldsymbol{\beta} \in \mathbb{R}^K$ is a column vector of regression coefficients. The link function $g(\cdot)$ used is $g(x) = 2 \arctan(x)$ [10, 12, 20].

The parameters to be estimated are β_0, $\boldsymbol{\delta} = [\delta_1, \ldots, \delta_J]$, $\boldsymbol{\beta} = [\beta_1, \ldots, \beta_P]$ and κ is the concentration of errors. A motivation for this model as well as extensive simulations to show the performance can be found in [25].

In a model with multiple groups and no continuous predictors, the circular intercept β_0 is the estimated mean direction of the reference group and $\boldsymbol{\delta}$ contains the estimated mean differences of the $J = K - 1$ other groups with this reference group. In a model with only continuous predictors, β_0 is the circular intercept and $\boldsymbol{\beta}$ contains the P slopes of the linear predictors $\boldsymbol{x}$.

Prior distributions need to be specified for the parameters β_0, κ, $\delta_1, \ldots, \delta_J$, and $\beta_1, \ldots, \beta_P$. We will use the prior specification as presented in [25]. First, the joint prior is factored as

$$p(\beta_0, \kappa, \boldsymbol{\delta}, \boldsymbol{\beta}) \propto p(\beta_0, \kappa \mid \boldsymbol{\delta}, \boldsymbol{\beta}) \prod_{j=1}^{J} p(\delta_j) \prod_{p=1}^{P} p(\beta_p). \qquad (10.4)$$

Then, for (β_0, κ) we use the conjugate prior as introduced by [14]

$$p(\beta_0, \kappa \mid \boldsymbol{\delta}, \boldsymbol{\beta}) \propto I_0(\kappa)^{-c} \exp \left[R_0 \kappa \cos(\beta_0 - \mu_0) \right],$$

with prior mean direction μ_0, prior resultant length R_0, and prior "sample size" c. In all our analyses, we use hyperparameters $c = 0$ and $R_0 = 0$ which makes the prior constant (and thus the value for μ_0 arbitrary). Further, for δ_j $(j = 1, \ldots, J)$, we use the circular uniform distribution $p(\delta_j) = 1/(2\pi)$, and for β_p $(p = 1, \ldots, P)$ we use the standard normal distribution $p(\beta_p) \sim N(0, 1)$. The choice of the variance of the latter is relevant and discussed and investigated in [25]. In both applications in this chapter, the standard normal distribution works well and [25] showed that it is generally a good choice for standardized predictors.

With MCMC sampling, a sample from the joint posterior is obtained. Using the algorithm of [25], this entails a Gibbs step to sample β_0, the rejection sampler introduced by [11] to sample κ, and Metropolis Hastings steps for the β_p and for the δ_j parameters.

10.3.2 Embedding

In Section 10.2.2, the $PN(\theta \mid \boldsymbol{\mu}, \boldsymbol{I})$ for a circular variable θ was provided. In the regression model, for observation i ($i = 1, \ldots, n$), the mean vector is $\boldsymbol{\mu}_i = \boldsymbol{B}^t \boldsymbol{x}_i$, with $\boldsymbol{B} = [\boldsymbol{\beta}^I, \boldsymbol{\beta}^{II}]$ and $\boldsymbol{x}_i$ a vector of predictors. Note that the first column of $\boldsymbol{x}_i$ is a vector of ones to estimate the intercept and subsequent columns can contain continuous predictors as well as dummy variables to represent categorical predictors. The dimension of $\boldsymbol{x}_i$ is therefore $1 + P + J$. Continuous predictors are centered before inclusion in the analysis.

Prior distributions need to be specified for the parameters in $\boldsymbol{B}$. We will use the prior specification as presented in [7]. For both components $\boldsymbol{\beta}^I = [\beta_1^I, \ldots, \beta_{1+P+J}^I]$ and $\boldsymbol{\beta}^{II} = [\beta_1^{II}, \ldots, \beta_{1+P+J}^{II}]$ a multivariate normal prior distribution is used, with mean vector $\boldsymbol{\beta}_0 = \boldsymbol{0}$ and precision matrix $\boldsymbol{\Lambda}_0$, with dimension $(1 + P + J) \times (1 + P + J)$ and diagonal values equal to 1×10^{-4} and off diagonal values zero. The latent lengths $\boldsymbol{r} - [r_i, \ldots, r_n]$ are defined on $[0, \infty)$.

With MCMC sampling, a sample from the joint posterior is obtained. Using the algorithm of [8], this entails Gibbs sampling for $\boldsymbol{B}$ and a slice sampling approach [15, 26] for $\boldsymbol{r}$. Note that, although not the parameters of main interest, the distribution of vector length r_i for each observation $i = 1, \ldots, n$ is obtained with the MCMC sampler as well. The parameters in $\boldsymbol{B}$ describe the group differences (in a design with multiple groups) and/or relations (in a design with continuous predictors) that we are primarily interested in.

10.4 The Development of Spatial Cognition

10.4.1 The Data

The data for this illustration come from Bullens et al. [5]. In their study they investigated the development of children's ability to locate objects in an environment. The test environment consisted of a circular arena where participants, under different conditions, were asked to retrieve hidden objects. The outcome variable of interest is the difference between the true location of the hidden object and the direction pointed at by the participant. Variations in the experiment were created by the presence of movable local landmarks (e.g., a traffic cone placed in the arena) and distal landmarks (e.g., lights or specific shapes surrounding the arena). Note that we will only use a subset of the data to illustrate the statistical procedures. For the full design, theoretical background, as well as the results and their (substantive) implications we refer to [5].

The data we use contain scores for three subgroups of participants: 5-year-olds, 7-year-olds, and adults. The data are plotted in Figure 10.4. The score per participant is the average over several trials within the same con-

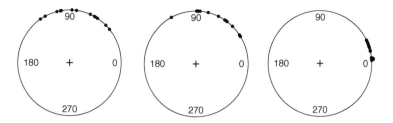

FIGURE 10.4
Data (angular errors) from the spatial memory experiment per subgroup (left: 5-year olds; middle: 7-year olds; right: adults).

dition. Note that, within this experiment, the locations of the hidden object, local landmarks, and distal landmarks are chosen such that a certain search strategy leads to pointing in the (near) correct direction and another search strategy leads to deviations in the counter clockwise direction. Of interest is to what extent, in the different age groups, respondents use the better strategy. Therefore, we will first estimate the mean angular errors for the three groups. Then, we will formulate our expectations in explicit hypotheses and show how these can be evaluated with Bayes factors.

10.4.2   Bayesian Inference

Intrinsic approach. In the formulation of the model, the adults were defined as the reference group, the 5-year-olds get $d_1 = 1$ and the 7-year-olds $d_2 = 1$ leading to

$$\mu_i = \beta_0 + \delta_1 d_{1i} + \delta_2 d_{2i}.$$

Application of the MCMC algorithm provides a sample of values for the model parameters β_0, δ_1, δ_2, and κ. A first set of iterations is needed as a burn-in phase and is therefore deleted. Subsequent iterations comprise a sample from the posterior distribution. It is important to inspect trace plots to determine the appropriate numbers of iterations required for burn-in and convergence of the chain. Convergence can further be checked by running multiple chains with different (arbitrary) starting values.

Convergence plots for all four parameters are provided in Figure 10.5. After deleting the first 500 iterations as burn-in, 50,000 iterations with thinning 5 provide the set of 10,000 iterations that are shown in the plots. We concluded that both the burn-in period and the total number of iterations are more than sufficient. All results in this section (estimates, model selection criterion, and Bayes factor) are based on this set of iterations.

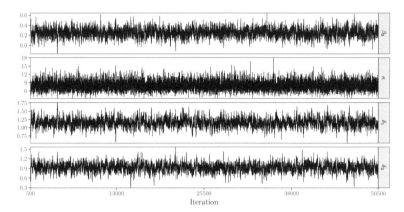

FIGURE 10.5

Trace plots for the model parameters (from top to bottom) β_0, κ, δ_1, δ_2 of the intrinsic approach.

In Table 10.1, we illustrate how the MCMC sample can be used to obtain any estimate of interest for each of the parameters presented in the columns β_0, δ_1, δ_2 and κ. First, we like to point out that row-wise, that is, per iteration, one can compute any function of parameters that is (also) of interest. In this example, we computed $\mu_{5yr} = \beta_0 + \delta_1$ and $\mu_{7yr} = \beta_0 + \delta_2$ within each iteration. Note that μ_{ad} equals the already available column β_0 but is repeated here to provide the three group means of interest together. This provides three additional columns that can be summarized in the same way as the original columns.

Column-wise summaries of the sampled values provide, for each parameter, estimates like posterior (circular) means, posterior (circular) standard deviations (SD), and lower bounds (LB) and upper bounds (UB) of 95% posterior credible intervals. Note that we computed *highest posterior density* (HPD) intervals (definition is provided below). All results are presented in the lower part of Table 10.1. Here, we will shortly outline how each estimate is computed.

First, it is important to note that summary statistics are computed differently for linear and circular variables. In this example, the only linear variable is κ. The posterior mean and posterior standard deviation are computed by applying the usual definitions for computing a mean and SD of a set of numbers, to the column of sampled values of κ. By sorting the κ values and denoting the resulting vector $\boldsymbol{\kappa}' = (\kappa^{(1)}, \ldots, \kappa^{(q)})$ for q iterations after burn-in, the LB and UB of the 95% HPD are obtained as

$$LB = \operatorname*{argmin}_{\kappa^{(i)} \in \kappa^{(1)}, \ldots, \kappa^{(.05q)}} \left\{ \kappa^{(i + \lfloor 0.95q \rfloor)} - \kappa^{(i)} \right\}, \qquad (10.5)$$

$$UB = \operatorname*{argmin}_{\kappa^{(i)} \in \kappa^{(.95q)}, \ldots, \kappa^{(q)}} \left\{ \kappa^{(i)} - \kappa^{(i - \lfloor 0.95q \rfloor)} \right\}, \qquad (10.6)$$

TABLE 10.1
Estimates for Bullens data using the intrinsic approach

It	β_0	δ_1	δ_2	κ	μ_{5yr}	μ_{7yr}	μ_{ad}
1	.	.	.	.	.	.	.
⋮	⋮	⋮	⋮	⋮	⋮	⋮	⋮
10000	.	.	.	.	.	.	.
Mean	0.26	1.14	0.93	7.92	1.40	1.19	0.26
Mode	0.26	1.11	0.93	7.72	1.40	1.21	0.26
SD	0.09	0.14	0.14	1.75	0.11	0.11	0.09
LB	0.08	0.87	0.65	4.59	1.19	0.98	0.08
UB	0.44	1.43	1.19	11.32	1.61	1.39	0.44

that is, the bounds define the shortest interval that contains 95% of the posterior samples. For the mode, the 10% HPD is computed adjusting the equation above accordingly, and the midpoint of the resulting interval is the estimated mode [37].

For the circular variables β_0, δ_1, and δ_2 as well as the three group mean directions μ_k, the summary statistics are computed differently. For instance, for β_0 the posterior mean is computed from the sampled values $\beta_0^{(1)}, \ldots, \beta_0^{(q)}$ as

$$\bar{\beta}_0 = \text{atan2}\left\{\sum_{i=1}^{q} \sin \beta_0^{(i)}, \sum_{i=1}^{q} \cos \beta_0^{(i)}\right\}, \tag{10.7}$$

where $\text{atan2}(a, b)$ computes the direction of vector (b, a) from the origin. It can be computed using

$$\text{atan2}(a, b) = \begin{cases} 2\arctan\left(\frac{a}{b+\sqrt{a^2+b^2}}\right) & \text{if } b > 0 \text{ or } a \neq 0, \\ \pi & \text{if } b < 0 \text{ and } a = 0, \\ \text{undefined} & \text{if } b = 0 \text{ and } a = 0. \end{cases}$$

The posterior standard deviation (SD) of β_0 is given in terms of the mean resultant length

$$\bar{R} = \sqrt{\left(\sum_{i=1}^{q} \sin \beta_0^{(i)}\right)^2 + \left(\sum_{i=1}^{q} \cos \beta_0^{(i)}\right)^2}. \tag{10.8}$$

The mean resultant length has a range from zero to one, where larger values refer to smaller spread. The circular SD is then computed as

$$SD_{circ} = \sqrt{-2\log\bar{R}}. \tag{10.9}$$

The circular HPD must also be computed differently because the shortest

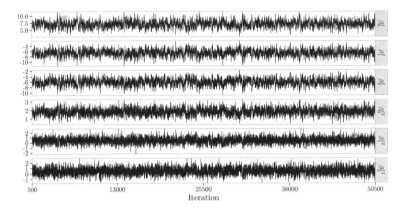

FIGURE 10.6

Trace plots for the model parameters (from top to bottom) β_0^I, β_1^I, β_2^I, β_0^{II}, β_1^{II}, β_2^{II} of the embedding approach.

arc containing 95% of the posterior samples might contain zero, which means it will not be identified correctly by the linear method described above. Therefore, we first ensure that the q sampled β_0 values (after burn-in) are within a range of 2π, e.g., from $0 - 2\pi$. From here, we treat the β_0 values as linear parameters. Sorting the β_0 values provides $\boldsymbol{\beta'_0} = (\beta_0^{(1)}, \ldots, \beta_0^{(q)})$ and this vector is augmented with $\boldsymbol{\beta'_0} + 2\pi$. Now, the shortest interval containing $.95q$ of the samples in $\{\boldsymbol{\beta'_0}, \boldsymbol{\beta'_0} + 2\pi\}$ provides the correct 95% circular HPD interval. Again, we use the midpoint of the 10% HPD to estimate the circular mode.

The same formulas provide the posterior mean, mode, SD, LB, and UB for the other circular variables.

Embedding approach. In the embedding approach we used the same dummy coding for the three groups as in the intrinsic approach, leading to

$$\mu_i^I = \beta_0^I + \beta_1^I d_{1i} + \beta_2^I d_{2i}$$

$$\mu_i^{II} = \beta_0^{II} + \beta_1^{II} d_{1i} + \beta_2^{II} d_{2i}$$

MCMC sampling in the embedding approach provides a sample of values for β_0^I, β_1^I, β_2^I, β_0^{II}, β_1^{II}, and β_2^{II}. Figure 10.6 shows convergence plots for 10,000 iterations (after thinning 5 and burn-in 500) for all parameters. They show that, also using the embedding approach, these numbers are sufficient. All results in this section are based on these 10,000 iterations.

As shown in the description of the intrinsic approach, from the MCMC output we can compute new columns representing functions of the sampled parameters. The original, sampled parameters are the separate component scores, and from these we can obtain the circular mean directions (μ_k), that

TABLE 10.2
Estimates for Bullens data using the embedding approach

		Mean	Mode	SD	LB	UB
First component:	β_0^I	7.02	6.77	1.23	4.58	9.32
	β_1^I	-6.62	-6.61	1.27	-9.11	-4.22
	β_2^I	-6.06	-5.83	1.27	-8.49	-3.56
Second component:	β_0^{II}	1.81	1.80	0.40	1.04	2.59
	β_1^{II}	0.43	0.57	0.60	-0.74	1.60
	β_2^{II}	0.51	0.40	0.61	-0.69	1.71
Mean directions:	μ_{5yr}	1.40	1.40	0.14	1.13	1.66
	μ_{7yr}	1.18	1.20	0.12	0.93	1.42
	μ_{ad}	0.25	0.25	0.04	0.18	0.32
Mean res. length:	ρ_{5yr}	0.89	0.93	0.05	0.79	0.96
	ρ_{7yr}	0.87	0.91	0.06	0.75	0.96
	ρ_{ad}	0.99	0.99	0.004	0.98	1.00

is, the parameters of primary interest. They are computed, for each iteration, as

$$\mu_{ad} = \text{atan2}(\beta_0^{II}, \beta_0^I)$$
$$\mu_{5yr} = \text{atan2}(\beta_0^{II} + \beta_1^{II}, \beta_0^I + \beta_1^I)$$
$$\mu_{7yr} = \text{atan2}(\beta_0^{II} + \beta_2^{II}, \beta_0^I + \beta_2^I).$$

Additionally, we can obtain estimates for the circular concentrations ρ_k by computing, for each iteration, $\bar{R}_k$. We do this using the following formula from [17]:

$$\bar{R}_k = \sqrt{\pi \xi_k/2} \exp\left(-\xi_k\right) \left[I_0\left(\xi_k\right) + I_1\left(\xi_k\right)\right]$$

where $\xi_k = ||\mu_k||^2/4$ and I_ν is the modified Bessel function of the first kind and order ν. For the three groups $||\mu_k||$ is given by

$$||\mu_{ad}|| = \sqrt{(\beta_0^I)^2 + (\beta_0^{II})^2}$$
$$||\mu_{5yr}|| = \sqrt{(\beta_0^I + \beta_1^I)^2 + (\beta_0^{II} + \beta_1^{II})^2}$$
$$||\mu_{7yr}|| = \sqrt{(\beta_0^I + \beta_2^I)^2 + (\beta_0^{II} + \beta_2^{II})^2}.$$

In Table 10.2, the first 6 rows present the posterior estimates for the β's of the two linear components. The lower half of the table provides the posterior estimates of interest, that is, the mean directions and the mean resultant lengths per subgroup. For the computation of the estimates, the same formulas for linear and circular parameters as in the intrinsic approach are used. The mean directions are circular parameters, whereas all other parameters are linear.

Comparison of approaches. Looking at the results in Tables 10.1 and 10.2 and comparing them, several observations are worth discussing. First of all, the results for the group mean directions μ_{5yr}, μ_{7yr}, and μ_{ad} are almost identical between approaches when we look at their posterior means or modes. The results are presented in radians but, for interpretation, could be expressed in degrees as well. The estimated mean directions are 80°, 68° and 15°, for the 5-year-olds, 7-year olds and adults, respectively, irrespective of which approach (intrinsic or embedding) is considered. These results are in line with a visual inspection of the data in Figure 10.4.

Differences between approaches are, however, observed when looking at the posterior SD and the LB and UB of the 95% HPD intervals. In the intrinsic approach the SD values for the three groups are almost equal (0.11, 0.11, 0.09), whereas with the embedding approach more variable results are obtained (0.14, 0.12, 0.04). This is a direct result of a difference in modeling. In the intrinsic approach we modeled one concentration parameter κ and thus enforced homogeneity of within group variances [29], that affects the resulting posterior standard deviations.

The embedding approach defines a fixed variance-covariance for the bivariate linear data. The estimation approach then estimates $\boldsymbol{B}$ and the latent lengths $\boldsymbol{r}$ freely, that is without a homogeneity assumption; the length $||\boldsymbol{\mu}_i|| = ||\boldsymbol{B}^t\boldsymbol{x}_i||$ is allowed to differ for each individual. In this case that means that the spread of the circular predicted values is allowed to differ for each group. This can be seen in the last three rows of Table 10.2, where estimated mean resultant lengths are provided for each subgroup. The mean resultant length of the *adult* group is clearly larger than the other two, which corresponds to a smaller spread of the data in this subgroup, as can be seen from Figure 10.4.

Also the differences in HPD intervals can be explained by this difference in modeling. The HPD for the mean direction for adults (expressed in degrees) is considerably smaller in the embedding approach (10°, 18°) compared to the intrinsic approach (4°, 25°), which can be explained by the substantially smaller spread in the data of this subgroup. Finally note that, although the numbers for the HPD intervals differ between the intrinsic and embedding approach, the same general pattern is observed; the 95% HPD intervals show considerable overlap for the 5-year-old and 7-year old groups, while neither of them overlap with the interval of the adult group.

Since the two approaches differ with respect to assuming homogeneity of variances (intrinsic) or not (embedding), an interesting question is which model fits best for these data. A popular measure for model comparison in Bayesian approaches is the DIC [35], which is computed as follows. Denote the set of all parameters of the model (either intrinsic or embedding) by ψ and the set of posterior means and mean directions by $\bar{\psi}$. The latter are estimated from the posterior sample $\psi^{(1)}, \ldots, \psi^{(q)}$. Then, for data $\boldsymbol{\theta}$, the DIC is computed as

$$DIC = -2\log p(\boldsymbol{\theta} \mid \bar{\psi}) + 4\left[\log p(\boldsymbol{\theta} \mid \bar{\psi}) - \frac{1}{Q}\sum_{i=1}^{Q}\log p(\boldsymbol{\theta} \mid \psi^{(q)})\right]. \quad (10.10)$$

Like other information criteria (e.g., AIC or BIC), the DIC contains a fit part (first term) and a penalty for models with more effective parameters (second term). The resulting DIC value is usually not interpreted directly, but used to compare the fit of different models against one another where lower values indicate better model fit. In our case we will compare the model that is based on the von Mises distribution and a homogeneity assumption with the model based on the assumption that the (heterogeneous) projected normal distribution fits the data.

The resulting DIC values are 42.01 for the intrinsic approach and 28.59 for the embedding approach showing a clear preference for the model used in the embedding approach. This seems to be a consequence of the observed heterogeneity in the data that is not modeled in the intrinsic approach as applied. For comparison, we rerun the analysis using the intrinsic approach without imposing a homogeneity assumption, that is, estimating κ for each group separately. The resulting DIC was 27.20, now showing a slight preference for the intrinsic approach over the embedding approach. This can be explained by the fact that the embedding approach estimates more parameters leading to a larger penalty term in the DIC.

10.4.3 Inequality Constrained Hypotheses

The second part of this illustration deals with the evaluation of inequality constrained hypotheses. To test inequality constrained hypotheses we will use the Bayes factor (BF), a Bayesian model selection criterion. A Bayes factor is the ratio of the marginal likelihoods of two models, e.g., representing two different hypotheses. Computation of marginal densities can be difficult in general. In the context of inequality constrained hypotheses, however, it is rather straightforward to compute the Bayes factor of a constrained model against the same model without order constraints. In a non-circular data context [18] showed that this can be done by computing the ratio between the proportion of the posterior samples in agreement with the order constraints, also called the fit of the constrained model, and the proportion of the prior in agreement with the order constraints, a measure of model size or complexity. Note that there are a few requirements for the prior specification for this result to hold (see [24]). In both our approaches (intrinsic and embedding) the priors comply with these criteria. This Bayes factor approach to order constrained hypotheses has been studied extensively for normal linear models (see, for instance, [16, 24]) and, more recently, has been extended to a broader class of models by [13]. In the context of circular data, this method is already described in [6] and [25]. Earlier work has presented (non-Bayesian) resampling tests for order constrained hypotheses for mean directions [2, 19]. Other related work focusses on estimation of parameters under order constraints using circular isotonic regression [3, 9, 31].

In this illustration, it can easily be imagined that the theoretical expectation about the mean directions is that they are ordered in the sense that the

TABLE 10.3
Bayes factors for the hypothesis tests on the Bullens data

	Homogeneity Intrinsic	Heterogeneity Intrinsic	Embedding
H_1 vs $H_{1'}$	59.09	39.76	39.64
H_2 vs $H_{2'}$	124.76	80.59	80.36
H_2 vs H_1	1.99	1.98	1.98

angular error decreases with age. This can be written as

$$H_1 : \mu_{ad} < \mu_{7yr} < \mu_{5yr},$$

and is formalized as a decrease of the circular distance between μ_k (for $k = 5yr, 7yr, ad$) and zero, that is, the shortest arc length between μ_k and zero. For $\mu_k \in [0, 2\pi)$ this can be computed as $\min(\mu_k, 2\pi - \mu_k)$.

A more specific hypothesis could be that additionally the difference in error between the 5- and 7-year-olds will be smaller than the difference in error between the 7-year-olds and adults. We will therefore also evaluate the hypothesis

$$H_2 : \mu_{ad} < \mu_{7yr} < \mu_{5yr} \quad \text{and} \quad (\mu_{5yr} - \mu_{7yr}) < (\mu_{7yr} - \mu_{ad}).$$

Each of the hypotheses will be evaluated against its complement, that is, any ordering other than the one hypothesized. We will use $H_{1'}$ to denote the complement of H_1, and $H_{2'}$ to denote the complement of H_2. The resulting BF is a measure of support for the constraints.

To obtain the BF we use the fact that all inequality constrained hypotheses (H_c) of interest are nested in the same unconstrained model (H_{unc}). This enables easy estimation of $BF_{c,unc}$ as the ratio of the fit and the complexity of H_c [18, 24]. The fit of a constrained hypothesis is the proportion of the posterior distribution in agreement with the constraints of H_c. This proportion is estimated as the proportion of the MCMC samples where the parameter values are in agreement with these constraints. The complexity reflects how precise the hypothesis is, or stated differently, what the size of the constrained parameter space is relative to the unconstrained parameter space. For H_1 this size is $1/6$ because 3 parameters can be ordered in 6 ways and H_1 specifies one such ordering. In a similar manner, the complexity of $H_{1'} = 5/6$, of $H_2 = 1/12$, and of $H_{2'} = 11/12$.

The BF for mutually comparing two constrained hypotheses is computed by

$$BF_{H_{c1}, H_{c2}} = \frac{BF_{H_{c1}, H_{unc}}}{BF_{H_{c2}, H_{unc}}}, \tag{10.11}$$

providing the amount of support for one constrained hypothesis (H_{c1}) against another (H_{c2}).

The results for comparing both H_1 and H_2 against its complement, and for comparing H_1 and H_2 mutually, are presented in Table 10.3. For comparison, we computed two sets of Bayes factors within the intrinsic approach, that is, using the model with the homogeneity assumption as well as a heterogeneous model where for the three subgroups different κ's are estimated.

The results show, again, that the homogeneity assumption has some effect on the results but that without this assumption the intrinsic and embedding approach provide results that are almost exactly equal. We conclude that the data provide rather strong evidence for both our hypotheses. Furthermore, the additional constraint in H_2 is supported as shown by the result $BF_{H_2,H_1} > 1$. The error in spatial navigation indeed decreases with age when comparing 5-year-olds, 7-year-olds and adults and the differences are small between 5 and 7 compared to differences between 7-year-olds and adults. These results are in line with our observations and interpretations of the estimated means and HPD intervals in the previous section.

10.5   Basic Human Values in the European Social Survey

The European Social Survey[1] (ESS) is administered every 2 years since 2002 in more than 30 nations. The ESS aims to measure, explain, and compare people's social values, cultural norms and behavior patterns across nations and years. To measure people's values the ESS includes the 21-item Basic Human Values Scale (BHV) by Schwartz [33, 34]. The BHV scale is grounded in theory and well-tested internationally. The theory discriminates 10 values (e.g., Self-direction, Universalism, and Security), measured by 10 subscales created from the original items. These values are viewed as organized along two dimensions, where one dimension contrasts *openness to change* and *conservation*, the other one contrasts *self-enhancement* and *self-transcendence*. Furthermore, the scales can be placed in fixed order on the circumplex with these two dimensions, as shown in Figure 10.7. Note that we decided to collapse two of the subscales, conformity and tradition, because according to Schwartz's model they do not differ with respect to their location on the circumference. From this point, we will therefore use 9 scales measuring 9 values that are theoretically placed on the circle as depicted in Figure 10.7. For a motivation of the theoretical model we refer to [33, 34]. Several scholars have further investigated properties of the scale but details go beyond the goals of this chapter.

As noted by [34] the distinction in 10 values is somewhat arbitrary and more or less fine-grained measures can be as reasonable, depending on the goals of the research. Also, conceiving human values as organized in a circular

---

[1] http://www.europeansocialsurvey.org/

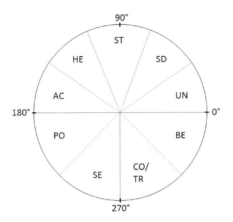

FIGURE 10.7
The Basic Human Values Scale with positioning of the values Stimulation (ST), Self-direction (SD), Universalism (UN), Benevolence (BE), Conformity/Tradition (CO/TR), Security (SE), Power (PO), Achievement (AC), Hedonism (HE).

structure implies that the whole set of values relates to other variables in an integrated manner. Analyzing the value scales separately does therefore no justice to the underlying theoretical framework of the BHV. By translating each person's score on the subscales into his or her position on the circle, that is, into an angular score, one tackles both shortcomings. The final score is an integration of the information on all values and no categorization takes place. Circular statistics can subsequently be used, for instance, to relate human values to other variables. We will use ESS data to illustrate Bayesian analysis of multiple regression models with a circular outcome. Again we will use both the intrinsic and embedding approach and compare results.

10.5.1   The Data

In this example, we will only use a small part of the ESS for illustration. Please note that data and variables are selected for this purpose and their selection was not based on theoretical models or with a specific interest in the substantive conclusions.

We used the data of 1690 Dutch respondents in the 2016 wave of data collection.[2] The distribution of the BHV scores in angles can be seen in Figure 10.8. For a set of predictors, we will investigate if they are related to BHV and how the relations can be interpreted. The included predictors are Age

---

[2]ESS Round 7: European Social Survey Round 7 Data (2014). Data file edition 2.1. NSD – Norwegian Centre for Research Data, Norway – Data Archive and distributor of ESS data for ESS ERIC.

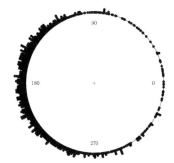

FIGURE 10.8
Distribution of scores of 1690 Dutch respondents on the BHV circumplex.

TABLE 10.4
Descriptives for ESS predictors used in the example

Variable	Mean	SD	Min	Max
Age (A)	51.4	17.6	14	92
Education (E)	9.2	5.2	1	18
Happy (H)	7.8	1.3	0	10
Religiousness (R)	4.2	3.1	0	10

(in years), Education (highest level of education on 18-point scale), and the two 10-point scales Happy (level of self-reported happiness from extremely unhappy to extremely happy) and Religiousness (self-reported level of how religious one is from not at all to very religious). Descriptive statistics for the predictors are provided in Table 10.4.

10.5.2 The Model

Intrinsic. As we have seen in Section 10.3.1, the mean direction of the vM model in a regression context is $\mu_i = \beta_0 + g(\boldsymbol{\beta}^t \boldsymbol{x}_i)$. Note that the dummy variables and corresponding parameters are now omitted since all predictors in this example are continuous. For the full model, containing all four candidate predictors, this leads to:

$$\mu_i = \beta_0 + 2\arctan(\beta_A x_{Ai} + \beta_E x_{Ei} + \beta_H x_{Hi} + \beta_R x_{Ri}).$$

With MCMC sampling, the marginal posteriors for each of the model parameters β_0, β_A, β_E, β_H, β_R, and κ are obtained and directly provide all desired estimates. The computation of posterior mean, mode, SD and HPD for both linear and circular variables is as explained in the previous example (see Section 10.4.2). Also DIC values for the comparison of models are computed using the MCMC sample.

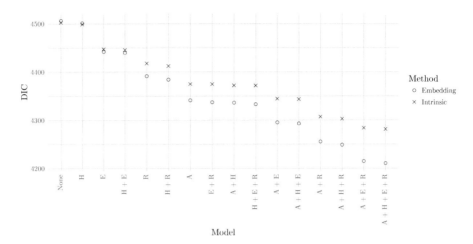

FIGURE 10.9
DIC values for all possible regression models for the intrinsic ($\times$) and the embedding ($\circ$) approach.

Embedding. As we have seen in Section 10.3.2, the mean vector in the embedding approach is $\boldsymbol{\mu}_i = \boldsymbol{B}^t \boldsymbol{x}_i$. For the model containing the four predictors this can be written as:

$$\mu_i^I \;\;= \beta_0^I + \beta_A^I x_{Ai} + \beta_E^I x_{Ei} + \beta_H^I x_{Hi} + \beta_R^I x_{Ri}$$
$$\mu_i^{II} \;= \beta_0^{II} + \beta_A^{II} x_{Ai} + \beta_E^{II} x_{Ei} + \beta_H^{II} x_{Hi} + \beta_R^{II} x_{Ri}.$$

With MCMC sampling, the marginal posteriors for each of the model parameters β_0^I, β_A^I, β_E^I, β_H^I, β_R^I, and β_0^{II}, β_A^{II}, β_E^{II}, β_H^{II}, β_R^{II} are obtained. However, to obtain estimates for the circular effects of the predictors we need to compute several functions of these parameters. These reparameterizations were introduced by [7] and provide more intuitive results than the parameters on the two linear components. This will be elaborated in Section 10.5.4.

Finally note that for both the intrinsic and embedding approach, trace plots were carefully monitored for convergence of the sampler. All results in the remainder of this section are based on summarizing 10,000 iterations from a sample of 50,500, where the first 500 were deleted as burn-in and every fifth iteration of the remaining 50,000 was saved. We will not provide the trace plots for parameters where the sample was clearly sufficient. In other cases, convergence issues will be discussed and illustrated.

10.5.3   Variable Selection

As a first step, all prediction models with either no, one, two, three, or all four predictors are estimated within the intrinsic as well as the embedding ap-

TABLE 10.5
Estimates for ESS data using the intrinsic approach

	Mean	Mode	SD	LB	UB
β_0	-3.030	-3.029	0.021	-3.070	-2.988
κ	1.986	2.006	0.070	1.856	2.125
β_A	-0.006	-0.006	0.001	-0.008	-0.005
β_E	0.010	0.010	0.002	0.006	0.014
β_H	0.017	0.018	0.008	0.002	0.032
β_R	-0.027	-0.027	0.003	-0.034	-0.021

proach. For all models, the model selection criterion DIC is computed and the best model (lowest DIC value) is detected. Note that considering all possible models is rather exploratory and therefore deserves some caution. Exploring many models for one data set is a way of capitalizing on chance and therefore increases the risk that results can not be replicated in new data. When the goal of the analysis is finding the best prediction model one should deal with this problem, for instance, by some form of cross validation. Since our goal is to demonstrate statistical tools and approaches, we will however ignore this issue in the remainder of the section.

The results are displayed in Figure 10.9. On the x-axis, the included predictors per model are listed and on the y-axis the resulting DIC values are plotted. Although the values differ between approaches, the order of fit of the different models is equal and both approaches agree that the best fitting model is the regression model with all four predictors. Therefore, we will show the estimated relations of this model in the next section.

10.5.4   Bayesian Inference

Intrinsic approach. The results of the intrinsic approach are presented in Table 10.5. All 95% HPD intervals (denoted as LB and UB in the table) exclude the value zero and therefore we conclude that the predictors have a relation with BHV. The predictors Education and Happy have positive coefficients, denoting that an increase in the predictor corresponds with an increase in the outcome, implying a counterclockwise move on the BHV scale. The other two predictors show negative and thus clockwise relations.

For the interpretation it is important to remember that all predictors are centered before inclusion in the analysis. Therefore, the intercept represents the predicted BHV score given average scores (i.e., zero after centering) on all predictors. The posterior mean ($\beta_0 = -3.030$) corresponds to the location 186° and falls in the category Power.

The interpretation of, for instance, the posterior mean of the regression coefficient for the predictor Age ($\beta_A = -0.0006$) is that this is the effect of an

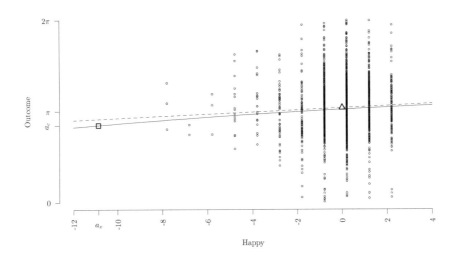

FIGURE 10.10
Regression curves for the marginal effect of Happy resulting from the intrinsic
(dashed line) and embedding approach (solid line), with the triangle at $x = 0$
and the square at $x = -10.7$ denoting the inflection points of the curves for
the intrinsic and embedding approaches, respectively.

increase of one year in Age on the circular outcome BHV conditional on all
other predictors being zero. Stated differently, the effect of Age on the BHV
score is a clockwise move of $0.34°$ per year, which seems a rather small effect.
   We can also examine prediction plots. As an example, we provide Figure
10.10, with Happy (centered) on the x-axis and the BHV scores on the y-axis.
The data are plotted together with the regression curve, represented by the
dashed line, of the marginal effect of Happy, that is, with other predictors
fixed at their means. The solid line in this figure resulted from the embedding
approach and is discussed later. At $x = 0$ a triangle through the dashed line is
plotted to denote the inflection point of the curve. In the intrinsic approach as
applied in this chapter, the inflection point is fixed at $x = 0$ and thus within
the range of the observed data. The posterior mean of the marginal effect of
Happy, $\beta_H = 0.017$, is the slope of the tangent line at this point.

Embedding approach. The results of the embedding approach are presented in
Table 10.6. On the first component, for the predictors Age and Religiousness
the HPD intervals exclude the value zero. This is not the case for predictors
Education and Happy. On the second component the HPD intervals for all four
predictors exclude the value zero. To combine the bivariate coefficients into one
circular coefficient, new measures were developed by [7]. Since these measures

TABLE 10.6
Estimates for ESS data using the embedding approach

First component	Mean	Mode	SD	LB	UB
β_0^I	-1.331	-1.334	0.033	-1.398	-1.270
β_A^I	-0.013	-0.013	0.002	-0.017	-0.010
β_E^I	-0.011	-0.011	0.006	-0.023	0.001
β_H^I	-0.032	-0.036	0.024	-0.079	0.014
β_R^I	-0.048	-0.051	0.010	-0.068	-0.028
Second component	Mean	Mode	SD	LB	UB
β_0^{II}	-0.080	-0.081	0.027	-0.131	-0.025
β_A^{II}	0.014	0.013	0.002	0.011	0.017
β_E^{II}	-0.034	-0.034	0.005	-0.044	-0.023
β_H^{II}	-0.051	-0.053	0.021	-0.091	-0.010
β_R^{II}	0.071	0.074	0.009	0.053	0.088

are functions of the bivariate coefficients they are easily computed from the MCMC output. The new interpretation tools allow us to assess whether there is an effect of a predictor on the mean of the circular outcome and how large this circular effect is.

First, predicted regression curves can be constructed to visualize the marginal circular effect of one predictor with the other predictors set at zero (i.e., due to the centering, to their mean). The circular predicted values at different x-values are computed using:

$$\hat{\theta} = \text{atan2}(\mu^{II}, \mu^I) = \text{atan2}(\beta_0^{II} + \beta_1^{II}x, \beta_0^I + \beta_1^I x).$$

The curve for the marginal effect of Happy is presented by the solid line in Figure 10.10.

Second, a reparametrization of the equation for $\hat{\theta}$ provides:

$$\hat{\theta} = a_c + \arctan[b_c(x - a_x)],$$

which is helpful for the interpretation of the effect of x on θ by focusing on the inflection point of the curve (in Figure 10.10 denoted by the square through the solid line, at $x = -10.7$). The position of the inflection point is given by a_x for the location on the x-axis and a_c for the location at the y-axis. The slope of the tangent line at the inflection point is denoted b_c and is a first quantification of the circular effect of x on θ. The formulas for the computation of a_x, a_c, and b_c are straightforward and provided in [7].

As can be seen in this example, the inflection point of the regression curve is not necessarily located near the observed data. Therefore it is not, in general, the best measure for the circular effect. Two better alternatives are the slope

TABLE 10.7
Estimates for the circular effects measures for ESS data

		Mean	Mode	SD	LB	UB
Age	b_c	-0.019	-0.019	0.002	-0.024	-0.014
	SAM	-0.011	-0.011	0.001	-0.013	-0.009
	AS	-0.011	-0.011	0.001	-0.013	-0.009
Educ	b_c	0.030	0.030	0.005	0.019	0.040
	SAM	0.025	0.024	0.004	0.017	0.033
	AS	0.025	0.024	0.004	0.017	0.032
Happy	b_c	0.068	0.053	0.758	0.012	0.172
	SAM	0.036	0.036	0.015	0.006	0.067
	AS	0.036	0.036	0.015	0.006	0.067
Relig	b_c	-0.077	-0.075	0.011	-0.098	-0.056
	SAM	-0.056	-0.055	0.007	-0.069	-0.043
	AS	-0.056	-0.055	0.006	-0.068	-0.043

at the mean $\bar{x}$ (SAM), and the average slope (AS) over all observed x_i [7]. Table 10.7 provides the circular effect measures for the marginal effects of each of the four predictors.

The results show that in the model with four predictors, the marginal effect of each predictor (evaluated at the mean of the other predictors) has a HPD that does not include zero. Therefore, we conclude that all predictors have an effect on the mean of BHV. In this illustration, the differences between b_c, SAM and AS are rather small, although for the predictor Happy the difference between b_c versus SAM and AS is more substantial. Also the SD of b_c is relatively large and as a consequence the interval width of the 95% HPD is large for this parameter (see LB and UB in the table) compared to the results in SAM and AS. The trace plots presented in Figure 10.11 illustrate these differences. The trace plots for SAM and AS show convergence, whereas the plot for b_c shows a pattern with many peaks of extreme values. Please note the differences in scale on the y-axis as well as the truncation at 0.5 and -0.5; the true observed maximum and minimum values for b_c were actually 18.7 and -64.9, respectively. An explanation for these extreme values and subsequent poor convergence is that b_c is estimated at the inflection point $x = a_x$, which is far away from the observed data as can be seen in Figure 10.10. In this case, we prefer the stability and interpretability (i.e., the slope of the curve within the range of the data) of the other measures (SAM/AS).

10.5.5   Comparison of Approaches

Also in this example, we observe results that are similar in many aspects but also show some differences. For instance, the DIC values are not the same between the two modeling approaches but the conclusions in terms of order

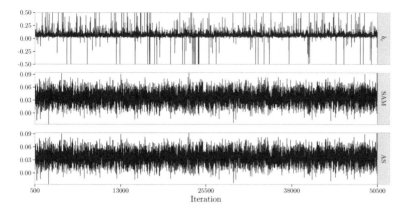

FIGURE 10.11

Trace plots for the parameters b_c, SAM and AS for the predictor Happy.

of best fitting models do not deviate from each other. The estimated relations between each predictor and the outcome in the regression model with all four predictors are, again, not exactly the same but lead to the same conclusions.

The differences can be explained by a difference in modeling assumptions. As before the intrinsic approach that we applied assumes homogeneity and conditional on this assumption measures to what extent the predictor affects the *location* of the outcome on the circle. In contrast, the embedding approach is a consensus model where effects on *location* and *spread* of the circular outcome are modeled simultaneously [7, 29]. For a predictor, the initial results on the two components β^I and β^{II} show that a predictor has any effect on the outcome variable when one or both β's are non-zero. It is, however, possible that the predictor has an effect on the spread of the circular outcome while the location on the circle is not affected. In [7] the terms *location and accuracy effects* are used to denote these effects and measures are introduced for distinguishing –to some extent– between the two.

One of these measures is the previously introduced b_c. When at least one of β^I and β^{II} of a predictor is non-zero, while $b_c = 0$, then the predictor affects the spread on the circle but not the location (only an accuracy effect). When $b_c \neq 0$, the predictor has an effect on the location but can also affect the spread on the circle (a location effect with or without an accuracy effect). The measures b_c, SAM, and AS represent the location effect.

For the ESS data analyzed in this section, the (small) differences between results are most likely due to a different fit of the data to the specified model in terms of the presence or absence of the homogeneity assumption. When we go back to Figure 10.9 and compare the DIC values of both approaches for the model with four predictors, we see a DIC of 4281 for the intrinsic and a DIC of 4210 for the embedding approach. We conclude that the four predictors also have an effect on the spread of the circular outcome and that this violation of the homogeneity assumption in the applied intrinsic approach causes the lower fit value.

10.6   Discussion

In this chapter we applied the intrinsic and embedding approaches on data sets from psychology and sociology. With these illustrations, we aimed to show that circular data are present in several branches of the social and behavioral sciences, and that the tools for the analysis of circular data are available and provide interpretable results. A third aim was to make a case for the Bayesian approach. An advantage of the Bayesian approach is its flexibility to perform inference, also in complex models. A second advantage is that by using a sampling based approach (i.e., MCMC) also the posterior distribution of any function of parameters is available. Using a frequentist approach, the computation of a new interpretation tool like the SAM in the embedding approach may be just as easy, but obtaining the corresponding standard error or confidence interval would be difficult if not impossible. For some Bayesians a third advantage would be the option to include (subjective) prior knowledge in the analysis, while others would see the need for specifying prior distributions as a disadvantage of the Bayesian approach. In this chapter, in both approaches, we worked with standard low-informative priors so that the impact of priors on results was minimal.

By comparing two modeling approaches, we gained insight in often implicit model assumptions and how they may affect the results. Some clarification needs to be made however. We more or less equated the intrinsic approach with assuming a von Mises likelihood with a homogeneity of variances assumption, but this assumption can be released by letting κ depend on predictors as in [20]. In contrast, the PN regression model has an innate heterogeneity of variance [29], so an assumption of homogeneity of variance, which is often used in social and behavioral sciences, is not possible. This can be seen as a limitation of the embedding approach.

In terms of ease of application, the intrinsic approach may look more appealing. By modeling a distribution directly on the circle, the model parameters represent the circular effects that we are interested in. This is not the case in the embedding approach, where linear β's on two components are estimated. However, the reparametrization into circular effects has solved this major disadvantage of the embedding approach.

An advantage of the embedding approach is the flexibility to extend to more complex models. Indeed, in the embedding approach, mixed-effects models have been proposed by [27] and spatial and spatio-temporal models by [22] and [38]. To what extent the new interpretation tools for the circular effects, as presented by [7], can be generalized to these models still needs to be investigated, though.

In conclusion, this chapter demonstrated two applications of Bayesian analyses in the case of circular data. We highlighted the flexibility of using a Bayesian MCMC approach as well as the differences between using an intrinsic or an embedding approach. In essence, they can be compared as follows. The intrinsic approach allows modeling location effects and variance effects separately and has directly interpretable parameters. The embedding approach models location and variance effects concurrently and allows for very flexible extensions.

Notes

1. The intrinsic circular regression models are available in the R package `CircGLMBayes` (https://github.com/keesmulder/CircGLMBayes).

2. The projected normal circular regression models are available in the R package `bpnreg` (https://github.com/joliencremers/bpnreg).

3. The authors are supported by grant 452-12-010 from the Netherlands Organization for Scientific Research (NWO), awarded to Irene Klugkist.

Bibliography

[1] M. Ashby and K. Bowers. A comparison of methods for temporal analysis of aoristic crime. *Crime Science*, 2(1):1, 2013.

[2] C. Baayen and I. Klugkist. Evaluating order-constrained hypotheses for circular data from a between-within subjects design. *Psychological Methods*, 19(3):398, 2014.

[3] S. Barragán, M. Fernández, C. Rueda, and S. Peddada. Isocir: An R package for constrained inference using isotonic regression for circular data, with an application to cell biology. *Journal of Statistical Software*, 54(4):i04, 2013.

[4] C. Brunsdon and J. Corcoran. Using circular statistics to analyze time patterns in crime incidence. *Computers, Environment and Urban Systems*, 30(3):300 – 319, 2006.

[5] J. Bullens, M. Nardini, C. Doeller, O. Braddick, A. Postma, and N. Burgess. The role of landmarks and boundaries in the development of spatial memory. *Developmental Science*, 13(1):170–180, 2010.

[6] J. Cremers and I. Klugkist. How to analyze circular data: A tutorial for projected normal regression models. Working paper.

[7] J. Cremers, K. Mulder, and I. Klugkist. Circular interpretation of regression coefficients. *British Journal of Mathematical and Statistical Psychology*, in press.

[8] J. Cremers, T. Mainhard, and I. Klugkist. Assessing a Bayesian embedding approach to circular regression models. *Methodology*, under review.

[9] M. Fernández, C. Rueda, and S. Peddada. Identification of a core set of signature cell cycle genes whose relative order of time to peak expression is conserved across species. *Nucleic Acids Research*, 40(7):2823–2832, 2012.

[10] N. Fisher and A. Lee. Regression models for an angular response. *Biometrics*, 48(3):665–677, 1992.

[11] P. Forbes and K. Mardia. A fast algorithm for sampling from the posterior of a von Mises distribution. *Journal of Statistical Computation and Simulation*, 85(13):2693–2701, 2015.

[12] J. Gill and D. Hangartner. Circular data in political science and how to handle it. *Political Analysis*, 18(3):316–336, 2010.

[13] X. Gu, J. Mulder, M. Deković, and H. Hoijtink. Bayesian evaluation of inequality constrained hypotheses. *Psychological Methods*, 19(4):511–527, 2014.

[14] P. Guttorp and R. Lockhart. Finding the location of a signal: A Bayesian analysis. *Journal of the American Statistical Association*, 83(402):322–330, 1988.

[15] D. Hernandez-Stumpfhauser, F. Breidt, and M. van der Woerd. The general projected normal distribution of arbitrary dimension: Modeling and Bayesian inference. *Bayesian Analysis*, 12(1):113–133, 2017.

[16] H. Hoijtink. *Informative Hypotheses: Theory and Practice for Behavioral and Social Scientists*. Boca Raton, FL: CRC Press, 2012.

[17] D. Kendall. Pole-seeking brownian motion and bird navigation. *Journal of the Royal Statistical Society: Series B (Statistical Methodology)*, 36(3):365–417, 1974.

[18] I. Klugkist, O. Laudy, and H. Hoijtink. Inequality constrained analysis of variance: a Bayesian approach. *Psychological Methods*, 10(4):477, 2005.

[19] I. Klugkist, J. Bullens, and A. Postma. Evaluating order-constrained hypotheses for circular data using permutation tests. *British Journal of Mathematical and Statistical Psychology*, 65(2):222–236, 2012.

[20] F. Lagona. Regression analysis of correlated circular data based on the multivariate von Mises distribution. *Environmental and Ecological Statistics*, 23(1):89–113, 2016.

[21] T. Leary. *Interpersonal Diagnosis of Personality*. New York: Ronald Press, 1957.

[22] G. Mastrantonio, G. Lasinio, and A. Gelfand. Spatio-temporal circular models with non-separable covariance structure. *TEST*, 25(2):331–350, 2016.

[23] F. Mechsner, P. Stenneken, J. Cole, G. Aschersleben, and W. Prinz. Bimanual circling in deafferented patients: Evidence for a role of visual forward models. *Journal of Neuropsychology*, 1(2):259–282, 2007.

[24] J. Mulder, H. Hoijtink, and I. Klugkist. Equality and inequality constrained multivariate linear models: Objective model selection using constrained posterior priors. *Journal of Statistical Planning and Inference*, 140(4):887–906, 2010.

[25] K. Mulder and I. Klugkist. Bayesian estimation and hypothesis tests for a circular generalized linear model. *Journal of Mathematical Psychology*, in press.

[26] R. Neal. Slice sampling. *The Annals of Statistics*, 31(3):705–767, 2003.

[27] G. Nuñez-Antonio and E. Gutiérrez-Peña. A Bayesian model for longitudinal circular data based on the projected normal distribution. *Computational Statistics & Data Analysis*, 71:506–519, 2014.

[28] A. Postma, S. Zuidhoek, M. Noordzij, and A. Kappers. Keep an eye on your hands: on the role of visual mechanisms in processing of haptic space. *Cognitive Processing*, 9(1):63–68, 2008.

[29] L. Rivest, T. Duchesne, A. Nicosia, and D. Fortin. A general angular regression model for the analysis of data on animal movement in ecology. *Journal of the Royal Statistical Society: Series C (Applied Statistics)*, 65 (3):445–463, 2016.

[30] M. Rocchi and C. Perlini. Is the time of suicide a random choice? A new statistical perspective. *Crisis*, 23(4):161–166, 2002.

[31] C. Rueda, M. Fernández, and S. Peddada. Estimation of parameters subject to order restrictions on a circle with application to estimation of phase angles of cell cycle genes. *Journal of the American Statistical Association*, 104(485):338–347, 2009.

[32] J. Russell. A circumplex model of affect. *Journal of Personality and Social Psychology*, 39(6):1161–1178, 1980.

[33] S. Schwartz. Are there universal aspects in the structure and contents of human values? *Journal of Social Issues*, 50(4):19–45, 1994.

[34] S. Schwartz. An overview of the Schwartz theory of basic values. *Online Readings in Psychology and Culture*, 2(1):1–20, 2012.

[35] D. Spiegelhalter, N. Best, B. Carlin, and A. Van Der Linde. Bayesian measures of model complexity and fit. *Journal of the Royal Statistical Society: Series B (Statistical Methodology)*, 64(4):583–639, 2002.

[36] T. Stoffregen, B. Bardy, O. Merhi, and O. Oullier. Postural responses to two technologies for generating optical flow. *Presence: Teleoperators and Virtual Environments*, 13(5):601–615, 2004.

[37] J. Venter. On estimation of the mode. *The Annals of Mathematical Statistics*, 38(5):1446–1455, 1967.

[38] F. Wang and A. Gelfand. Modeling space and space-time directional data using projected Gaussian processes. *Journal of the American Statistical Association*, 109(508):1565–1580, 2014.

[39] T. Wubbels, P. den Brok, I. Veldman, and J. van Tartwijk. Teacher interpersonal competence for Dutch secondary multicultural classrooms. *Teachers and Teaching*, 12(4):407–433, 2006.

11

Nonparametric Classification for Circular Data

Marco Di Marzio

DISFPEQ, Chieti-Pescara University

Stefania Fensore

DISFPEQ, Chieti-Pescara University

Agnese Panzera

DISIA, Florence University

Charles C. Taylor

Department of Statistics, University of Leeds

CONTENTS

11.1	Density Estimation on the Circle	243
11.2	Classification via Density Estimation	245
11.3	Local Logistic Regression	246
	11.3.1 Binary Regression via Density Estimation	246
	11.3.2 Local Polynomial Binary Regression	248
11.4	Numerical Examples	250
11.5	Classification of Earth's Surface	251
11.6	Conclusion	253
	Bibliography	256

Circular data occur when the sample space is the unit circle. The peculiarity of a circular measurement scale is that its beginning and its end coincide. After both an origin and an orientation have been chosen, a circular observation can be measured, in radians, by an angle $\theta \in [-\pi, \pi)$. Circular data often arise in biology (migration paths, flight directions of animals), meteorology (wind and marine current directions), and geology (orientations of joints and faults, landforms, oriented stones). For comprehensive accounts see, for example, [9] and [1].

There are some proposals in the literature for classifying circular data. For example, SenGupta and Roy [10] proposed a classification rule using a chord-length based distance between a new observation and the observations of two known circular populations. Recently, SenGupta and Ugwuowo [11] introduced a new method involving a generalized likelihood ratio test for classifying toroidal and cylindrical data in two populations, assuming that one of the probabilities of misclassification is known. [8] extended the naive Bayes classifier to the case where the conditional probability distribution of the predictor is assumed to be von Mises-Fisher. Fernandes and Cardoso [5] proposed a classifier by introducing a directional version of logistic regression, also presenting several real data examples.

We discuss some nonparametric classification methods for circular data which are based on kernel estimation of the population densities and on local logistic regression. In particular we consider the case when there are two sub-populations. This field seems unexplored, although the need for flexible methods is easily illustrated by the toy examples shown in Figure 11.1.

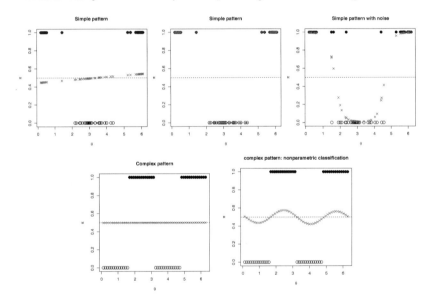

FIGURE 11.1
Examples of binary regression. The two groups are labelled as filled dots (group 1) and open circles (group 2), with the y-axis depicting the (estimated) probability (shown in red) of belonging to group 1. In each case we quote the misclassification rate in brackets. Top panel from left: Linear binary regression (0.42); Circular binary regression (0); Circular binary regression with overlapping groups (0.06). Bottom panel from left: Circular binary regression (0.5); Kernel binary regression (0.04).

In the first sub-figure, we observe that a linear model could give very poor performance (depending on where the circle is "cut") in classifying circular observations even when the pattern is simple. Using a periodic model the global logistic regression works perfectly for the same pattern (second panel) and its performance is slightly worse when there is some noise (third panel). Coming to a more complex pattern, the parametric logistic regression is no longer successful in classifying the two groups (bottom row), whereas fitting a nonparametric model (final panel) leads to good classification.

The chapter is organized as follows. Section 11.1 collects some basic results on circular density estimation. Section 11.2 discusses discrimination using density estimation. Section 11.3 deals with two different approaches to nonparametric regression with a circular predictor and a binary response. Finally, Section 11.4 presents some simulation examples and Section 11.5 shows a real-data case study.

11.1 Density Estimation on the Circle

A circular kernel K_κ with concentration parameter κ is a periodic function such that

i) for $\theta \in [-\pi, \pi)$, it admits a uniformly convergent Fourier series

$$\frac{1 + 2\sum_{j=1}^{\infty} \gamma_j(\kappa) \cos(j\theta)}{2\pi},$$

where $\gamma_j(\kappa)$ is a strictly monotonic function of κ;

ii) $\int_{-\pi}^{\pi} K_\kappa(\theta)\, d\theta = 1$, and, if K_κ takes negative values, there exists $0 < M < \infty$ such that, for all $\kappa > 0$

$$\int_{-\pi}^{\pi} |K_\kappa(\theta)|\, d\theta \leq M\,;$$

iii) for all $0 < \delta < \pi$,

$$\lim_{\kappa \to \infty} \int_{\delta \leq |\theta| \leq \pi} |K_\kappa(\theta)|\, d\theta = 0.$$

Typical examples of circular kernels include: the *von Mises density* where $\gamma_j(\kappa) = \mathcal{I}_j(\kappa)/\mathcal{I}_0(\kappa)$, with $\mathcal{I}_j(\kappa)$ being the modified Bessel function of the first kind and order $j \in \mathbb{N}$, the density of the *Wrapped Normal* distribution in which $\gamma_j(\kappa) = \kappa^{j^2}$, and the density of the *Wrapped Cauchy* distribution in which $\gamma_j(\kappa) = \kappa^j$.

Given a random sample of angles $\Theta_1, \ldots, \Theta_n$ from an unknown circular density f, the kernel estimator of f at $\theta \in [-\pi, \pi)$ can then be defined as

$$\hat{f}(\theta; \kappa) := \frac{1}{n} \sum_{i=1}^{n} K_\kappa(\Theta_i - \theta). \tag{11.1}$$

As in the *linear* setting, the role of the kernel function is to emphasize, in the estimation process, the contribution of the observations which are in a neighbourhood of θ. The concentration parameter κ controls the width of that neighbourhood in such a way that its role is the inverse of the bandwidth in the linear case, in the sense that smaller values of κ give wider neighborhoods.

Now, for $j \in \mathbb{N}$ and a circular kernel K_κ, we set

$$\eta_j(K_\kappa) := \int_{-\pi}^{\pi} K_\kappa(\alpha) \sin^j(\alpha) d\alpha \qquad \text{and} \qquad \nu(K_\kappa) := \int_{-\pi}^{\pi} K_\kappa^2(\alpha) d\alpha, \quad (11.2)$$

so that K_κ is said to be a r-th sin-order kernel if $\eta_0(K_\kappa) = 1$, $\eta_j(K_\kappa) = 0$ for $j < r$ and $\eta_r(K_\kappa) \neq 0$. In the sequel we will assume that the concentration parameter κ depends on the sample size n in such a way that $\lim_{n \to \infty} \eta_2(K_\kappa) = 0$ and $\lim_{n \to \infty} n^{-1} \nu(K_\kappa) = 0$. Moreover, we also assume that K_κ is a second sin-order circular kernel, satisfying $\eta_j(K_\kappa) = o(\eta_2(K_\kappa))$ for each $j > 2$. Now, if f'' is continuous at $\theta \in [-\pi, \pi)$, the asymptotic bias and variance of $\hat{f}(\theta; \kappa)$ respectively are

$$\mathsf{E}[\hat{f}(\theta; \kappa)] - f(\theta) = \frac{\eta_2(K_\kappa)}{2} f''(\theta) + o(\eta_2(K_\kappa)), \tag{11.3}$$

and

$$\mathsf{Var}[\hat{f}(\theta; \kappa)] = \frac{\nu(K_\kappa)}{n} f(\theta) + o\left(\frac{\nu(K_\kappa)}{n}\right). \tag{11.4}$$

For the case of a von Mises kernel, i.e., $K_\kappa(\theta) := \{2\pi \mathcal{I}_0(\kappa)\}^{-1} \exp(\kappa \cos(\theta))$, it holds that

$$\lim_{\kappa \to \infty} \eta_2(K_\kappa) = \frac{1}{\kappa}, \qquad \text{and} \qquad \lim_{\kappa \to \infty} \nu(K_\kappa) = \frac{\kappa^{1/2}}{2\pi^{1/2}}. \tag{11.5}$$

Then, the above assumptions on the concentration parameter imply that $\kappa \to \infty$ and $n^{-1} \kappa^{1/2} \to 0$ as $n \to \infty$. This shows a trade-off between the asymptotic bias and variance, with the following asymptotic mean squared error for $\hat{f}(\theta; \kappa)$

$$\mathsf{AMSE}[\hat{f}(\theta; \kappa)] = \left\{ \frac{f''(\theta)}{2\kappa} \right\}^2 + \frac{\kappa^{1/2} f(\theta)}{2\pi^{1/2} n}.$$

It is then easy to show that the value of κ which minimizes $\mathsf{AMSE}[\hat{f}(\theta; \kappa)]$ is

$$\hat{\kappa} = \left\{ \frac{2n\pi^{1/2} (f''(\theta))^2}{f(\theta)} \right\}^{2/5}, \tag{11.6}$$

which gives a convergence rate of $n^{-4/5}$. Note that this corresponds to the rate attained by a kernel density estimator, equipped with a second-order kernel, in the linear setting. Practical selection rules for κ can be found, for example, in [13], [3], and [6].

11.2   Classification via Density Estimation

Kernel density estimation is commonly used also for classification purposes. Consider two populations $\mathcal{P}_1$ and $\mathcal{P}_2$, respectively, described by the circular densities f_1 and f_2. Let π_i, $i \in (1, 2)$, be the prior probability that a new datum, say θ, is drawn from $\mathcal{P}_i$. An *optimal* classification rule, in the sense of minimising the Bayes risk, assigns θ to $\mathcal{P}_1$ if $\pi_1 f_1(\theta) - \pi_2 f_2(\theta) \geq 0$.

When f_1 and f_2 are unknown, and random samples of sizes n_1 and n_2 are respectively available from $\mathcal{P}_1$ and $\mathcal{P}_2$, an empirical version of the above rule can be based on the kernel estimates of $f_1(\theta)$ and $f_2(\theta)$. Hence, letting $\hat{f}_1(\theta; \kappa_1)$ and $\hat{f}_2(\theta; \kappa_2)$ be the kernel estimators of $f_1(\theta)$ and $f_2(\theta)$, respectively, and setting $\hat{h}(\theta; \kappa_1, \kappa_2) := \hat{f}_1(\theta; \kappa_1) - \hat{f}_2(\theta; \kappa_2)$, according to the above classification rule, in the special case $\pi_1 = \pi_2$, an observation θ will be classified as coming from $\mathcal{P}_1$ if $\hat{h}(\theta; \kappa_1, \kappa_2) \geq 0$. More generally, if the cases were selected at random from the (pooled) population, then we would classify θ to $\mathcal{P}_1$ if $n_1 \hat{f}_1(\theta; \kappa_1) > n_2 \hat{f}_2(\theta; \kappa_2)$. For the linear counterpart of this rule, its L_2 properties and some ways to select the appropriate bandwidths, see [4] and [7].

Now, recalling the assumptions on the circular kernel discussed in the previous section, and assuming the continuity of $f_j''(\theta)$, $j \in (1, 2)$, by virtue of results (11.3) and (11.4) we have that

$$
\begin{aligned}
\mathsf{E}[\hat{h}(\theta; \kappa_1, \kappa_2)] \;=\; & f_1(\theta) - f_2(\theta) + \frac{1}{2} \left\{ \eta_2(K_{\kappa_1}) f_1''(\theta) - \eta_2(K_{\kappa_2}) f_2''(\theta) \right\} \\
& + o(\eta_2(K_{\kappa_1}) + \eta_2(K_{\kappa_2})),
\end{aligned}
$$

and

$$
\mathsf{Var}[\hat{h}(\theta; \kappa_1, \kappa_2)] = \frac{\nu(K_{\kappa_1})}{n_1} f_1(\theta) + \frac{\nu(K_{\kappa_2})}{n_2} f_2(\theta) + o\left(\frac{\nu(K_{\kappa_1})}{n_1} + \frac{\nu(K_{\kappa_2})}{n_2} \right).
$$

When K_{κ_1} and K_{κ_2} are both von Mises kernels, using the approximations in (11.5), the asymptotic mean squared error of $\hat{h}(\theta; \kappa_1, \kappa_2)$, $\theta \in [-\pi, \pi)$, can be easily obtained as

$$
\mathsf{AMSE}[\hat{h}(\theta; \kappa_1, \kappa_2)] = \frac{1}{4} \left\{ \frac{f_1''(\theta)}{\kappa_1} - \frac{f_2''(\theta)}{\kappa_2} \right\}^2 + \frac{1}{2\pi^{1/2}} \left\{ \frac{f_1(\theta)\kappa_1^{1/2}}{n_1} + \frac{f_2(\theta)\kappa_2^{1/2}}{n_2} \right\}.
$$

The values of κ_1 and κ_2 minimizing $\mathsf{AMSE}[\hat{h}(\theta; \kappa_1, \kappa_2)]$ are (see [4]):

$$
\hat{\kappa}_1 = \left\{ \left[2\pi^{1/2} n_1 (f_1''(\theta))^2 - (2\sqrt{\pi} n_1 f_1''(\theta))^{5/3} (2\pi^{1/2} n_2)^{-2/3} f_2''(\theta)^{1/3} \right] / f(\theta) \right\}^{2/5},
$$

and

$$
\hat{\kappa}_2 = \left\{ \left[2\pi^{1/2} n_2 (f_2''(\theta))^2 - (2\pi^{1/2} n_2 f_2''(\theta))^{5/3} (2\pi^{1/2} n_1)^{-2/3} f_1''(\theta)^{1/3} \right] / f(\theta) \right\}^{2/5}.
$$

By inspecting the result for $\hat{\kappa}_1$ (or $\hat{\kappa}_2$) we can note that it depends on *both* f_1 and f_2, and on *both* n_1 and n_2. When n_1 (n_2, resp.) is fixed and n_2 (n_1, resp.) goes to infinity, the optimal value of κ_1 (κ_2, resp.) increases to the optimal value of κ obtained for the case of a single population in (11.6).

Different selection rules can be used for the practical choice of κ_1 and κ_2. The simplest one can be considered the *von Mises reference rule*, where we replace the unknown quantities appearing in the formulation of $\hat{\kappa}_1$ and $\hat{\kappa}_2$ by the corresponding quantities obtained for a von Mises density (see [13] for the case of a single population).

11.3   Local Logistic Regression

One of the possible regression models for dealing with dichotomous data is logistic regression. The goal is to find the best fit to describe the relationship between a binary outcome and a set of independent predictors. Logistic regression determines the membership degree of each individual to one of the two groups by fitting a continuous function taking values in the interval $[0, 1]$. Logistic regression can be also used for classification purposes.

Let Y and Θ be a binary response and a circular predictor, respectively, and set $\lambda(\theta) := P(Y = 1 \mid \Theta = \theta)$. Denote the density functions in the circular covariate space for the successes ($Y = 1$) and for the failures ($Y = 0$) by f_1 and f_2, respectively, and let π_1 be the proportion of successes in the population, and $\pi_2 = 1 - \pi_1$. Then, for $\theta \in [-\pi, \pi)$, we have

$$\lambda(\theta) = \frac{\pi_1 f_1(\theta)}{\pi_1 f_1(\theta) + \pi_2 f_2(\theta)}. \tag{11.7}$$

Given a point (θ, y), a classification rule predicts $\hat{y} = 1$ if the estimate of $\lambda(\theta)$, say $\hat{\lambda}(\theta)$, satisfies $\hat{\lambda}(\theta) \geq 0.5$, and $\hat{y} = 0$ otherwise.

In what follows we discuss two different approaches to nonparametrically estimate $\lambda(\theta)$. Specifically, the first one is based on the kernel estimation of the densities appearing in (11.7), while the second one is a local likelihood approach which leads to a whole class of nonparametric estimators of $\lambda(\theta)$.

11.3.1   Binary Regression via Density Estimation

Given n independent copies of (Θ, Y): $(\Theta_1, Y_1), \ldots, (\Theta_n, Y_n)$, assume that the sample has been reordered so that the first n_1 pairs are successes and the last $n_2 = n - n_1$ ones are failures. Replacing π_j in (11.7) with n_j/n, $j \in (1, 2)$, a kernel estimator of $\lambda(\theta)$, $\theta \in [-\pi, \pi)$, can be defined as

$$\hat{\lambda}(\theta; \kappa_1, \kappa_2) = \frac{n_1 \hat{f}_1(\theta; \kappa_1)}{n_1 \hat{f}_1(\theta; \kappa_1) + n_2 \hat{f}_2(\theta; \kappa_2)}, \tag{11.8}$$

where $\hat{f}_j$, $j \in (1,2)$, are kernel estimators of the f_js. Estimators like the above one, which satisfy $0 \leq \hat{\lambda}(\theta; \kappa_1, \kappa_2) \leq 1$, have been studied in the Euclidean setting by [12].

When we use the same kernel function for both $\hat{f}_1$ and $\hat{f}_2$ in (11.8), and, in addition, a single smoothing parameter, i.e. $\kappa_1 = \kappa_2 = \kappa$, we obtain the Nadaraya-Watson estimator for circular-linear regression

$$\hat{\lambda}(\theta; \kappa) = \frac{\sum_{i=1}^{n} Y_i K_\kappa(\Theta_i - \theta)}{\sum_{i=1}^{n} K_\kappa(\Theta_i - \theta)},$$

that is a weighted average of the observations of the response variable with weights given by circular kernels, see [2] for details.

Note that the more general classification rule for which $\kappa_1 \neq \kappa_2$ is equivalent to the kernel density classification rule discussed in the previous section. Reasoning as in the Euclidean setting (see [12]), since we focus on the estimation of the two densities, we ignore the error of magnitude $O(n^{-1/2})$ implied by the replacement of π_i, $i \in (1,2)$, by n_i/n.

Now, we assume that n_1 and n_2 go to infinity in such a way that $n_1/n_2 \to \pi_1/\pi_2$. Letting $m(\theta) := \pi_1 f_1(\theta) + \pi_2 f_2(\theta)$ and $\hat{m}(\theta; \kappa_1, \kappa_2) := \pi_1 \hat{f}_1(\theta; \kappa_1) + \pi_2 \hat{f}_2(\theta; \kappa_2)$, we have

$$\hat{\lambda}(\theta; \kappa_1, \kappa_2) = \frac{\pi_1 \hat{f}_1(\theta; \kappa_1)}{\hat{m}(\theta; \kappa_1, \kappa_2)},$$

which can be re-written as

$$\hat{\lambda}(\theta; \kappa_1, \kappa_2) = \frac{\pi_1 f_1(\theta) \left(\frac{\hat{f}_1(\theta; \kappa_1) - f_1(\theta)}{f_1(\theta)} + 1 \right)}{m(\theta) \left(\frac{\hat{m}(\theta; \kappa_1, \kappa_2) - m(\theta)}{m(\theta)} + 1 \right)}.$$

Since $\hat{f}_1(\theta; \kappa_1) - f_1(\theta)$ and $\hat{m}(\theta; \kappa_1, \kappa_2) - m(\theta)$ are asymptotically *small*, expanding $\hat{f}_1(\theta, \kappa_1)$ and $\hat{m}(\theta; \kappa_1, \kappa_2)$ respectively around $f_1(\theta)$ and $m(\theta)$, it results

$$\hat{\lambda}(\theta; \kappa_1, \kappa_2) = \lambda(\theta) \left(1 + \frac{\hat{f}_1(\theta; \kappa_1) - f_1(\theta)}{m(\theta)} - \frac{\hat{m}(\theta; \kappa_1, \kappa_2) - m(\theta)}{m(\theta)} \right)$$
$$+ O_p(\{\hat{m}(\theta; \kappa_1, \kappa_2) - m(\theta)\}^2) + O_p(\{\hat{f}_1(\theta; \kappa_1) - f_1(\theta)\}^2),$$

and recalling the definition of $\hat{m}(\theta; \kappa_1, \kappa_2)$, one has

$$\hat{\lambda}(\theta; \kappa_1, \kappa_2) = \lambda(\theta) + \frac{\pi_1(1 - \lambda(\theta))(\hat{f}_1(\theta; \kappa_1) - f_1(\theta)) - \pi_2 \lambda(\theta)(\hat{f}_2(\theta; \kappa_2) - f_2(\theta))}{m(\theta)}$$
$$+ O_p(\{\hat{m}(\theta; \kappa_1, \kappa_2) - m(\theta)\}^2) + O_p(\{\hat{f}_1(\theta; \kappa_1) - f_1(\theta)\}^2).$$

Now suppose that: both f_1 and f_2 admit continuous derivatives up to the second order; that K_{κ_1} and K_{κ_2} both satisfy the assumptions for circular kernels in Section 11.1; and that $\eta_2(K_{\kappa_1}) = O(\eta_2(K_{\kappa_2}))$ and $\nu(K_{\kappa_1}) =$

$O(\nu(K_{\kappa_2}))$. Then, starting from above approximation, by virtue of results (11.3) and (11.4), we obtain

$$
\begin{aligned}
E[\hat{\lambda}(\theta; \kappa_1, \kappa_2)] - \lambda(\theta) \;=\;& \frac{\pi_1 \pi_2}{2m^2(\theta)} \left(\eta_2(K_{\kappa_1}) f_2(\theta) f_1''(\theta) - \eta_2(K_{\kappa_2}) f_1(\theta) f_2''(\theta) \right) \\
& + o(\eta_2(K_{\kappa_1})),
\end{aligned}
$$

and

$$
\begin{aligned}
\mathsf{Var}[\hat{\lambda}(\theta; \kappa_1, \kappa_2)] \;=\;& \frac{\lambda(\theta)(1 - \lambda(\theta))}{n\, m(\theta)} [(1 - \lambda(\theta))\nu(K_{\kappa_1}) + \lambda(\theta)\nu(K_{\kappa_2})] \\
& + o\left(n^{-1}\nu(K_{\kappa_1}) \right).
\end{aligned}
$$

In the special case when K_{κ_1} and K_{κ_2} are both von Mises kernels, with $\kappa_1 \approx \kappa_2$, using approximations (11.5), we have

$$
E[\hat{\lambda}(\theta; \kappa_1, \kappa_2)] - \lambda(\theta) = \frac{\pi_1 \pi_2}{2m^2(\theta)} \left(\frac{f_2(\theta) f_1''(\theta)}{\kappa_1} - \frac{f_1(\theta) f_2''(\theta)}{\kappa_2} \right) + o\left(\frac{1}{\kappa_1} \right),
$$

and

$$
\mathsf{Var}[\hat{\lambda}(\theta; \kappa_1, \kappa_2)] = \frac{\lambda(\theta)(1 - \lambda(\theta))}{2n\pi^{1/2} m(\theta)} \left[(1 - \lambda(\theta))\kappa_1^{1/2} + \lambda(\theta)\kappa_2^{1/2} \right] + o\left(\frac{\kappa_1^{1/2}}{n} \right).
$$

Concerning optimal smoothing selection the standard approach in the Euclidean setting is to consider a weighted version of the mean squared error. In a practical implementation the smoothing parameters can be selected by minimizing an empirical version of the weighted mean squared error (for details see [12]).

11.3.2   Local Polynomial Binary Regression

A different way to nonparametrically estimate $\lambda(\theta)$ is based on the local likelihood approach. Using a circular kernel K_κ as the weight function, the local log-likelihood at $\theta \in [-\pi, \pi)$ is

$$
\sum_{i=1}^{n} \left\{ Y_i \log \left(\frac{\lambda(\Theta_i)}{1 - \lambda(\Theta_i)} \right) + \log(1 - \lambda(\Theta_i)) \right\} K_\kappa(\Theta_i - \theta),
$$

which, using $\delta = \log(\lambda/(1 - \lambda))$, can be rewritten on the logistic scale as

$$
\sum_{i=1}^{n} \left\{ Y_i \delta(\Theta_i) - \log(1 + \exp(\delta(\Theta_i))) \right\} K_\kappa(\Theta_i - \theta).
$$

Now, let g be a generic periodic function which admits continuous derivatives up to a given order p, in a neighborhood of $\theta \in [-\pi, \pi)$. Then, a pth degree series expansion for g at a point α in a neighborhood of θ is

$$
g(\alpha) \approx \sum_{j=0}^{p} \frac{\sin^j(\alpha - \theta) g^{(j)}(\theta)}{j!}.
$$

The above *sine-polynomial*, which has been introduced by [2] in the context of nonparametric circular-linear regression, is essentially the leading part of a Taylor series expansion with the increments replaced by their sines: this assures that both the sign and the periodicity of the increments are preserved.

Now, using the above sine-polynomial to approximate $\delta(\Theta_i)$ around θ in the local log-likelihood function yields a class of nonparametric estimators for $\lambda(\theta)$. Specifically, for $\boldsymbol{\beta} = (\beta_0, \dots, \beta_p)$, where β_j stands for the derivative of order j of δ at θ, we can rewrite the local log-likelihood as

$$\sum_{i=1}^{n} \{Y_i q_p(\Theta_i; \boldsymbol{\beta}) - \log(1 + \exp(q_p(\Theta_i; \boldsymbol{\beta})))\} K_\kappa(\Theta_i - \theta), \tag{11.9}$$

where $q_p(\Theta_i; \boldsymbol{\beta}) = \sum_{j=0}^{p} \beta_j \sin^j(\Theta_i - \theta)/(j!)$. Letting $\hat{\beta}_0$ be the solution for β_0 of the maximization of (11.9) with respect to $\boldsymbol{\beta}$, we obtain

$$\hat{\lambda}(\theta; \kappa) = \frac{\exp(\hat{\beta}_0)}{1 + \exp(\hat{\beta}_0)},$$

as a local polynomial estimator for λ at θ.

Clearly, different values of p will give different estimators. When $p = 0$, the resulting estimator is the Nadaraya–Watson (or local constant) one discussed in the previous section, while, when $p = 1$, we have a *local linear* logistic estimator. The Euclidean version of the latter estimator has been introduced by [12].

Alternatively, we can use different weights, K_{κ_1} and K_{κ_2}, for successes and failures in local likelihood (11.9) obtaining

$$\sum_{i=1}^{n} Y_i q_p(\Theta_i; \boldsymbol{\beta}) K_{\kappa_1}(\Theta_i - \theta) - \log(1 + \exp(q_p(\Theta_i; \boldsymbol{\beta}))) K_{\kappa_2}(\Theta_i - \theta).$$

In this case, using $p = 0$ in the sine-polynomial approximation, we obtain the estimator previously given by (11.8).

Concerning asymptotic properties, when a von Mises kernel is employed as the weight function, under suitable regularity conditions, assuming that $\kappa_1 = \kappa_2 = \kappa$ with $\kappa \to \infty$ and $\kappa/n \to 0$ as $n \to \infty$, then, when $p = 0$, we have

$$E[\hat{\lambda}(\theta; \kappa)] - \lambda(\theta) = \frac{1}{2\kappa} \left(\lambda''(\theta) + \frac{2\lambda'(\theta)m'(\theta)}{m(\theta)} \right) + o\left(\frac{1}{\kappa}\right),$$

and when $p = 1$, we obtain

$$E[\hat{\lambda}(\theta; \kappa)] - \lambda(\theta) = \frac{1}{2\kappa}\delta''(\theta)\lambda(\theta)(1 - \lambda(\theta)) + o\left(\frac{1}{\kappa}\right),$$

and in both cases

$$\mathsf{Var}[\hat{\lambda}(\theta; \kappa)] = \frac{\kappa^{1/2}}{n} \frac{\lambda(\theta)(1 - \lambda(\theta))}{2\pi^{1/2}m(\theta)} + o\left(\frac{\kappa^{1/2}}{n}\right).$$

Asymptotic properties for the more general case of another circular kernel, which involve the quantities in Equation (11.5), can also be easily obtained.

Additionally, notice that, as expected, the local constant estimator for a binary response has the same asymptotic bias and variance obtained by [2] for the general case of a real-valued response. Also, for the logistic local linear function, the asymptotic bias, depending on the second derivative of δ (and thus on λ) but not on m, exhibits a structure which is reminiscent of the asymptotic bias of the local linear estimator with real-valued response, as given by [2].

For both local constant and local linear estimators, the value of κ which minimizes the asymptotic mean squared error is $O(n^{2/5})$ and gives a convergence rate of magnitude $O(n^{-4/5})$.

Finally, concerning the practical selection of κ, a possible way is to start from a least-squares objective function, and choose the value of κ which minimizes

$$\sum_{i=1}^{n}\left(Y_i - \hat{\lambda}_{-i}(\Theta_i;\kappa)\right)^2, \tag{11.10}$$

where $\hat{\lambda}_{-i}(\Theta_i;\kappa)$ stands for the estimate of λ at Θ_i with the ith sample observation removed. A more natural way is to start from the leave-one-out version of the local log-likelihood, i.e., to select the value of κ which maximizes

$$\sum_{i=1}^{n}\left\{Y_i \log\left(\frac{\hat{\lambda}_{-i}(\Theta_i;\kappa)}{1-\hat{\lambda}_{-i}(\Theta_i;\kappa)}\right) + \log\left(1-\hat{\lambda}_{-i}(\Theta_i;\kappa)\right)\right\}. \tag{11.11}$$

11.4   Numerical Examples

In this section we compare the performance of the proposed classification methods using artificial data. In the first simulation experiment we test the performance of the kernel density classification method. The notation $vM(\mu,\kappa)$ refers to the von Mises density function with mean direction μ and concentration parameter κ. In particular, we consider two samples of size $n_1 = n_2 = 100$ from von Mises populations with different mean directions and equal concentration parameter: the first sample is drawn from $vM(2,4)$; the second one from $vM(4,4)$. We will use two circular kernel density estimates taking the von Mises distribution as kernel and selecting the concentration parameters using the von Mises reference rule. The classifier then assigns each observation to the population having the highest corresponding density estimate. The misclassification rate over 100 simulated datasets is 0.03252. The same experiment is performed considering samples drawn from not well separated populations described by $vM(4,4)$ and $vM(\pi,2)$ densities. The misclassification rate over 100 simulations is 0.2541. Figure 11.2 illustrates the estimates obtained for one dataset in each of the described experiments.

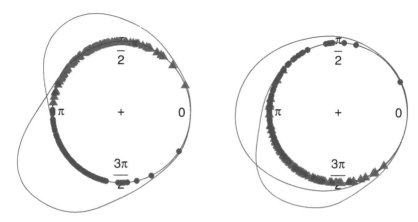

FIGURE 11.2
Left: samples from vM(2,4) (red) and vM(4,4) (blue); right: samples from vM(4,4) (red) and vM(π,2) (blue) with the corresponding density estimates.

In a second simulation we explore the behaviour of local polynomial binary regression. We compare results for local constant ($p = 0$) and local linear ($p = 1$) estimators in terms of average squared prediction error (MSPE) and misclassification rate, when the smoothing parameters are selected by cross-validation. The results are shown in Table 11.1. Here we consider the case of overlapping groups using 100 samples with $n_1 = 100$ observations drawn from $vM(4,4)$ and $n_2 = 60$ from $vM(\pi,2)$. For both the accuracy indicators we can see that the local linear estimator performs slightly better than the local constant one. We also compare medians of the values of κ obtained by the least squares cross-validation selector minimizing (11.10) and the likelihood cross-validation one maximizing (11.11). Resulting values are respectively denoted by $\hat{\kappa}_{MSPE}$ and $\hat{\kappa}_{MR}$. The bandwidths for the cases $p = 0$ and $p = 1$ are not directly comparable, but it is interesting to note that the bandwidths chosen by likelihood cross-validation are less than those chosen by least squares. We can note that, for $p = 0$, the limiting case ($\kappa \to \infty$) gives a classifier which is equivalent to the nearest neighbour (NN) classifier. For this simulation setting the average leave-one-out error rate for the NN classifier is 0.311 which is somewhat greater than the kernel methods, although the two are not strictly comparable.

11.5 Classification of Earth's Surface

Using Google maps (https://maps.googleapis.com/maps/api/staticmap) we have collected color images for (3×3) pixels which are centered at randomly

TABLE 11.1
Estimate of the mean squared prediction errors (MSPE) and misclassification rates (MR) (along with median of smoothing parameters selected by
least squares cross-validation (11.10) and likelihood cross-validation (11.11),
respectively) for local binary regression with p=0 and p=1, using 100 samples with $n_1 = 100$ observations from vM(4,4) and $n_2 = 60$ observations from
vM(π,2).

Polynomial degree	MSPE $(\hat{\kappa}_{MSPE})$	MR $(\hat{\kappa}_{MR})$
$p = 0$	0.1572	0.2147
	(12.5)	(6)
$p = 1$	0.1557	0.2128
	(3.75)	(2.5)

located positions on the sphere (and then converted to latitude, longitude).
These blocks, which are at high resolution (about 70 m×70 m), contain 9
pixels, with each pixel having an RGB color vector (integers on a 0–255 scale)
associated with it. Using "training" pixels from the middle of the ocean, we
identified water as having RGB values of $(163, 204, 255)$, respectively. The
randomly sampled (3×3) pixel blocks were then classified as "water" or not
according to whether the average of these pixels (for each color) were equal
to this identified blue color. Note that this would include very small lakes,
estuaries, and wider rivers. These small features will result in a challenging
classification problem. In all we sampled 2000 points on the sphere, converted
these to latitude and longitude, and then classified 1435 as water, and 565 as
not water. These points are displayed in Figure 11.3, in which the outlines
of the continents are visible. Data which were "missing" near the poles were
allocated as water near the North pole, and land near the South pole. For
the circular data analysis, we can consider only the longitude of these data.
Given that the "default" classifier to allocate every case as water is very hard
to beat, we consider only observations lying between 30° and 60° north. This
leaves 379 observations, of which 196 are water.

We start with kernel density classification using von Mises kernels with
concentration parameters selected according to the von Mises reference rule
($\hat{\kappa}_1$=1.74, $\hat{\kappa}_2$=2.55); see Figure 11.4. Using this classification rule the resulting
MR is 0.19528. The same result is also obtained using the kernel estimator
for binary regression as discussed in Section 11.3.1. The nearest neighbour
classifier (equivalent to kernel density classification with $\kappa_i \to \infty, i = 1, 2$)
gives a leave-one-out misclassification rate of 0.15040.

Coming to the likelihood approach we use a parametric model for the logit
as the benchmark. In particular, we define as the link function the periodic
model $\alpha_0 + \alpha_1 \sin(\Theta_i) + \alpha_2 \cos(\Theta_i)$. Then we classify the data using local
polynomial binary regression discussed in Section 11.3.2 with $p = 0$, $p =$
1 and $p = 2$. The results are collected in Table 11.2. We can see that all

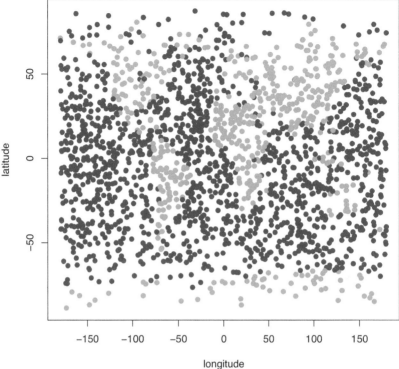

FIGURE 11.3
"Water" (blue) and "not water" (green) observations.

the nonparametric estimators beat the benchmark. In particular we note a decreasing mean prediction squared error (MSPE) when the polynomial degree increases. In terms of misclassification rate local linear wins, and we can see that the improvement on the parametric model is almost 63% while on the local constant and local quadratic side it is 5% and 3%, respectively. See Figures 11.5 and 11.6 for an illustration of the behavior of the estimators.

11.6 Conclusion

This chapter collects different nonparametric approaches for classifying circular data, exploring both theoretical properties and practical behaviour. The first classification rule is based on the difference between two kernel estimates of the circular densities which describe the sub-populations. The second rule is based on the nonparametric estimate of the logistic regression using the kernel estimates of the circular densities in the covariate spaces of successes

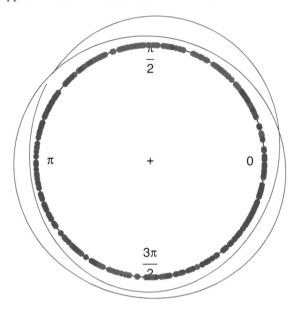

FIGURE 11.4
Kernel density estimation for classifying "water" (blue) from "not water" (red)
observations.

TABLE 11.2
MSPE and MR (along with smoothing parameters $\hat{\kappa}_{MSPE}$ and $\hat{\kappa}_{MR}$ respectively selected by least squares cross-validation (11.10) and likelihood cross-validation (11.11)) comparing classification of longitude data between 30 and 60 degrees North by using: (a) kernel density classification, (b) a circular model for binary regression, (c) local polynomial binary regression with p=0, p=1 and p=2.

Classifier	MSPE $(\hat{\kappa}_{MSPE})$	MR $(\hat{\kappa}_{MR})$
KDE	- (-)	0.19528 $(\hat{\kappa}_1 = 1.74; \hat{\kappa}_2 = 2.55)$
Parametric model	0.20237 (-)	0.25593 (-)
$p = 0$ estimator	0.08632 (1.5)	0.10026 (7.5)
$p = 1$ estimator	0.07843 (1.5)	0.09499 (7.5)
$p = 2$ estimator	0.07710 (5)	0.09763 (3.5)

and failures. After noting that estimator of the difference of densities and the binary regression one are equivalent, it is shown, as expected, that they

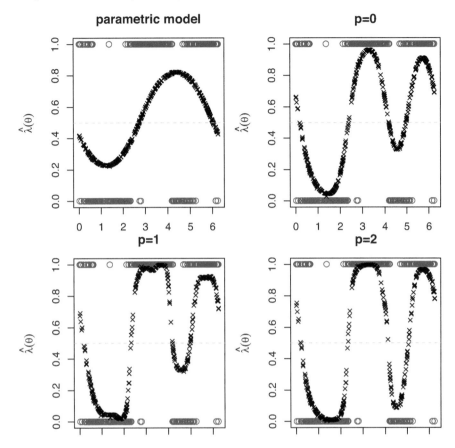

FIGURE 11.5
Top line, from left: parametric and local polynomial ($p \in (0, 1, 2)$) estimates of λ using "water" (blue) from "not water" (red) observations. The bandwidths have been selected using likelihood cross-validation.

share the convergence rate. The third approach lies on the local likelihood idea. The logit function is approximated by a pth degree periodic polynomial, whose coefficients are the *arguments* of the local likelihood function. For lower polynomial orders, i.e. $p \in (0, 1)$, the rate of convergence is the same of the other discussed classification approaches, while higher order polynomial would lead to faster rates. As for the practical performances, if compared to the classical parametric methods, the proposed estimators are preferable, as expected, especially when the data scenarios are rather complex. Finally, the discussed methods are applied to a real dataset for discriminating between water and land locations using the longitude of the observations.

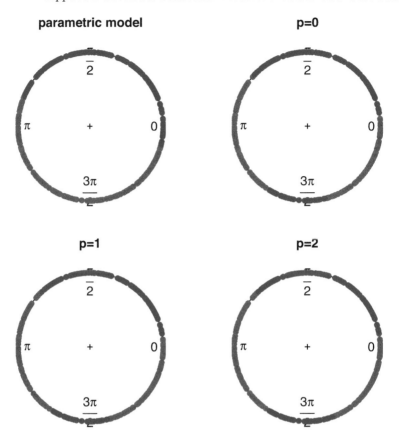

FIGURE 11.6

Top line, from left: classification of "water" (blue) – "not water" (red) observations using parametric and local polynomial ($p \in (0, 1, 2)$) estimators.

Bibliography

[1] Lee A. Circular data. *Computational Statistics*, 2:477–486, 2010.

[2] M. Di Marzio, A. Panzera, and C.C. Taylor. Local polynomial regression for circular predictors. *Statistics & Probability Letters*, 79:2066–2075, 2009.

[3] M. Di Marzio, A. Panzera, and C.C. Taylor. Kernel density estimation on the torus. *Journal of Statistical Planning and Inference*, 141:2156–2173, 2011.

[4] M. Di Marzio and C.C. Taylor. Kernel density classification and boosting: an L_2 analysis. *Statistics and Computing*, 15:113–123, 2005.

[5] K. Fernandes and J.S. Cardoso. Discriminative directional classifiers. *Neurocomputing*, 207:141–149, 2016.

[6] E. García-Portugués. Exact risk improvement of bandwidth selectors for kernel density estimation with directional data. *Electronic Journal of Statistics*, 7:1655–1685, 2013.

[7] P. Hall and K.H. Kang. Bandwidth choice for nonparametric classification. *The Annals of Statistics*, 33:284–306, 2005.

[8] P.L. Lopez-Cruz, C. Bielza, and Larranaga P. Directional naive Bayes classifiers. *Pattern Analysis and Applications*, 18:225–246, 2015.

[9] K.V. Mardia and P.E. Jupp. *Directional Statistics*. Chichester: J. Wiley, 2000.

[10] A. SenGupta and S. Roy. *A Simple Classification Rule for Directional Data*, pages 81–90. Springer, 2005.

[11] A. SenGupta and F.I. Ugwuowo. A classification method for directional data with application to the human skull. *Communication in Statistics – Theory Methods*, 40:457–466, 2011.

[12] D.F. Signorini and M.C. Jones. Kernel Estimators for Univariate Binary Regression. *Journal of the American Statistical Association*, 99:119–126, 2004.

[13] C.C. Taylor. Automatic bandwidth selection for circular density estimation. *Computational Statistics and Data Analysis*, 52:3493–3500, 2008.

12

Directional Statistics in Machine Learning: A Brief Review

Suvrit Sra

Massachusetts Institute of Technology

CONTENTS

12.1	Introduction	259
12.2	Basic Directional Distributions	260
	12.2.1 Uniform Distribution	260
	12.2.2 The von Mises-Fisher Distribution	261
	12.2.3 Watson Distribution	261
	12.2.4 Other Distributions	262
12.3	Related Work and Applications	263
12.4	Modeling Directional Data: Maximum-Likelihood Estimation ..	263
	12.4.1 Maximum-Likelihood Estimation for vMF	264
	12.4.2 Maximum-Likelihood Estimation for Watson	265
12.5	Mixture Models	266
	12.5.1 EM Algorithm	267
	12.5.2 Limiting Versions	268
	12.5.3 Application: Clustering Using movMF	269
	12.5.4 Application: Clustering Using moW	271
12.6	Conclusion	272
	Bibliography	272

12.1 Introduction

Data are often represented as vectors in a Euclidean space $\mathbb{R}^p$. But frequently they possess more structure and treating them as Euclidean vectors may be inappropriate. A simple example of this instance is when data are normalized to have unit norm, and thereby put on the surface of the *unit sphere* $\mathbb{S}^{p-1}$. Such data are better viewed as objects on a manifold, and when building mathematical models for them, it is often advantageous to exploit the geometry of the manifold (here $\mathbb{S}^{p-1}$).

For example, in information retrieval it has been long known that for high-dimensional vectors cosine similarity is a more effective measure of similarity than just Euclidean distances. There is substantial empirical evidence that normalizing the data vectors helps us remove the biases induced by the length of a document and obtain superior results [39, 40]. On a related note, the spherical k-means (`spkmeans`) algorithm [16], which runs k-means with cosine similarity for clustering unit norm vectors, has been found to work well for text clustering and a variety of other data. Another widely used similarity measure is *Pearson correlation*: given $x, y \in \mathbb{R}^d$ this is defined as $\rho(x,y) := \frac{\sum_i (x_i - \bar{x})(y_i - \bar{y})}{\sqrt{\sum_i (x_i - \bar{x})^2} \times \sqrt{\sum_i (y_i - \bar{y})^2}}$, where $\bar{x} = \frac{1}{d}\sum_i x_i$ and $\bar{y} = \frac{1}{d}\sum_i y_i$. Mapping $x \mapsto \tilde{x}$ with $\tilde{x}_i = \frac{x_i - \bar{x}}{\sqrt{\sum_i (x_i - \bar{x})^2}}$ (similarly define $\tilde{y}$), we obtain the inner-product $\rho(x,y) = \langle \tilde{x}, \tilde{y} \rangle$. Moreover, $\|\tilde{x}\| = \|\tilde{y}\| = 1$. Thus, the Pearson correlation is exactly the cosine similarity between $\tilde{x}$ and $\tilde{y}$. More broadly, domains where similarity measures such as cosine, Jaccard or Dice [36] are more effective than measures derived from Mahalanobis type distances, possess intrinsic "directional" characteristics, and are hence better modeled as directional data [30].

This chapter recaps basic statistical models for *directional data*, which herein refers to unit norm vectors for which "direction" is more important than "magnitude." In particular, we recall some basic distributions on the unit hypersphere, and then discuss two of the most commonly used ones: the von Mises-Fisher and Watson distributions. For these distributions, we describe maximum likelihood estimation as well as mixture modeling via the Expectation Maximization (EM) algorithm. In addition, we include a brief pointer to recent literature on applications of directional statistics within machine learning and related areas.

We warn the advanced reader that no new theory is developed in this chapter, and our aim herein is to merely provide an easy introduction. The material of this chapter is based on the author's thesis [41], and the three papers [6, 42, 43], and the reader is referred to these works for a more detailed development and additional experiments.

12.2 Basic Directional Distributions

12.2.1 Uniform Distribution

The probability element of the uniform distribution on $\mathbb{S}^{p-1}$ equals $c_p d\mathbb{S}^{p-1}$. The normalization constant c_p ensures that $\int_{\mathbb{S}^{p-1}} c_p d\mathbb{S}^{p-1} = 1$, from which it follows that

$$c_p = \Gamma(p/2)/2\pi^{p/2},$$

where $\Gamma(s) := \int_0^\infty t^{s-1} e^{-t} dt$ is the well-known Gamma function.

12.2.2 The von Mises-Fisher Distribution

The vMF distribution (see, e.g., [27, §2.3.1]) is one of the simplest distributions for directional data and it has properties analogous to those of the multivariate Gaussian on $\mathbb{R}^p$. For instance, the maximum entropy density on $\mathbb{S}^{p-1}$ subject to $E[x]$ being fixed is a vMF density (see, e.g., [35, pp. 172–174] and [29]).

A unit norm vector x has the von Mises-Fisher (vMF) distribution[1] if its density is

$$p_{\text{vmf}}(x; \mu, \kappa) := c_p(\kappa) e^{\kappa \mu^T x},$$

where $\|\mu\| = 1$ and $\kappa \geq 0$. Integrating using polar coordinates, it can be shown [41, App. B.4.2] that the normalizing constant is given by

$$c_p(\kappa) = \frac{\kappa^{p/2-1}}{(2\pi)^{p/2} I_{p/2-1}(\kappa)},$$

where $I_s(\kappa)$ denotes the modified Bessel function of the first kind [1].[2]

The vMF density p_{vmf} is parametrized by the mean direction μ, and the *concentration* parameter κ, so-called because it characterizes how strongly the unit vectors drawn according to p_{vmf} are concentrated about the mean direction μ. Larger values of κ imply stronger concentration about the mean direction. In particular when $\kappa = 0$, p_{vmf} reduces to the uniform density on $\mathbb{S}^{p-1}$, and as $\kappa \to \infty$, p_{vmf} tends to a point density.

12.2.3 Watson Distribution

The uniform and the vMF distributions are defined over *directions*. However, sometimes the observations are *axes* of direction, i.e., the vectors $\pm x \in \mathbb{S}^{p-1}$ are equivalent. This constraint is also denoted by $x \in \mathbb{P}^{p-1}$, where $\mathbb{P}^{p-1}$ is the projective hyperplane of dimension $p-1$. The multivariate Watson distribution [27, 31] models such data; it is parametrized by a *mean-direction* $\mu \in \mathbb{P}^{p-1}$, and a *concentration* parameter $\kappa \in \mathbb{R}$, and has probability density

$$p_{\text{wat}}(x; \mu, \kappa) := d_p(\kappa) e^{\kappa (\mu^T x)^2}, \qquad x \in \mathbb{P}^{p-1}. \tag{12.1}$$

The normalization constant $d_p(\kappa)$ is given by

$$d_p(\kappa) = \frac{\Gamma(p/2)}{2\pi^{p/2} M(\frac{1}{2}, \frac{p}{2}, \kappa)}, \tag{12.2}$$

where M is the confluent hypergeometric function [20, formula 6.1(1)]

$$M(a, c, \kappa) = \sum_{j \geq 0} \frac{a^{\bar{j}}}{c^{\bar{j}}} \frac{\kappa^j}{j!}, \qquad a, c, \kappa \in \mathbb{R}, \tag{12.3}$$

---

[1]Some authors, e.g., [27] call it the Fisher-von Mises-Langevin (FvML) distribution; we retain the shorter name von Mises-Fisher in this chapter.

[2]Note that sometimes in directional statistics literature, the integration measure is normalized by the uniform measure, so that instead of $c_p(\kappa)$, one uses $c_p(\kappa)2\pi^{p/2}/\Gamma(p/2)$.

and $a^{\bar{0}} = 1$, $a^{\bar{j}} = a(a+1)\cdots(a+j-1)$, $j \geq 1$, denotes the *rising-factorial*.

Observe that for $\kappa > 0$, the density concentrates around μ as κ increases, whereas for $\kappa < 0$, it concentrates around the great circle orthogonal to μ as κ decreases.

12.2.4 Other Distributions

We briefly summarize a few other interesting directional distributions, and refer the reader to [27, 31] for a more thorough development.

Bingham Distribution

Some axial data do not exhibit the rotational symmetry of Watson distributions. Such data could be potentially modeled using Bingham distributions, where the density at a point x is $B_p(x; K) := c_p(K)e^{x^T K x}$, where the normalizing constant can be shown to be $c_p(K) = \dfrac{\Gamma(p/2)}{2\pi^{p/2}M(\frac{1}{2},\frac{p}{2},K)}$, where $M(\cdot,\cdot,K)$ denotes the confluent hypergeometric function of matrix argument [34].

Note that since $x^T(K + \delta I_p)x = x^T K x + \delta$, the Bingham density is identifiable only up to a constant diagonal shift. Thus, one can assume $\mathrm{Tr}(K) = 0$, or that the smallest eigenvalue of K is zero [31]. Intuitively, one can see that the eigenvalues of K determine the axes around which the data clusters, e.g., greatest clustering will be around the axis corresponding to the leading eigenvector of K.

Bingham-Mardia Distribution

Certain problems require rotationally symmetric distributions that have a "modal ridge" rather than just a mode at a single point. To model data with such characteristics [31] suggest a density of the form

$$p(x; \mu, \kappa, \nu) = c_p(\kappa)e^{\kappa(\mu^T x - \nu)^2}, \tag{12.4}$$

where as usual $c_p(\kappa)$ denotes the normalization constant.

Fisher–Watson Distributions

This distribution is a simpler version of the more general Fisher–Bingham distribution [31]. The density is

$$p(x; \mu, \mu_0, \kappa, \kappa_0) = c_p(\kappa_0, \kappa, \mu_0^T \mu)e^{\kappa_0 \mu_0^T x + \kappa(\mu^T x)^2}. \tag{12.5}$$

Fisher–Bingham

This is a more general directional distribution; its density is

$$p(x; \mu, \kappa, A) = c_p(\kappa, A)e^{\kappa \mu^T x + x^T A x}. \tag{12.6}$$

There does not seem to exist an easy integral representation of the normalizing constant, and in an actual application one needs to resort to some sort of approximation for it (such as a saddle-point approximation). Kent distributions arise by putting an additional constraint $A\mu = 0$ in (12.6).

12.3 Related Work and Applications

The classical references on directional statistics are [29, 30, 31, 49]; a more recent reference is [27]. Additionally, for readers interested in statistics on manifolds, a good starting point is [12]. To our knowledge, the first work focusing on high-dimensional directional statistics was [5], where the key application was clustering of text and gene expression data using mixtures of vMFs. Another early reference on high-dimensional directional data is [18]. There exist a vast number of applications and settings where hyperspherical or manifold data arise. Summarizing all of these is clearly beyond the scope of this chapter. We mention below a smattering of applications that are directly related to this chapter.

We note a work on feature extraction based on correlation in [21]. Classical data mining applications such as topic modeling for normalized data are studied in [4, 37]. A semi-parametric setting using Dirichlet process mixtures for spherical data is [44]. Several directional data clustering settings include: depth images using Watson mixtures [22]; a k-means++ [3] style procedure for mixture of vMFs [32]; clustering on orthogonal manifolds [11]; mixtures of Gaussian and vMFs [25]. Directional data has also been used in several biomedical (imaging) applications, for example [33], fMRI [26], white matter supervoxel segmentation [10], and brain imaging [38]. In signal processing there are applications to spatial fading using vMF mixtures [28] and speaker modeling [47]. Finally, beyond vMF and Watson, it is worthwhile to consider the Angular Gaussian distribution [48], which has been applied to model natural images for instance in [24].

12.4 Modeling Directional Data: Maximum-Likelihood Estimation

In this section we briefly recap data models involving vMF and Watson distributions. In particular, we describe maximum-likelihood estimation for both distributions. As is well-known by now, for these distributions estimating the mean μ is simpler than estimating the concentration parameter κ.

12.4.1   Maximum-Likelihood Estimation for vMF

Let $\mathcal{X} = \{x_1, \ldots, x_n\}$ be a set of points drawn from $p_{\mathrm{vmf}}(x; \mu, \kappa)$. We wish to estimate μ and κ by solving the optimization problem

$$\max \ell(\mathcal{X}; \mu, \kappa) := n \log c_p(\kappa) + \sum_{i=1}^{n} \kappa \mu^T x_i, \quad \text{s.t. } \|\mu\| = 1, \ \kappa \geq 0. \quad (12.1)$$

Writing $\frac{\|\sum_i x_i\|}{n} = \bar{r}$, a brief calculation shows that the optimal solution satisfies

$$\mu = \frac{1}{n\bar{r}} \sum_{i=1} x_i, \quad \kappa = A_p^{-1}(\bar{r}), \quad (12.2)$$

where the nonlinear map A_p is defined as

$$A_p(\kappa) = \frac{-c_p'(\kappa)}{c_p(\kappa)} = \frac{I_{p/2}(\kappa)}{I_{p/2-1}(\kappa)} = \bar{r}. \quad (12.3)$$

The challenge is to solve (12.3) for κ. For small values of p (e.g., $p = 2, 3$) the simple estimates provided in [31] suffice. But for machine learning problems, where p is typically very large, these estimates do not suffice. In [6], the authors provided efficient numerical estimates for κ that were obtained by truncating the continued fraction representation of $A_p(\kappa)$ and solving the resulting equation. These estimates were then corrected to yield the approximation

$$\hat{\kappa} = \frac{\bar{r}(p - \bar{r}^2)}{1 - \bar{r}^2}, \quad (12.4)$$

which turns out to be remarkably accurate in practice.

Subsequently, [46] showed simple bounds for κ by exploiting inequalities about the Bessel ratio $A_p(\kappa)$—this ratio possesses several nice properties, and is very amenable to analytic treatment [2]. The work of [46] lends theoretical support to the empirically determined approximation (12.4), by essentially showing this approximation lies in the "correct" range. Tanabe et al. [46] also presented a fixed-point iteration based algorithm to compute an approximate solution κ.

The *critical* difference between this approximation and the next two is that it does not involve any Bessel functions (or their ratio). That is, not a single evaluation of $A_p(\kappa)$ is needed—an advantage that can be significant in high-dimensions where it can be computationally expensive to compute $A_p(\kappa)$. Naturally, one can try to compute $\log I_s(\kappa)$ ($s = p/2$) to avoid overflows (or underflows as the case may be), though doing so introduces yet another approximation. Therefore, when running time and simplicity are of the essence, approximation (12.4) is preferable.

Approximation (12.4) can be made more exact by performing a few iterations of Newton's method. To save runtime, Sra [42] recommends only two iterations of Newton's method, which amounts to computing κ_0 using (12.4), followed by

$$\kappa_{s+1} = \kappa_s - \frac{A_p(\kappa_s) - \bar{R}}{1 - A_p(\kappa_s)^2 - \frac{(p-1)}{\kappa_s} A_p(\kappa_s)}, \quad s = 0, 1. \quad (12.5)$$

Approximation (12.5) was shown in [42] to be competitive in running time with the method of [46], and was seen to be overall more accurate. Approximating κ remains a topic of research interest, as can be seen from the recent works [13, 23].

12.4.2 Maximum-Likelihood Estimation for Watson

Let $\mathcal{X} = \{x_1, \ldots, x_n\}$ be i.i.d. observations drawn from $p_{\text{wat}}(x; \mu, \kappa)$. We wish to estimate μ and κ by maximizing the log-likelihood

$$\ell(\mathcal{X}; \mu, \kappa) = n\left(\kappa\mu^\top S\mu - \ln M(1/2, p/2, \kappa) + \gamma\right), \qquad (12.6)$$

subject to $\mu^T\mu = 1$, where $S = \frac{1}{n}\sum_{i=1}^n x_i x_i^\top$ is the sample *scatter matrix*, and γ is a constant. Considering the first-order optimality conditions of (12.6) leads to the following parameter estimates [31, Sec. 10.3.2]

$$\hat{\mu} = \pm s_1 \quad \text{if} \quad \hat{\kappa} > 0, \qquad \hat{\mu} = \pm s_p \quad \text{if} \quad \hat{\kappa} < 0, \qquad (12.7)$$

where $s_1, s_2, \ldots, s_p$ are (normalized) eigenvectors of the scatter matrix S corresponding to the eigenvalues $\lambda_1 \geq \lambda_2 \geq \cdots \geq \lambda_p$.

To estimate the concentration parameter $\hat{\kappa}$ we must solve:[3]

$$g\left(\tfrac{1}{2}, \tfrac{p}{2}; \hat{\kappa}\right) := \frac{\frac{\partial}{\partial\kappa}M\left(\frac{1}{2}, \frac{p}{2}, \kappa\right)|_{\kappa=\hat{\kappa}}}{M\left(\frac{1}{2}, \frac{p}{2}, \hat{\kappa}\right)} = \hat{\mu}^\top S\hat{\mu} := r \qquad (0 \leq r \leq 1). \quad (12.8)$$

Notice that (12.7) and (12.8) are coupled—so we simply solve both $g(1/2, p/2; \hat{\kappa}) = \lambda_1$ and $g(1/2, p/2; \hat{\kappa}) = \lambda_p$, and pick the solution that yields a higher log-likelihood.

The hard part is to solve (12.8). One could use a root-finding method (e.g., Newton–Raphson), but similar to the vMF case, an out-of-the-box root-finding approach can be unduly slow or numerically hard as data dimensionality increases. The authors of [43] consider the following more general equation:

$$g(a, c; \kappa) := \frac{M'(a, c; \kappa)}{M(a, c; \kappa)} = r$$
$$c > a > 0, \quad 0 \leq r \leq 1, \qquad (12.9)$$

and derive for it high-quality closed form numerical approximations. These approximations improve upon two previous approaches, that of [8] and [41]. Bijral et al. [8] followed the continued-fraction approach of [6] to obtain the heuristic approximation

$$BBG(r) := \frac{cr - a}{r(1 - r)} + \frac{r}{2c(1 - r)}. \qquad (12.10)$$

This approximation turns out to be less accurate than the functional approximations cited in Theorem 1, which was obtained in [43].

---

[3] We need $\lambda_1 > \lambda_2$ to ensure a unique estimate for positive κ, and $\lambda_{p-1} > \lambda_p$, for negative κ.

Theorem 1 ([43]) *Let the solution to $g(a, c; \kappa) = r$ be denoted by $\kappa(r)$. Consider the following three bounds:*

$$(\textit{lower bound}) \qquad L(r) = \frac{rc - a}{r(1-r)}\left(1 + \frac{1-r}{c-a}\right), \qquad (12.11)$$

$$(\textit{bound}) \qquad B(r) = \frac{rc - a}{2r(1-r)}\left(1 + \sqrt{1 + \frac{4(c+1)r(1-r)}{a(c-a)}}\right), \qquad (12.12)$$

$$(\textit{upper bound}) \qquad U(r) = \frac{rc - a}{r(1-r)}\left(1 + \frac{r}{a}\right). \qquad (12.13)$$

Let $c > a > 0$, and $\kappa(r)$ be the solution (12.9). Then, we have

1. *for $a/c < r < 1$,*

$$L(r) < \kappa(r) < B(r) < U(r), \qquad (12.14)$$

2. *for $0 < r < a/c$,*

$$L(r) < B(r) < \kappa(r) < U(r). \qquad (12.15)$$

3. *and if $r = a/c$, then $\kappa(r) = L(a/c) = B(a/c) = U(a/c) = 0$.*

12.5 Mixture Models

Many times a single vMF or Watson distribution is insufficient to model data. In these cases, a richer model (e.g., for clustering, or as a generative model, etc.) such as a mixture model may be more useful. We summarize in this section mixtures of vMF (movMF) and mixtures of Watson (moW) distributions. The former was originally applied to high-dimensional text and gene expression data in [6], and since then it has been used in a large number of applications (see also Section 12.3). The latter has been applied to genetic data [43], as well as in other data mining applications [8].

Let $p(x; \mu, \kappa)$ denote either a vMF density or a Watson density. We consider mixture models of K different vMF densities or K different Watson densities. Thus, a given unit norm observation vector x has the *mixture density*

$$f\left(x; \{\mu_j\}_{j=1}^{K}, \{\kappa_j\}_{j=1}^{K}\right) := \sum_{j=1}^{K} \pi_j p(x; \mu_j, \kappa_j). \qquad (12.1)$$

Suppose we observe the set $\mathcal{X} = \{x_1, \ldots, x_n \in \mathbb{P}^{p-1}\}$ of i.i.d. observations drawn from (12.1). Our aim is to infer the mixture parameters $(\pi_j, \mu_j, \kappa_j)_{j=1}^{K}$, where $\sum_j \pi_j = 1$, $\pi_j \geq 0$, $\|\mu_j\| = 1$, and $\kappa_j \geq 0$ for an movMF and $\kappa_j \in \mathbb{R}$ for a moW.

12.5.1 EM Algorithm

A standard, practical approach to estimating the mixture parameters is via the Expectation Maximization (EM) algorithm [15] applied to maximize the mixture log-likelihood for $\mathcal{X}$. Specifically, we seek to maximize

$$\ell(\mathcal{X}; \{\pi_j, \mu_j, \kappa_j\}_{j=1}^K) := \sum_{i=1}^n \ln\Big(\sum_{j=1}^K \pi_j p(x_i; \mu_j, \kappa_j)\Big). \qquad (12.2)$$

First, we use Jensen's inequality on (12.2) to obtain the lower bound

$$\ell(\mathcal{X}; \{\pi_j, \mu_j, \kappa_j\}_{j=1}^K) \geq \sum_{ij} \beta_{ij} \ln\left(\pi_j p(x_i|\mu_j, \kappa_j)/\beta_{ij}\right). \qquad (12.3)$$

Then, the E-Step sets β_{ij} to the *posterior* probability (for x_i given μ_j and κ_j):

$$\beta_{ij} := \frac{\pi_j p(x_i|\mu_j, \kappa_j)}{\sum_l \pi_l p(x_i|\mu_l, \kappa_l)}. \qquad (12.4)$$

With this choice of β_{ij}, the M-Step involves maximizing the bound (12.3) over the mixture parameters. In particular, this results in the updates
M-Step for movMF:

$$\mu_j = \frac{r_j}{\|r_j\|}, \quad r_j = \sum_{i=1}^n \beta_{ij} x_i, \qquad (12.5)$$

$$\kappa_j = A_p^{-1}(\bar{r}_j), \quad \bar{r}_j = \frac{\|r_j\|}{\sum_{i=1}^n \beta_{ij}}. \qquad (12.6)$$

M-Step for moW:

$$\mu_j = s_1^j \;\; \text{if} \;\; \kappa_j > 0, \qquad \mu_j = s_p^j \;\; \text{if} \;\; \kappa_j < 0, \qquad (12.7)$$

$$\kappa_j = g^{-1}(1/2, p/2, r_j), \quad \text{where} \quad r_j = \mu_j^\top S^j \mu_j \qquad (12.8)$$

where s_1^j denotes the top eigenvector corresponding to eigenvalue $\lambda_1(S^j)$ of the *weighted-scatter matrix*

$$S^j = \frac{1}{\sum_{i=1}^n \beta_{ij}} \sum_{i=1}^n \beta_{ij} x_i x_i^T.$$

For both movMF and moW, the component probabilities are as usual $\pi_j = \frac{1}{n}\sum_i \beta_{ij}$. Iterating between (12.4) and the M-Steps we obtain an EM algorithm. Pseudo-code for such a procedure is shown as Algorithm 1.

Hard Assignments. To speed up EM, we can replace the E-Step (12.4) by the standard *hard-assignment* rule [6]:

$$\beta_{ij} = \begin{cases} 1, & \text{if } j = \text{argmax}_{j'} \ln \pi_l + \ln p(x_i|\mu_l, \kappa_l), \\ 0, & \text{otherwise.} \end{cases} \qquad (12.9)$$

The corresponding M-Step also simplifies considerably. Such hard-assignments maximize a lower-bound on the incomplete log-likelihood and yield *partitional-clustering* algorithms.

Initialization. For movMF, typically an initialization using spherical k-means (spkmeans) [16] can be used. The next section presents arguments that explain why this initialization is natural for movMF. Similarly, for moW, an initialization based on diametrical k-means [17] can be used, though sometimes even an spkmeans initialization suffices [43].

Input: $x = \{x_1, \dots, x_n :$ where each $\|x_i\| = 1\}$, K
Output: Parameter estimates π_j, μ_j, and κ_j, for $1 \leq j \leq K$
Initialize π_j, μ_j, κ_j for $1 \leq j \leq K$
while *not converged* **do**
$\quad$ {*Perform the E-step of EM*}
$\quad$ **foreach** *i and j* **do**
$\quad\quad$ | Compute β_{ij} using (12.5.4) (or via (12.5.9))
$\quad$ **end**
$\quad$ {*Perform the M-Step of EM*}
$\quad$ **for** $j = 1$ *to* K **do**
$\quad\quad$ $\pi_j \leftarrow \frac{1}{n} \sum_{i=1}^{n} \beta_{ij}$
$\quad\quad$ For movMF: compute μ_j and κ_j using (12.5.5) and (12.5.6)
$\quad\quad$ For moW: compute μ_j and κ_j using (12.5.7) and (12.5.8)
$\quad$ **end**
end

Algorithm 1: EM Algorithm for movMF and moW

12.5.2 Limiting Versions

It is well-known that the famous k-means algorithm may be obtained as a limiting case of the EM algorithm applied to a mixture of Gaussians. Analogously, the spherical k-means algorithm of [16] that clusters unit norm vectors and finds unit norm means (hence "spherical") can be viewed as the limiting case of a movMF. Indeed, assume that the priors of all mixture components are equal. Furthermore, assume that all the mixture components have equal concentration parameters κ and let $\kappa \to \infty$. Under these assumptions, the E-Step (12.9) reduces to assigning a point x_i to the cluster nearest to it, which here is given by the cluster with whose centroid the given point has largest dot product. In other words, a point x_i is assigned to cluster $k = \text{argmax}_j \, x_i^T \mu_j$ because $\beta_{ik} \to 1$ and $\beta_{ij} \to 0$ for $j \neq k$ in (12.9).

In a similar manner, the diametrical clustering algorithm of [17] also may be viewed as a limiting case of EM applied to a moW. Recall that the diametrical clustering algorithm groups together correlated and anti-correlated unit norm data vectors into the same cluster, i.e., it treats diametrically opposite points equivalently. Remarkably, it turns out that the diametrical clustering algorithm of [17] can be obtained as follows: Let $\kappa_j \to \infty$, so that for each i, the corresponding posterior probabilities $\beta_{ij} \to \{0, 1\}$; the particular β_{ij} that tends to 1 is the one for which $(\mu_j^\top x_i)^2$ is maximized in the E-Step; subsequently the M-Step (12.7), (12.8) also simplifies, and yields the same updates as made by the diametrical clustering algorithm.

Alternatively, we can obtain diametrical clustering from the hard-assignment heuristic of EM applied to a moW where all mixture components have the same (positive) concentration parameter κ. Then, in the E-Step (12.9), we can ignore κ altogether, which reduces Algorithm 12.5.1 to the diametrical clustering procedure.

12.5.3 Application: Clustering Using movMF

Mixtures of vMFs have been successfully used in text clustering; see [7] for a detailed overview. We recall below results of two main experiments: (i) simulated data; and (ii) Slashdot news articles.

The key characteristic of text data is its high dimensionality. And for modeling clusters of such data using a movMF, the approximate computation of the concentration parameter κ as discussed in Section 12.4.1 is of great importance: without this approximation, the computation breaks down due to floating point difficulties.

For (i), we simulate a mixture of 4 vMFs, each with $p = 1000$, and draw a sample of 5000 data points. The clusters are chosen to be roughly of the same size and their relative mixing proportions are $(0.25, 0.24, 0.25, 0.26)$, with concentration parameters (to one digit) $(651.0, 267.8, 267.8, 612.9)$, and random unit vectors as means. This is the same data as the `big-mix` data in [7]. We generated the samples using the `vmfsamp` code (available from the author upon request).

For (ii), we recall a part of the results of [7] on news articles from the Slashdot website. These articles are tagged and cleaned to retain 1000 articles that more clearly belong to a primary category / cluster. We report results on 'Slash-7' and 'Slash-6'; the first contains 6714 articles in 7 primary categories: business, education, entertainment, games, music, science, and internet; while the second contains 5182 articles in 6 categories: biotech, Microsoft, privacy, Google, security, and space.

Performance Evaluation

There are several ways to evaluate performance of a clustering method. For the simulated data, we know the true parameters from which the dataset was simulated, hence we can compare the error in estimated parameter values. For the Slashdot data, we have knowledge of "ground truth" labels, so we can use the *normalized mutual information (NMI)* [45] (a measure that was also previously used to assess movMF based clustering [6, 7]) as an external measure of cluster quality. Suppose the predicted cluster labels are $\hat{Y}$ and the true labels are Y, then the NMI between Y and $\hat{Y}$ is defined as

$$\mathrm{NMI}(Y,\hat{Y}) := \frac{I(Y,\hat{Y})}{\sqrt{H(Y)H(\hat{Y})}}, \tag{12.10}$$

where $I(\cdot,\cdot)$ denotes the usual mutual information and H denotes the entropy [14]. When the predicted labels agree perfectly with the true labels, then NMI equals 1; thus higher values of NMI are better.

Results on "Bigsim"

The results of the first experiment are drawn from [6], and are reported in Table 12.1. From the results it is clear that on this particular simulated data, EM manages to recover the true parameters to quite a high degree of accuracy. Part of this reason is due to the high values of the concentration parameter: as κ increases, the probability mass concentrates, which makes it easier to separate the clusters using EM. To compute the values in the table, we ran EM with soft-assignments and then after convergence used assignment (12.9).

TABLE 12.1
Accuracy of parameter estimation via EM for movMF on the "bigsim" data.

$\min \mu^T \hat{\mu}$	$\max \frac{\lvert\kappa-\hat{\kappa}\rvert}{\kappa}$	$\max \frac{\lvert\pi-\hat{\pi}\rvert}{\pi}$
0.994	0.006	0.002

We report the worst values (the averages were better) seen across 20 different runs. For the estimated mean, we report the worst inner product with the true mean; for the concentration and mixture weights we report worst case relative errors.

Results on Slashdot

These results are drawn from [7]. The results here are reported merely as an illustration, and we refer the reader to [7] for more extensive results. We report performance of our implementation of Algorithm 12.5.1 (EM for movMF) against Latent Dirichlet Allocation (LDA) [9] and a Exponential-family Dirichlet compound multinomial model (EDCM) [19].

   Table 12.2 reports results of comparing Algorithm 12.5.1 specialized for movMFs against LDA and EDCM. As can be seen from the results, the vMF mixture leads to much higher quality clustering than the other two competing approaches. We did not test an optimized implementation (and used our own MATLAB implementation), but note anecdotally that the EM procedure was 3–5 times faster than the others.

TABLE 12.2
Comparison of NMI values of moVMF versus LDA and ECDM (derived from [7]).

Dataset	moVMF	LDA	ECDM
Slash-7	0.39	0.22	0.31
Slash-6	0.65	0.36	0.46

The NMI values obtained by the moVMF model are substantially higher than LDA as well as ECDM, suggesting that the clusters obtained via moVMF are of higher quality.

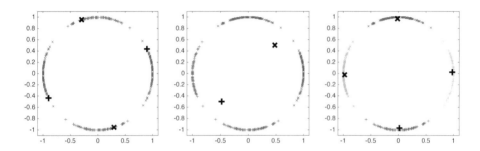

FIGURE 12.1
The left panel shows axially symmetric data that has two clusters (centroids are indicated by '+' and 'x'). The middle and right panel shows clustering yielded by (Euclidean) K-means (note that the centroids fail to lie on the circle in this case) with $K = 2$ and $K = 4$, respectively. Diametrical clustering recovers the true clusters in the left panel.

12.5.4 Application: Clustering Using moW

Figure 12.1 shows a toy example of axial data. Here, the original data has two clusters (leftmost panel of Figure 12.1). If we cluster this data into two clusters using Euclidean k-means, we obtain the plot in the middle panel; clustering into 4 groups using Euclidean k-means yields the rightmost panel. As is clear from the figure, Euclidean k-means cannot discover the desired structure, if the true clusters are on axial data. The true clusters (left panel) are recovered by diametrical clustering (right panel) [17], which also shows the mean vectors $\pm\mu$ for each cluster (marked with a '+' and 'x'). Recall that as mentioned above, the diametrical clustering method is obtained as the limiting case of EM on moW.

12.6 Conclusion

We summarized a few distributions from directional statistics that are useful for modeling normalized data. We focused in particular on the von Mises-Fisher distribution (the "Gaussian" of the hypersphere) and the Watson distribution (for axially symmetric data). For both of these distributions, we recapped maximum likelihood parameter estimation as well as mixture modeling using the EM algorithm. For extensive numerical results on clustering using mixtures of vMFs, we refer the reader to the original paper [6]; similarly, for mixtures of Watsons please see [43]. The latter paper also describes asymptotic estimates of the concentration parameter in detail.

Now directional distributions are widely used in machine learning (Section 12.3 provides some pointers to related work), and we hope the brief summary provided in this chapter helps promote wider understanding about these. In particular, we hope to see more exploration of directional models in the following important subareas: Bayesian models, Hidden Markov Models using directional models, and deep generative models.

Bibliography

[1] M. Abramowitz and I. A. Stegun. *Handbook of Mathematical Functions, With Formulas, Graphs, and Mathematical Tables.* Dover, New York, 1974.

[2] D. E. Amos. Computation of modified Bessel functions and their ratios. *Mathematics of Computation*, 28(125):235–251, 1974.

[3] D. Arthur and S. Vassilvitskii. k-means++: The advantages of careful seeding. In *Proceedings of the eighteenth annual ACM-SIAM Symposium on Discrete Algorithms*, pages 1027–1035. Society for Industrial and Applied Mathematics, 2007.

[4] A. Banerjee and S. Basu. Topic models over text streams: A study of batch and online unsupervised learning. In *SIAM International Conference on Data Mining*, volume 7, pages 437–442. SIAM, 2007.

[5] A. Banerjee, I. Dhillon, J. Ghosh, and S. Sra. Generative model-based clustering of directional data. In *Proceedings of the ninth ACM SIGKDD International Conference on Knowledge Discovery and Data Mining*, pages 19–28. ACM, 2003.

[6] A. Banerjee, I. S. Dhillon, J. Ghosh, and S. Sra. Clustering on the unit

hypersphere using von Mises-Fisher distributions. *Journal of Machine Learning Research*, 6:1345–1382, 2005.

[7] A. Banerjee, I. S. Dhillon, J. Ghosh, and S. Sra. Text Clustering with Mixture of von Mises-Fisher Distributions. In A. N. Srivastava and M. Sahami, editors, *Text Mining: Theory, Applications, and Visualization*. CRC Press, 2009.

[8] A. Bijral, M. Breitenbach, and G. Z. Grudic. Mixture of Watson distributions: a generative model for hyperspherical embeddings. In *Artificial Intelligence and Statistics (AISTATS), Journal of Machine Learning Research* - Proceedings Track 2, pages 35–42, 2007.

[9] D. M. Blei, A. Y. Ng, and M. I. Jordan. Latent Dirichlet Allocation. *Journal of Machine Learning Research*, 3:993–1022, 2003.

[10] R. P. Cabeen and D. H. Laidlaw. White matter supervoxel segmentation by axial DP-means clustering. In *Medical Computer Vision. Large Data in Medical Imaging*, pages 95–104. Springer, 2014.

[11] H. Cetingul and R. Vidal. Intrinsic mean shift for clustering on stiefel and Grassmann manifolds. In *IEEE Conference on Computer Vision and Pattern Recognition*, pages 1896–1902. IEEE, 2009.

[12] Y. Chikuse. *Statistics on special manifolds*, volume 174. Springer Science & Business Media, 2012.

[13] D. Christie. Efficient von Mises-Fisher concentration parameter estimation using Taylor series. *Journal of Statistical Computation and Simulation*, 85(16):1–7, 2015.

[14] T. Cover and J. Thomas. *Elements of Information Theory*. John Wiley & Sons, New York, USA, 1991.

[15] A. Dempster, N. Laird, and D. Rubin. Maximum Likelihood from incomplete data via the EM algorithm. *Journal of the Royal Statistical Society*, 39, 1977.

[16] I. S. Dhillon and D. S. Modha. Concept decompositions for large sparse text data using clustering. *Machine Learning*, 42(1):143–175, 2001.

[17] I. S. Dhillon, E. M. Marcotte, and U. Roshan. Diametrical clustering for identifying anti-correlated gene clusters. *Bioinformatics*, 19(13):1612–1619, 2003.

[18] I. L. Dryden et al. Statistical analysis on high-dimensional spheres and shape spaces. *The Annals of Statistics*, 33(4):1643–1665, 2005.

[19] C. Elkan. Clustering documents with an exponential-family approximation of the dirichlet compound multinomial distribution. In *Proceedings of the 23rd International Conference on Machine Learning*, pages 289–296. ACM, 2006.

[20] A. Erdélyi, W. Magnus, F. Oberhettinger, and F. G. Tricomi. *Higher transcendental functions*, volume 1. McGraw Hill, 1953.

[21] Y. Fu, S. Yan, and T. S. Huang. Correlation metric for generalized feature extraction. *IEEE Transactions on Pattern Analysis and Machine Intelligence*, 30(12):2229–2235, 2008.

[22] M. A. Hasnat, O. Alata, and A. Trémeau. Unsupervised clustering of depth images using Watson mixture model. In *22nd International Conference on Pattern Recognition (ICPR)*, pages 214–219. IEEE, 2014.

[23] K. Hornik and B. Grün. On maximum likelihood estimation of the concentration parameter of von Mises-Fisher distributions. *Computational statistics*, 29(5):945–957, 2014.

[24] R. Hosseini. *Natural Image Modelling using Mixture Models with compression as an application*. PhD thesis, Berlin, Technische Universtität Berlin, Diss., 2012, 2012.

[25] P. Kasarapu and L. Allison. Minimum message length estimation of mixtures of multivariate Gaussian and von Mises-Fisher distributions. *Machine Learning*, pages 1–46, 2015.

[26] D. Lashkari, E. Vul, N. Kanwisher, and P. Golland. Discovering structure in the space of fmri selectivity profiles. *Neuroimage*, 50(3):1085–1098, 2010.

[27] C. Ley and T. Verdebout. *Modern Directional Statistics*. Chapman & Hall/CRC, 2017.

[28] K. Mammasis, R. W. Stewart, and J. S. Thompson. Spatial fading correlation model using mixtures of von Mises-Fisher distributions. *IEEE Transactions on Wireless Communications*, 8(4):2046–2055, 2009.

[29] K. V. Mardia. Statistics of directional data. *Journal Royal Statistical Society, Series B (Methodological)*, 37(3):349–393, 1975.

[30] K. V. Mardia. *Statistical Distributions in Scientific Work*, volume 3, chapter Characteristics of directional distributions, pages 365–385. Reidel, Dordrecht, 1975.

[31] K. V. Mardia and P. Jupp. *Directional Statistics*. John Wiley & Sons Ltd., second edition, 2000.

[32] M. Mash'al and R. Hosseini. K-means++ for mixtures of von Mises-Fisher distributions. In *Conference on Information and Knowledge Technology (IKT)*, pages 1–6. IEEE, 2015.

[33] T. McGraw, B. C. Vemuri, B. Yezierski, and T. Mareci. Von Mises-Fisher mixture model of the diffusion ODF. In *3rd IEEE International Symposium on Biomedical Imaging: Nano to Macro*, pages 65–68. IEEE, 2006.

[34] R. J. Muirhead. *Aspects of Multivariate Statistical Theory*. John Wiley, 1982.

[35] C. R. Rao. *Linear Statistical Inference and its Applications*. Wiley, New York, 2nd edition, 1973.

[36] E. Rasmussen. Clustering algorithms. In W. Frakes and R. Baeza-Yates, editors, *Information Retrieval: Data Structures and Algorithms*, pages 419–442. Prentice Hall, New Jersey, 1992.

[37] J. Reisinger, A. Waters, B. Silverthorn, and R. J. Mooney. Spherical topic models. In *Proceedings of the 27th International Conference on Machine Learning (ICML)*, pages 903–910, 2010.

[38] S. Ryali, T. Chen, K. Supekar, and V. Menon. A parcellation scheme based on von Mises-Fisher distributions and markov random fields for segmenting brain regions using resting-state fMRI. *NeuroImage*, 65:83–96, 2013.

[39] G. Salton. *Automatic Text Processing: The Transformation, Analysis, and Retrieval of Information by Computer*. Addison-Wesley (Reading MA), 1989.

[40] G. Salton and M. J. McGill. *Introduction to Modern Retrieval*. McGraw-Hill Book Company, 1983.

[41] S. Sra. *Matrix Nearness Problems in Data Mining*. PhD thesis, University of Texas at Austin, 2007.

[42] S. Sra. A short note on parameter approximation for von Mises-Fisher distributions: and a fast implementation of $i_s(x)$. *Computational Statistics*, 27(1):177–190, 2012.

[43] S. Sra and D. Karp. The multivariate Watson distribution: maximum-likelihood estimation and other aspects. *Journal of Multivariate Analysis*, 114:256–269, 2013.

[44] J. Straub, J. Chang, O. Freifeld, and J. W. Fisher III. A dirichlet process mixture model for spherical data. In *Proceedings of the Eighteenth International Conference on Artificial Intelligence and Statistics*, pages 930–938, 2015.

[45] A. Strehl and J. Ghosh. Cluster Ensembles – A Knowledge Reuse Framework for Combining Multiple Partitions. *Journal of Machine Learning Research*, 3:583–617, 2002.

[46] A. Tanabe, K. Fukumizu, S. Oba, T. Takenouchi, and S. Ishii. Parameter estimation for von Mises-Fisher distributions. *Computational Statistics*, 22(1):145–157, 2007.

[47] H. Tang, S. M. Chu, and T. S. Huang. Generative model-based speaker clustering via mixture of von Mises-Fisher distributions. In *IEEE International Conference on Acoustics, Speech and Signal Processing.*, pages 4101–4104. IEEE, 2009.

[48] D. E. Tyler. Statistical analysis for the angular central gaussian distribution on the sphere. *Biometrika*, 74(3):579–589, 1987.

[49] G. S. Watson. The statistics of orientation data. *The Journal of Geology*, 74(5, Part 2):786–797, 1966.

13

Applied Directional Statistics with R: An Overview

Arthur Pewsey

University of Extremadura Cáceres, Spain

CONTENTS

13.1 Introduction 277
13.2 The Circular Package 278
13.3 Packages that Use the Circular Package 280
13.4 Other Packages for Circular Statistics 281
13.5 The Directional Package 281
13.6 Other Packages for Directional Statistics 283
13.7 Unsupported Directional Statistics Methodologies 284
13.8 Conclusions 285
Bibliography 285

13.1 Introduction

In its widest definition, directional statistics is that branch of statistics that deals with data for which Riemannian manifolds, such as the unit circle, torus, cylinder, sphere and extensions thereof, are the natural supports. The geometries and topologies of such supports generally disqualify the use of the more familiar statistical techniques developed for data observed on supports such as $\mathbb{R}$, $\mathbb{R}^2$, etc.

As implied in the prefaces of the seminal text [40] as well as its heavily revised more recent incarnation, [42], historically a major impediment to the application of Directional Statistics has been a lack of software implementing the methodology particular to it. In recent years, the advent of the R statistical computing environment [53] and its myriad contributed packages has partially addressed that dearth.

In this chapter we provide a quick reference guide identifying and commenting on the functionality of those R packages relevant to the analysis of directional data. The two main ones are the `circular` and `Directional` packages. The functionality of the `circular` package is primarily for the analysis of data whose natural support is the unit circle; such as directions or times of the day, week, month or year. Others of its functions deal with data distributed on extensions of the unit circle such as the torus (circular-circular data) or cylinder (circular-linear data). The `Directional` package includes functions for use with circular, toroidal and cylindrical data as well as data distributed on the unit sphere or hypersphere.

Section 13.2 focuses on the `circular` package, and Section 13.3 on those packages that make use of it. Section 13.4 considers other packages relevant to the analysis of circular data. Details of the `Directional` package are provided in Section 13.5. Other packages available for analyzing directional data are considered in Section 13.6. Four important methodologies that, as yet, are not supported by dedicated R packages are discussed in Section 13.7. Conclusions are drawn in Section 13.8.

13.2   The Circular Package

The `circular` package [5] has evolved from the S-Plus `CircStats` package written by Ulrich Lund and ported to R by Claudio Agostinelli. The functionality of the `CircStats` package was originally based upon the content of [28], whereas that of the most recent version of the `circular` package includes techniques from a wider range of sources.

The `circular` package has basic functions with which to handle, summarize, simulate, graphically represent and model circular data. Numerous circular data sets, taken mainly from [19], are also available from within it. More specifically, the package includes functions to:

- Perform kernel density estimation;

- Compute densities for classical circular distributions such as the circular uniform, cardioid, von Mises, wrapped normal, wrapped Cauchy, projected normal, Cartwright, asymmetric triangular, axial von Mises, two-component generalized von Mises and two-component von Mises mixture, as well as the more recently proposed symmetric Jones–Pewsey family [30] and the highly flexible four-parameter Kato–Jones family [34];

- Simulate random variates from the circular uniform, cardioid, von Mises, wrapped normal, wrapped Cauchy, wrapped stable, Kato–Jones and two-component von Mises mixture distributions;

- Carry out maximum likelihood estimation for the parameters of the von Mises, wrapped Cauchy and wrapped normal distributions, and calculate confidence intervals for the parameters of the von Mises distribution;

- Apply tests for uniformity, homogeneity, goodness-of-fit, change points and constant concentration;

- Perform circular-circular and circular-linear regression and one-way circular ANOVA under the assumption of von Mises populations.

Many of the `circular` package's capabilities are illustrated in [51]. The latter's companion `CircStatsInR` workspace, available from `http://circstatinr.st-andrews.ac.uk/`, includes over 150 additional functions for techniques not covered by the `circular` package. Amongst them are functions to:

- Compute the densities of Jones–Pewsey, Batschelet [10, Section 15.7], [52], sine-skewed Jones–Pewsey [1], asymmetric extended Jones–Pewsey [2] and inverse Batschelet [31] distributions;

- Calculate distribution and quantile functions for, and simulate variates from, the cardioid, Cartwright, wrapped Cauchy, von Mises, Jones–Pewsey, Batschelet, sine-skewed Jones–Pewsey, asymmetric extended Jones–Pewsey and inverse Batschelet distributions;

- Apply bootstrap based tests of uniformity and reflective symmetry;

- Carry out inference for key distributional characteristics such as the mean direction, mean resultant length and coefficients of circular asymmetry and kurtosis;

- Perform maximum likelihood estimation for the parameters of, investigate model reduction and carry out goodness-of-fit testing for, the Jones–Pewsey and inverse Batschelet distributions;

- Apply a bootstrap version of Watson's test for a common mean direction [65], a randomization version of Fisher's test for a common median direction [19, Section 5.3.2], large-sample and bootstrap versions of Fisher's test for a common concentration [19, Section 5.4.4], large-sample and bootstrap versions of the Mardia-Watson-Wheeler test for a common distribution [67], [40], a randomization version of Watson's test for a common distribution of two samples [64], Moore's test for a common distribution for two paired samples [44];

- Compute the Fisher–Lee correlation coefficient for rotational dependence [20] and bootstrap based confidence intervals for it, perform randomization based tests for circular-circular independence based on the correlation coefficient of Jammalamadaka and Sarma [27] and the test statistic of Rothman [55] and for circular-linear independence based on the rank correlation coefficient of Mardia [41].

13.3   Packages that Use the Circular Package

R packages that make use of the functionality of the `circular` package include, in alphabetical order:

`bcpa`, used to perform behavioural change point analysis ([22]) with the aim of identifying hidden changes in the underlying parameters of a time series, developed specifically for irregularly sampled animal movement data;

`Bios2cor`, with functions for the computation and analysis of correlation in multiple sequence alignments and in side-chain motions during molecular dynamics simulations ([50]);

`CircMLE`, providing a series of wrapper functions to fit, via maximum likelihood, the 10 animal orientation models considered in [60];

`CircOutlier`, designed to detect outliers in simple circular-circular regression models as described in [3];

`depth`, used to apply Euclidian and spherical depth function methodologies in multivariate analysis;

`FLightR`, developed to identify the position of an animal from solar geolocation data, based on a hidden Markov model;

`isocir`, which offers a set of functions for analyzing circular data under order restrictions as described in [57], [9], [56];

`KarsTS`, a graphical user interface for karstic time series analysis based on the `tcltk` package. Such series typically exhibit strong non-linear behaviour and missing values;

`kernplus`, a machine learning based tool for estimating multivariate power curves and predicting wind power output as proposed in [38];

`kineticF`, providing data cleaning, processing, visualisation and analysis functions for kinetic visual field data;

`localdepth`, with functions to calculate simplicial, Mahalanobis and elliptical local and global depths;

`monogeneaGM`, for performing geometric morphometric and evolutionary biology analyses of anchor shape from four anchored monogeneans, as expounded in [35];

`monographaR`, with functions to aid the production of plant taxonomic monographs as described in [54];

`move`, an extensive package for ecologists offering functions to access, visualize and analyze animal movement data, including fitting dynamic Brownian bridge movement models as proposed in [36];

`movMF`, for model based clustering of unit vectors using mixtures of von Mises–Fisher distributions as described in [25]. Important applications include the clustering of text and gene expression data;

NPCirc, which provides functions to apply non-parametric kernel methods for density and regression estimation as discussed in [46], [47], [48] and [49];

picante, offering a range of functions for use in the integration of phylogenies and ecology as proposed in [66];

sharpshootR, a toolkit providing soil survey supporting functions to handle, summarize and visualize soil data as described in [11];

SOPIE, providing functions to perform non-parametric estimation of the off-pulse interval of a source function originating from a pulsar, as proposed in [61];

wle, including functions to compute minimum distance and weighted likelihood robust estimates of the parameters of the von Mises and wrapped normal distributions, as considered in [4].

13.4   Other Packages for Circular Statistics

Other R packages with functions relevant to the analysis of circular data include, in alphabetical order:

CircNNTSR, which offers functions with which to fit non-negative trigonometric sum models to circular, multivariate circular, and spherical data as proposed in [16], [17], and [18];

OmicCircos, for generating high quality circular plots for "omics" (such as genomics, proteomics or metabolomics) data as proposed in [26];

plotrix, a general purpose graphical package with various options for circular displays;

psych, which implements techniques for psychological, psychometric and personality research including functions for the analysis of circadian and diurnal data;

season, a package based on [8] and [19] for analyzing seasonal health data which includes functions for plotting circular data and performing stationary and non-stationary cosinor regression modelling to detect seasonality in yearly data or circadian patterns in hourly data;

spatstat, a vast package for the analysis of spatial point patterns which includes kernel density estimation for circular data.

13.5   The Directional Package

The Directional package implements a range of techniques from [42] and [51] as well as other more recent proposals, offering functions to manage, transform, simulate, visualize and analyze data distributed on the circle, torus, cylinder, sphere and their extensions. For circular data analysis, it includes functions to:

- Produce summary statistics for circular and grouped circular data;

- Perform kernel density estimation;

- Compute the density of the von Mises distribution;

- Simulate random variates from the von Mises distribution and a mixture of von Mises distributions;

- Estimate the parameters of the bivariate angular Gaussian, wrapped Cauchy and generalized von Mises distributions using maximum likelihood;

- Apply tests of uniformity, a specified mean direction and the equality of von Mises concentration parameters;

- Calculate circular-circular and circular-linear correlation coefficients;

- Perform ANOVA and circular regression;

- Carry out goodness-of-fit for grouped data.

For data distributed on the (hyper-)sphere, it has functions to:

- Generate, check, transform and manipulate key matrices;

- Compute saddlepoint approximations of the normalizing constant of the Fisher–Bingham distribution;

- Simulate random matrices and vectors from the von Mises–Fisher, angular central Gaussian, Bingham, matrix Fisher, spherical Fisher-Bingham, spherical Kent and elliptically symmetric angular Gaussian (ESAG) distributions;

- Perform kernel density estimation;

- Summarize spherical data;

- Compute the density of the ESAG distribution;

- Produce contour plots of von Mises–Fisher, Kent, and mixtures of von Mises–Fisher distributions;

- Estimate the parameters of the von Mises–Fisher, Kent, matrix Fisher, angular central Gaussian, spherical projected normal, ESAG and Wood bimodal distributions by maximum likelihood;

- Test for the equality of von Mises–Fisher concentration parameters;

- Compute spherical-spherical correlation and perform spherical-spherical regression;

- Perform ANOVA, regression, discriminant and cluster analysis;

- Compare the goodness-of-fit of the Kent and von Mises–Fisher distributions.

13.6   Other Packages for Directional Statistics

Other packages relevant to the analysis of directional data include, in alphabetical order:

bReeze, with functions to analyze, visualize and interpret wind data and to calculate the potential energy production of wind turbines;

geosphere, for computing distances and related measures on the sphere for geographic applications, as proposed in [33];

globe, for plotting 2D and 3D views of spheres including the Earth with major coastlines;

mgcv, for fitting smooth functions including splines on the sphere, as proposed in [63];

misc3d, a graphics package to define, manipulate and plot meshes on simplices, spheres, balls, rectangles and tubes, as well as to produce directional and other multivariate histograms;

moveHMM, a package described in [43] with tools for modelling animal movement data using the hidden Markov methodology ([68]);

NHMSAR, for the calibration, simulation, fitting and validation of Markov switching autoregressive models with Gaussian or von Mises innovations, which can be used to fit an extension of the von Mises process of [13] as described in [6] for wind time series;

rgl, for producing 3D interactive graphics;

rstiefel, for simulating random orthonormal matrices from linear and quadratic exponential family distributions on the Stiefel manifold, particularly the matrix-variate Bingham-von Mises-Fisher distribution as proposed in [23];

shape, a general purpose graphics package for plotting circles, cylinders and other shapes;

skmeans, to perform spherical k-means clustering of unit vectors as described in [24], important applications being the clustering of text and gene expression data;

sm, with functions to implement the smoothing methodology considered in [12], in particular for kernel density estimation on the sphere;

sphereplot, offering functions to create spherical coordinate system plots using extensions to the rgl package;

SphericalCubature, providing functions with which to perform numerical integration over spheres and balls in d dimensions;

13.7   Unsupported Directional Statistics Methodologies

In this final section of the chapter we consider four important directional statistics methodologies that are not presently supported by dedicated R packages.

Principal component analysis (PCA) is a data-reduction technique that is often applied in the analysis of multivariate linear data. In recent years, analogous techniques have been developed for data distributed on a hypersphere or torus. A decomposition for hyperspherical data, referred to as principal nested spheres (PNS), was introduced by [32] and discussed in [39, Section 7.6]. An approach to PCA for toroidal data, incorporating a variant of PNS, was proposed recently in [15]. Although Matlab and Python implementations are available, we are unaware of any R packages presently supporting these methodologies.

The KarsTS, moveHMM, NHMSAR and season packages, referred to above, include functions supporting various approaches to directional time series analysis. However, as far as we are aware, no package is presently available which specifically implements all four modelling approaches discussed in [21].

In recent years, considerable attention has been given to modelling spatial as well as spatio-temporal directional data. Wave direction over space and time is an example of the latter type of data. Many recent developments are discussed in two of the other chapters of this book [29], [37]. Although no specific package presently exists to fit the various models described there, a package designed to fit some of them, with the tentative title of CircComplexMod, is presently under development and due to be released in November 2017. We note that the CircSpatial package, developed during the production of [45] for visualizing, simulating and kriging circular-spatial data, has been removed from the CRAN repository. A 2009 version of it is however available from the archive.

Compositional data analysis deals with data for which the response is generally defined as a vector of non-negative proportions summing to one. The predominant approach to analyzing such data is that based upon Aitchison's methodology [7]. In R, the `compositions` and `Compositional` packages support such techniques. Nevertheless, alternative approaches have been proposed in the literature [59]. One such approach, based on the square-root transformation from a unit simplex to a unit hypersphere, was discussed in [62] and has received renewed attention more recently in the Directional Statistics literature [58]. The application of the square-root transformation based approach to compositional data analysis would be very much enhanced by an R package supporting its use. Drawing together extensions of two of the ideas referred to in this section, PCA for functional directional data was proposed recently in [14]. As shown there, it can be used to analyze square-root transformed longitudinal compositional data.

13.8 Conclusions

The aim of this chapter has been to provide an overview of the R packages presently available for the analysis of directional data. As we have seen, the `circular` and `Directional` packages are the basic ones providing functions with which to handle, summarize, simulate, visualize, model and perform basic forms of analysis such as ANOVA and regression for directional data distributed on the circle, torus, cylinder, sphere and their extensions. Other packages are available for general methods such as graphics production, numerical integration, depth calculation, non-parametric kernel estimation, spline fitting, outlier detection, discriminant analysis, cluster analysis, time series analysis, and the analysis of circular data under order restrictions. Packages are also available for more specific tasks such as the analysis of animal movement, kinetic visual field, soil, circadian, diurnal, omics and seasonal health data, molecular dynamics, karstic time series, geometric morphometric and evolutionary biology, as well as for integrating phylogenies and ecology, modelling wind data and predicting wind power output, and the preparation of plant taxonomic monographs. Other important developing methodologies such as PCA for hyperspherical and toroidal data, spatial and spatio-temporal directional data analysis and compositional data analysis based on the square-root transformation are not presently supported by dedicated R packages. However, the R software undergoes constant evolution and it is reasonable to assume that in the coming years more packages will become available with which to perform these and other forms of statistical analysis for directional data.

Bibliography

[1] T. Abe and A. Pewsey. Sine-skewed circular distributions. *Statistical Papers*, 52:683 707, 2011.

[2] T. Abe, A. Pewsey, and K. Shimizu. Extending circular distributions through transformation of argument. *Annals of the Institute of Statistical Mathematics*, 65:833–858, 2013.

[3] A. H. Abuzaid, A. G. Hussin, and I. B. Mohamed. Detection of outliers in simple circular regression models using the mean circular error statistic. *Journal of Statistical Computation and Simulation*, 83:269–277, 2013.

[4] C. Agostinelli. Robust estimation for circular data. *Computational Statistics and Data Analysis*, 51:5867–5875, 2007.

[5] C. Agostinelli and U. Lund. *R Package Circular: Circular Statistics (version 0.4-93)*, 2017.

[6] P. Ailliot, J. Bessac, V. Monbet, and F. Pène. Non-homogeneous hidden Markov-switching models for wind time series. *Journal of Statistical Planning and Inference*, 160:75–88, 2015.

[7] J. Aitchison. *The Statistical Analysis of Compositional Data*. Chapman and Hall, London, 1986.

[8] A. G. Barnett and A. J. Dobson. *Analysing Seasonal Health Data*. Springer, Heidelberg, 2010.

[9] S. Barragán, M. A. Fernández, C. Rueda, and S. D. Peddada. isocir: an R package for constrained inference using isotonic regression for circular data, with an application to cell biology. *Journal of Statistical Software*, 54:1–17, 2013.

[10] E. Batschelet. *Circular Statistics in Biology*. Academic Press, New York, 1981.

[11] D. E. Beaudette, P. Roudier, and A. T. O'Geen. Algorithms for quantitative pedology: a toolkit for soil scientists. *Computers and Geosciences*, 52:258–268, 2013.

[12] A. W. Bowman and A. Azzalini. *Applied Smoothing Techniques for Data Analysis: the Kernel Approach with S-Plus Illustrations*. Oxford University Press, Oxford, 1997.

[13] J. Breckling. *The Analysis of Directional Time Series: Applications to Wind Speed and Direction*. Springer, London, 1989.

[14] X. Dai and H.-G. Müller. Principal component analysis for functional data on Riemannian manifolds and spheres. *arXiv:1705.06226*, 2017.

[15] B. Eltzner, S. Huckemann, and K. V. Mardia. Torus principal component analysis with an application to RNA structures. *arXiv:1511.04993*, 2015.

[16] J. J. Fernández-Durán. Circular distributions based on nonnegative trigonometric sums. *Biometrics*, 60:499–503, 2004.

[17] J. J. Fernández-Durán. Models for circular-linear and circular-circular data constructed from circular distributions based on nonnegative trigonometric sums. *Biometrics*, 63:579–585, 2007.

[18] J. J. Fernández-Durán and M. M. Gregorio-Domínguez. CircNNTSR: an R package for the statistical analysis of circular, multivariate circular, and spherical data using nonnegative trigonometric sums. *Journal of Statistical Software*, 70:1–19, 2016.

[19] N. I. Fisher. *Statistical Analysis of Circular Data*. Cambridge University Press, Cambridge, 1993.

[20] N. I. Fisher and A. J. Lee. A correlation coefficient for circular data. *Biometrika*, 70:327–332, 1983.

[21] N. I. Fisher and A. J. Lee. Time series analysis of circular data. *Journal of the Royal Statistical Society, Series B*, 56:327–339, 1994.

[22] E. Gurarie, R. D. Andrews, and K. L. Laidre. A novel method for identifying behavioural changes in animal movement data. *Ecology Letters*, 12:395–408, 2009.

[23] P. D. Hoff. Simulation of the matrix Bingham–von Mises–Fisher distribution, with applications to multivariate and relational data. *Journal of Computational and Graphical Statistics*, 18:438–456, 2009.

[24] K. Hornik, I. Feinerer, M. Kober, and C. Buchta. Spherical k-means clustering. *Journal of Statistical Software*, 50:1–22, 2012.

[25] K. Hornik and B. Grün. movMF: an R package for fitting mixtures of von Mises-Fisher distributions. *Journal of Statistical Software*, 58:1–31, 2014.

[26] Y. Hu, C. Yan, C.-H. Hsu, Q.-R. Chen, K. Niu, G. A. Komatsoulis, and D. Meerzaman. OmicCircos: a simple-to-use R package for the circular visualization of multidimensional omics data. *Cancer Informatics*, 13:13, 2014.

[27] S. R. Jammalamadaka and Y. Sarma. A correlation coefficient for angular variable. In K. Matusita, editor, *Statistical Theory and Data Analysis II*, pages 349–364. North Holland, Amsterdam, 1988.

[28] S. R. Jammalamadaka and A. SenGupta. *Topics in Circular Statistics*. World Scientific, Singapore, 2001.

[29] G. Jona-Lasinio, A. E. Gelfand, and G. Mastrantonio. Spatial and spatio-temporal circular processes with application to wave directions. In C. Ley and T. Verdebout, editors, *Applied Directional Statistics*. CRC Press, Boca Raton, FL, 2018.

[30] M. C. Jones and A. Pewsey. A family of symmetric distributions on the circle. *Journal of the American Statistical Association*, 100:1422–1428, 2005.

[31] M. C. Jones and A. Pewsey. Inverse Batschelet distributions for circular data. *Biometrics*, 68:183–193, 2012.

[32] S. Jung, I. L. Dryden, and J. S. Marron. Analysis of principal nested spheres. *Biometrika*, 99:551–568, 2012.

[33] C. F. F. Karney. Algorithms for geodesics. *Journal of Geodesy*, 87:43–55, 2013.

[34] S. Kato and M. C. Jones. A tractable and interpretable four-parameter family of unimodal distributions on the circle. *Biometrika*, 102:181–190, 2015.

[35] T. F. Khang, O. Y. M. Soo, W. B. Tan, and L. H. S. Lim. Monogenean anchor morphometry: systematic value, phylogenetic signal, and evolution. *PeerJ*, 4:e1668, 2016.

[36] B. Kranstauber, R. Kays, S. D. LaPoint, M. Wikelski, and K. Safi. A dynamic Brownian bridge movement model to estimate utilization distributions for heterogeneous animal movement. *Journal of Animal Ecology*, 81:738–746, 2012.

[37] F. Lagona. Correlated cylindrical data. In C. Ley and T. Verdebout, editors, *Applied Directional Statistics*. CRC Press, Boca Raton, FL, 2018.

[38] G. Lee, Y. Ding, M. G. Genton, and L. Xie. Power curve estimation with multivariate environmental factors for inland and offshore wind farms. *Journal of the American Statistical Association*, 110:56–67, 2015.

[39] C. Ley and T. Verdebout. *Modern Directional Statistics*. CRC Press, Boca Raton, FL, 2017.

[40] K. V. Mardia. *Statistics of Directional Data*. Academic Press, London, 1972.

[41] K. V. Mardia. Linear-circular correlation coefficients and rhythmometry. *Biometrika*, 63:403–405, 1976.

[42] K. V. Mardia and P. E. Jupp. *Directional Statistics*. Wiley, Chichester, 1999.

[43] T. Michelot, R. Langrock, and T. A. Patterson. movehmm: an R package for the statistical modelling of animal movement data using hidden Markov models. *Methods in Ecology and Evolution*, 7:1308–1315, 2016.

[44] B. R. Moore. A modification of the Rayleigh test for vector data. *Biometrika*, 67:175–180, 1980.

[45] W. J. Morphet. *Simulation, kriging, and visualization of circular-spatial data*. PhD thesis, Utah State University, 2009.

[46] M. Oliveira, R. M. Crujeiras, and A. Rodríguez-Casal. A plug-in rule for bandwidth selection in circular density estimation. *Computational Statistics and Data Analysis*, 56:3898–3908, 2012.

[47] M. Oliveira, R. M. Crujeiras, and A. Rodríguez-Casal. Nonparametric circular methods for exploring environmental data. *Environmental and Ecological Statistics*, 20:1–17, 2013.

[48] M. Oliveira, R. M. Crujeiras, and A. Rodríguez-Casal. CircSiZer: an exploratory tool for circular data. *Environmental and Ecological Statistics*, 21:143–159, 2014.

[49] M. Oliveira, R. M. Crujeiras, and A. Rodríguez-Casal. NPCirc: an R package for nonparametric circular methods. *Journal of Statistical Software*, 61:1–26, 2014.

[50] J. Pelé, M. Moreau, H. Abdi, P. Rodien, H. Castel, and M. Chabbert. Comparative analysis of sequence covariation methods to mine evolutionary hubs: examples from selected GPCR families. *Proteins: Structure, Function, and Bioinformatics*, 82:2141–2156, 2014.

[51] A. Pewsey, M. Neuhäuser, and G. D. Ruxton. *Circular Statistics in R*. Oxford University Press, Oxford, 2013.

[52] A. Pewsey, K. Shimizu, and R. de la Cruz. On an extension of the von Mises distribution due to Batschelet. *Journal of Applied Statistics*, 38:1073–1085, 2011.

[53] R Core Team. *R: A Language and Environment for Statistical Computing*. R Foundation for Statistical Computing, Vienna, Austria, 2014.

[54] M. Reginato. monographaR: an R package to facilitate the production of plant taxonomic monographs. *Brittonia*, 68:212–216, 2016.

[55] E. D. Rothman. Tests of coordinate independence for a bivariate sample on a torus. *The Annals of Mathematical Statistics*, 42:1962–1969, 1971.

[56] C. Rueda, M. A. Fernández, S. Barragán, and S. D. Peddada. Some advances in constrained inference for ordered circular parameters in oscillatory systems. In I. L. Dryden and J. T. Kent, editors, *Geometry Driven Statistics*, pages 97–114. Wiley, Chichester, 2015.

[57] C. Rueda, M. A. Fernández, and S. D. Peddada. Estimation of parameters subject to order restrictions on a circle with application to estimation of phase angles of cell cycle genes. *Journal of the American Statistical Association*, 104:338–347, 2009.

[58] J. L. Scealy and A. H. Welsh. Regression for compositional data by using distributions defined on the hypersphere. *Journal of the Royal Statistical Society, Series B*, 73:351–375, 2011.

[59] J. L. Scealy and A. H. Welsh. Colours and cocktails: compositional data analysis: 2013 Lancaster lecture. *Australian and New Zealand Journal of Statistics*, 56:145–169, 2014.

[60] J. T. Schnute and K. Groot. Statistical analysis of animal orientation data. *Animal Behaviour*, 43:15–33, 1992.

[61] W. D. Schutte. *Nonparametric estimation of the off-pulse interval(s) of a pulsar light curve*. PhD thesis, North-West University, 2014.

[62] M. A. Stephens. Use of the von Mises distribution to analyse continuous proportions. *Biometrika*, 69:197–203, 1982.

[63] G. Wahba. Spline interpolation and smoothing on the sphere. *SIAM Journal on Scientific and Statistical Computing*, 2:5–16, 1981.

[64] G. S. Watson. Goodness-of-fit tests on a circle. II. *Biometrika*, 49:57–63, 1962.

[65] G. S. Watson. *Statistics on Spheres*. Wiley, New York, 1983.

[66] C. O. Webb, D. D. Ackerly, and S. W. Kembel. Phylocom: software for the analysis of phylogenetic community structure and trait evolution. *Bioinformatics*, 24:2098–2100, 2008.

[67] S. Wheeler and G. S. Watson. A distribution-free two-sample test on the circle. *Biometrika*, 51:256–257, 1964.

[68] W. Zucchini, I. L. MacDonald, and R. Langrock. *Hidden Markov Models for Time Series: An Introduction Using R*. CRC Press, London, 2nd edition, 2016.

Index

A

Abe–Ley density, 46, 49
Akaike Information Criteria (AIC), 180, 184, 191, 225
Ambiguous rotations, 29–33
  analysis of examples, 42–43
  distributions, 38–40
  examples, 26–29
  orthogonal axial frames, 30–31
  summary statistics, 35–36
  symmetric arrays, 33–35
  symmetric frames, 32–33
  symmetry groups, 31
  testing uniformity, 36–37
  tests of location, 40–41
Amino acid residues, 4
Amino acids, 2, 63
Amino acid side chains, 2, 4, 15–17
Artificial neural nets, 6
Astrophysical application, 108–109

B

Basic Human Values (BHV) scale, 212, 227–236
BASILISK, 15–17
Bayesian information criterion (BIC), 52, 54, 82, 225
Bayesian modeling, 216
  advantages, 236
  DIC measure for model comparison, 224–225, 229–231, 234–235
  inequality constrained hypotheses, 225–227
  intrinsic and embedding circular data modeling approaches, 214–218, 219–225, 229–237

kriging with projected Gaussian processes, 146
protein structure modeling (TorusDBN), 10, 12–15, 67
R package tools, 237n
social and behavioral sciences applications, 211–237, *See also* Social and behavioral sciences applications
spatial and spatio-temporal modeling for wave directions, 130, 133
toroidal diffusions and protein structure evolution modeling, 62, 77–90, *See also* ETBN
wind direction bias correction and ensemble calibration, 131
See also Markov chain Monte Carlo (MCMC) modeling
Bessel ratio, 264
BIC statistic (Bayesian information criterion), 52, 54, 82, 225
Binary regression and circular data classification, 242–243, 246–250
Bingham distribution, 38–39, 262
Bingham-Mardia distribution, 262
Bird flight trajectories, 165, 173–174
Boltzmann distribution, 8
Bootstrap methods
  cylindrical data modeling, 56–57
  software tools, 279
  wildfire modeling applications, 195, 202, 204–205

C

Cardioid distributions, 167

Cayley distributions, 39

CircSiZer, 193–194, 205–206

Circular data classification, 241–243
  binary regression examples,
    242–243
  density estimation, 243–246
  earth surface classification,
    251–253
  local logistic regression, 246–250
  performance evaluation, 250–251

Circular data modeling, 129–131,
    149–150, 212, 241
  Bayesian framework, 130,
    216–218, *See also* Bayesian
    modeling
  Bayesian social and behavioral
    sciences applications,
    211–237, *See also* Social
    and behavioral sciences
    applications
  circumplex model, 212, 227
  classification, 241–256, *See also*
    Circular data classification
  comparing wrapped and
    projected Gaussian process,
    151–153
  continuous ranked probability
    score (CRPS), 134
  intrinsic and embedding
    approaches, 213–218,
    219–225, 229–237
  joint modeling of wave height
    and direction, 153–157
  kriging and forecasting, 133,
    146–148
  probability distributions, 164
  projected Gaussian process,
    141–153, 157
  projected normal distribution,
    215–216
  software tools, 278–281
  wildfire modeling applications,

190–191, 198–200, *See also*
    Wildfire modeling
  wrapped spatial and
    spatio-temporal process,
    131–141, 151–153, 157
  See also Circular distributions;
    Cylindrical distributions;
    specific distributions

Circular density estimation, 243–246

Circular distributions, 167–168
  sine-skewed perturbation, 168
  wildfire modeling applications,
    190–191, *See also* Wildfire
    modeling
  See also Cardioid distributions;
    Circular data modeling;
    Cylindrical distributions;
    von Mises distributions

Circular package, R software,
    277–280

Circumplex model, 212, 227

Cluster-related directional statistics,
    260, 269–271

Compositional data analysis,
    284–285

Continuous ranked probability score
    (CRPS), 134

Continuous-Time Markov Chains
    (CTMC), 79

Correlated cylindrical data, 45–57
  hidden Markov models, 46,
    49–57
  identification of sea regimes, 47,
    52–54
  segmentation of current fields,
    54–55

Cosine similarity, 260

Cosmic rays, 108–109

Crown growth asymmetry, 174–184

Crystal orientations, 26–27, 42

Cylindrical data classification, 242

Cylindrical data modeling, 45–57,
    163–164
  bird flight trajectory example,
    165, 173–174

hidden Markov models, 49–57
identification of sea regimes, 47, 52–54
mixture models, 46
segmentation of current fields, 54–55
time series and spatial series examples, 45
tree crown growth application, 174–184
See also Cylindrical distributions
Cylindrical distributions, 164–165, 168, 180–184
bird flight trajectory example, 165, 173–174
Gamma-von Mises, 170–171, 174, 180
generalized Gamma-von Mises, 171
Johnson-Wehrly, 168–169, 180
parameter estimation, 173
review of univariate probability distributions, 165–167
sine-skewed Weibull-von Mises distribution, 172–173
tree crown growth, 174–184
Weibull-von Mises, 169–170
wildfire modeling applications, 202–204
wrapped Cauchy, 167, 168–169

D
Demagnetization, 107
Deviance information criterion (DIC), 224–225, 229–231, 234–235
DIC measure for model comparison, 224–225, 229–231, 234–235
Diopside crystals, 26, 42
Directional package, R software, 277–278, 281–283
Directional statistics, 25–26
basic distributions, 260–263, *See also specific distributions*

Bayesian approaches, *See* Bayesian modeling
circular data, *See* Circular data modeling
cylindrical data, *See* Cylindrical data modeling
limitations of vector representations, 259–260
noisy data, *See* Noisy directional data, goodness-of-fit testing in the spherical convolution model
related work and applications, 263
rotational data, *See* Ambiguous rotations; SO(3)
software tools, 277–285

E
Earthquake focal mechanisms, 27–29, 42–43
Earth surface classification, 251–253
Embedding approach, 33–35, 215–216, 218, 222–225, 230, 232–237
ETDBN (Evolutionary Torus Dynamic Bayesian Network), 62, 77–90
case study, 87–90
hidden Markov model structure, 77–80
model training, 81–83
performance evaluation, 83–87
site-classes, 80–81
time-reversibility, 65–66, 79, 80
Euler pseudo-tpd, 72–73
European Social Survey (ESS) data, 227–229
Evolutionary hidden states, 87–90
Evolution of protein structure, *See* Protein structure evolution
Expectation-maximization (EM), 82
mixture models, 260, 268
wildfire modeling applications, 191

Exponential distribution, 165, 169
Exponential-family Dirichlet
        compound multinomial
        model (EDCM), 270

F
Fire regimes modeling, 187–207, *See
        also* Wildfire modeling
Fisher-Bingham distribution, 262
Fisher-Watson distribution, 262
Flying bird trajectory model, 165,
        173–174
Forecasting and kriging, 133,
        146–148
Forward filtering backward sampling
        (FFBS) algorithm, 80, 82
Fourier analysis on $SO(3)$ and S^2,
        96–98

G
Gait analysis, 112, 121–123
Gamma distribution, 166–167
Gamma function, 260–263
Gamma-von Mises distribution,
        170–171, 174, 180
Generalized Gamma distribution,
        167
Generalized Gamma-von Mises
        distribution, 171
Generalized von Mises (mGvM)
        regression model, 17
Goodness-of-fit testing for noisy
        directional data, 95–109,
        See also Noisy directional
        data, goodness-of-fit testing
        in the spherical convolution
        model

H
Helmholtz free energy, 8
Hidden Markov models (HMMs),
        55–57
    Abe–Ley density, 46, 49
    Bayesian protein structure

modeling (TorusDBN), 10,
        12–15
    correlated cylindrical data, 46,
        49–57
    cylindrical spatial series, 51–52
    cylindrical time series, 49–50
    Evolutionary Torus Dynamic
        Bayesian Network (ETBN)
        and protein structure
        evolution modeling, 77–80
    independence assumptions and
        protein structure modeling,
        15–17
    sample size requirements, 56–57
Histogram-based protein structure
        modeling, 11–12
Homology modeling, 7, 65

I
Ilmenite crystals, 26–27, 42
Indels, 66
Input output hidden Markov model
        (IOHMM), 15–17
Intrinsic approach for circular data
        modeling, 214–215, 229,
        231–232, 234–237

J
Johnson-Wehrly distribution,
        168–169, 180

K
Kent distributions, 263
Kernel density estimation (KDE),
        192–193, 243–246, 278
K-frames, 32–33
K-means algorithm, 268
Knee motion modeling, 112, 121–123
Kriging, 133, 146–148

L
Langevin diffusions, 70
Laplace distribution, 99
Latent Dirichlet allocation (LDA),
        270

Likelihood cross validation (LCV), 202

M

Machine learning applications, 259, 272
  basic directional distributions, 260–263
  mixture models, 266–271
  MLE for von Mises distribution, 264–265
  MLE for Watson distribution, 265–266
  text clustering, 269–271
Magnetic fields and mineral properties, 107
Markov chain modeling, 50, *See also* Hidden Markov models; Markov chain Monte Carlo (MCMC) modeling
Markov chain Monte Carlo (MCMC) modeling
  circular data modeling for social and behavioral sciences applications, 212, 215, 217–218, 219–225, 229–230
  protein structure modeling, 8–10
  spatial and spatio-temporal modeling for wave directions, 130, 133, 135
Markov chains, time-reversibility, 79
Maximum likelihood estimation (MLE), 263
  cylindrical data modeling example, 180
  parameter estimation for SE(3) models, 116–117
  software tools, 278, 279
  toroidal diffusions and protein structure evolution, 72–75
  for von Mises distribution, 263–264
Metropolis algorithm, 9
Metropolis-Hastings (MH) algorithm, 9, 82, 83, 217

Mixture models, 12, 46, 260, 266–271
Molecular dynamics (MD), 8
Molecular structure evolution, *See* Protein structure evolution
Monoclinic crystal orientations, 26
Monte Carlo principle, 8–9, *See also* Markov chain Monte Carlo (MCMC) modeling
Monte Carlo simulation, SE(3) models, 121–122

N

Nadaraya-Watson estimator, 247, 249, 260, 261–262, 265–266
Newton's method, 264
Neyman-Scott paradox, 116
Noisy directional data, goodness-of-fit testing in the spherical convolution model, 95–109
  harmonic analysis on SO(3) and S^2, 96–98
  real data examples, 107–108
  simulations, 104–106
  test constructions, 100–101
  testing procedures, 102–104

O

Ornstein-Uhlenbeck toroidal diffusion analogues, 62, 68, 70–72
Orthogonal axial frames, 30–31

P

Paleomagnetism, 107
Pearson correlation, 260
Peptide bonds, 4–5
PHAISTOS, 18
Point groups of the first kind, 31
Polypeptide backbones of proteins, 4, 10–15
Potts model, 51–52
Prediction and kriging, 133, 146–148
Primary structure of proteins, 2

Principal component analysis (PCA), 284, 285
Principal nested spheres (PNS), 284
Projected Gaussian process, 141–150, 157
    comparing wrapped and projected Gaussian process, 151–153
    kriging and forecasting, 146–148
    model fitting and inference, 145
    separable space-time wave direction example, 149–150
    space-time process, 149
Projected normal distribution for circular data, 215–216
Protein structure, 1–6, 61–64
    determination and prediction, 6–10
    dynamical Bayesian network model (TorusDBN), 10, 12–15, 67
    generative models, amino acid side chains, 15–17, 67
    generative models, polypeptide backbone, 10–15
    Markov chain Monte Carlo simulations, 8–10
    toroidal diffusions, 62, 67–70
Protein structure evolution, 61–62, 64–66
    case study, 87–90
    evolutionary hidden states, 87–90
    generative model, 62, 66–67, 77–90, *See also* ETDBN
    model performance evaluation, 75–77, 83–87
    time-reversibility, 65–66, 79, 80
    toroidal Ornstein-Uhlenbeck analogues, 62, 68, 70–72

Q
Quaternary structure of proteins, 3

R
Radially symmetric distributions, 38
Ramachandran distribution, 12, 80
Random number generation algorithm, 172–173
Regression-based circular data classifications, 242–243, 246–250
Rotational group SE(3), *See* SE(3) modeling
Rotational harmonics, 97
Rotational Laplace distribution, 99
Rotation group SO(3), *See* SO(3)
Rotation matrices, 111–113
    Cardan angles and SO(3) objects, 113
    one axis model, 113–115
    See also SE(3) modeling
R package tools, 277–285
    Bayesian circular data modeling, 237n
    circular package, 277–280
    compositional data analysis, 285
    Directional package, 277–278, 281–283
    unsupported directional statistics methodologies, 284–285

S
Sample size requirements, 56–57
Satellite imaging data, 189
SE(3) modeling, 111–113
    data analysis, 121–123
    data modeling, 115–116
    MLE, 116–117
    one axis model, 113–115
    open problems, 124
    parameter estimation, 116–119
    probabilistic models, 112
    simulations, 119–120
Sea currents, 48–49, 54–55
Sea wave directions and heights, *See* Wave directions or heights

Secondary structure of proteins, 3, 6, 63–64
Seismology
    earthquake focal mechanisms, 27–29, 42–43
    orthogonal axial frames, 31
Shoji-Ozaki pseudo tpd, 73
Similarity measures, 260
Sine-skewed circular distribution, 168
Sine-skewed Weibull-von Mises distribution, 172–173
Skewed Gaussian process, 136–141
SO(3)
    ambiguous rotations, 29–33, *See also* Ambiguous rotations
    distributions, 38–40, *See also* Cardioid distributions; von Mises distributions; *specific types*
    embedding approach, 33–35
    goodness-of-fit testing for noisy directional data, 95–109
    harmonic analysis on, 96–98
    K-frames, 32–33
    point groups of the first kind, 31
    summary statistics, 35–36
    symmetric arrays, 33–35
    testing uniformity, 36–37
Social and behavioral sciences applications, 211–213
    Basic Human Values (BHV) scale, 212, 227–236
    Bayesian modeling, 216–228
    circumplex model, 212, 227
    DIC measure for model comparison, 224–225, 229–231, 234–235
    inequality constrained hypotheses, 225–227
    intrinsic and embedding circular data modeling approaches, 213–218, 219–225
    spatial cognition development, 218–227

Software tools for directional statistics, 277–285
Space-time projected Gaussian process, 149–150
Spatial and spatio-temporal circular processes, 129–158, *See also* Circular data modeling; Wave directions or heights
Spatial cognition development, 218–227
Spherical data, goodness-of-fit testing for noisy directional data, 95–109, 259
Spherical k-means (spkmeans) algorithm, 260
Stochastic differential equations (SDEs), 68
Stochastic expectation-maximization (StEM), 82
Symmetric arrays, 33–35
Symmetric frames, 32–33
Symmetry groups, 31

T
Template-based modeling, 7
Tertiary structure of proteins, 3, 7
Text clustering, 260, 269–271
Time-reversibility in molecular structure evolution modeling, 65–66, 79, 80
Toroidal data classification, 242
Toroidal diffusions, 61–62, 67–70
    Bayesian modeling and protein structure evolution (ETBDN), 62, 77–90, *See also* ETDBN
    estimation, 72–75
    Ornstein-Uhlenbeck analogues, 62, 68, 70–72
    performance evaluation, 75–77
Toroidal probability distribution, 164
TorusDBN, 10, 12–15, 67
Tree crown growth, 174–184
Trigonal crystal orientations, 26–27

U

Ultra high energy cosmic rays (UHECR), 108–109
Uniform axis-random spin distributions, 38
Uniform distribution, 260
Univariate probability distributions, 165–167

V

Vector representations of data, 259–260
von Mises distributions, 68, 130, 167, 250–251, 269–271
   circular density estimation, 243–244
   Gamma-von Mises, 170–171, 174, 180
   generalized Gamma, 171
   intrinsic approach for circular data modeling, 214–215
   mixture models, 12, 266–271
   MLE for, 263–264
   protein structure modeling, 11, 12
   text clustering application, 269–271
   Weibull-von Mises distribution, 169–170
   wildfire modeling applications, 191–192, 196, 199–201
   See also Cardioid distributions
von Mises-Fisher distribution, 199–201, 260, 261
von Mises reference rule, 246, 250

W

Water geographical classification application, 251–253
Watson distribution, 39, 103–104, 105, 260, 261–262
   clustering application, 271
   mixture models, 266–271
Watson's test for common distributions, 279

Watson-Williams test, 122
Wave directions or heights, 45, 129–131, 149–150
   comparing wrapped and projected Gaussian process, 151–153
   cylindrical hidden Markov modeling, 52–54
   joint modeling of wave height and direction, 153–157
   projected Gaussian process, 141–150, 157
   separable space-time wave direction example, 149–150
   software tools, 284
   wrapped skewed Gaussian process, 136–141
   wrapped spatial and spatio-temporal process, 131–141, 157
Weibull distribution, 166
Weibull-von Mises distribution, 169–170, 172–173
Wildfire modeling, 187–207
   global scale, 188, 195–197
   local or landscape scale, 188, 189–195, 197–198
   open problems, 206–207
   orientation, 197–206
   orientation-size joint distribution, 200–205
   seasonality, 188–197
Wind direction or speed
   bias correction and ensemble calibration, 131
   cylindrical data modeling, 46, 47
   wildfire modeling applications, 198
Wrapped Cauchy distribution, 167, 168–169, 243
Wrapped Langevin diffusions, 70
Wrapped normal distributions, 243
Wrapped skew normal (WSN) distribution, 139

Wrapped spatial and spatio-temporal
　　process, 131, 157
　　comparing wrapped and
　　　projected Gaussian process,
　　　151–153
　　kriging and forecasting, 133

skewed Gaussian process,
　　136–141
spatial Gaussian process, 132
spatio-temporal process, 133
wave data for examples,
　　134–136